Mechanical Engineering Series

Frederick F. Ling
Series Editor

Mechanical Engineering Series

J. Angeles, **Fundamentals of Robotic Mechanical Systems: Theory, Methods, and Algorithms, 2nd ed.**

P. Basu, C. Kefa, and L. Jestin, **Boilers and Burners: Design and Theory**

J.M. Berthelot, **Composite Materials: Mechanical Behavior and Structural Analysis**

I.J. Busch-Vishniac, **Electromechanical Sensors and Actuators**

J. Chakrabarty, **Applied Plasticity**

G. Chryssolouris, **Laser Machining: Theory and Practice**

V.N. Constantinescu, **Laminar Viscous Flow**

G.A. Costello, **Theory of Wire Rope, 2nd ed.**

K. Czolczynski, Rotordynamics of Gas-Lubricated Journal Bearing Systems

M.S. Darlow, **Balancing of High-Speed Machinery**

J.F. Doyle, **Nonlinear Analysis of Thin-Walled Structures: Statics, Dynamics, and Stability**

J.F. Doyle, **Wave Propagation in Structures: Spectral Analysis Using Fast Discrete Fourier Transforms, 2nd ed.**

P.A. Engel, **Structural Analysis of Printed Circuit Board Systems**

A.C. Fischer-Cripps, Introduction to Contact Mechanics

A.C. Fischer-Cripps, Nanoindentation, 2nd ed.

J. García de Jalón and E. Bayo, **Kinematic and Dynamic Simulation of Multibody Systems: The Real-Time Challenge**

W.K. Gawronski, **Advanced Structural Dynamics and Active Control of Structures**

W.K. Gawronski, Dynamics and Control of Structures: A Modal Approach

K.C. Gupta, **Mechanics and Control of Robots**

J. Ida and J.P.A. Bastos, **Electromagnetics and Calculations of Fields**

M. Kaviany, **Principles of Convective Heat Transfer, 2nd ed.**

M. Kaviany, **Principles of Heat Transfer in Porous Media, 2nd ed.**

(continued after index)

Pierre Ladevèze Jean-Pierre Pelle

Mastering Calculations in Linear and Nonlinear Mechanics

Translated by Theofanis Strouboulis

With 143 Figures

Springer

Pierre Ladevèze
Lab. Mecanique et Technologie
Ecole Normale Superieure de Cachan, France
61, avenue du President Wilson
Cachan Cedex 94235, France
ladeveze@lmt.ens-cachan.fr

Jean-Pierre Pelle
Lab. Mecanique et Technologie
Ecole Normale Superieure de Cachan, France
61, avenue du President Wilson
Cachan Cedex 94235, France

Series Editor
Frederick F. Ling
Ernest F. Gloyna Regents Chair in Engineering, Emeritus
Department of Mechanical Engineering
The University of Texas at Austin
Austin, TX 78712-1063, USA
and
William Howard Hart Professor Emeritus
Department of Mechanical Engineering, Aeronautical Engineering and Mechanics
Rensselaer Polytechnic Institute
Troy, NY 12180-3590, USA

Library of Congress Cataloging-in-Publication Data
Ladevèze, Pierre, 1945–
[Maîtrise du calcul en mécanique linéaire et non linéaire. English]
Mastering calculations in linear and nonlinear mechanics/Pierre Ladevèze and Jean-Pierre Pelle; translated by Theofanis Strouboulis.
p. cm. — (Mechanical engineering series)
Includes bibliographical references and index.
ISBN 0-387-21294-9 (alk. paper)
1. Mechanics, Applied—Mathematical models. I. Pelle, Jean-Pierre. II.Title. III. Mechanical engineering series (Berlin, Germany)
TA350.L2213 2004
620.1′001′51—dc22 2004048201

Translated from the French *Maîtrise du calcul en mécanique linéaire et non linéaire*, by Pierre Ladevèze and Jean-Pierre Pelle, © 2001 Hermes-Lavoisier Science Publishers, Paris, France.

ISBN 0-387-21294-9 Printed on acid-free paper.

Printed in the United States of America. (SBA)

9 8 7 6 5 4 3 2 1 SPIN 10959008

springeronline.com

Series Preface

Mechanical engineering, an engineering discipline forged and shaped by the needs of the industrial revolution, is once again asked to do its substantial share in the call for industrial renewal. The general call is urgent as we face profound issues of productivity and competitiveness that require engineering solutions. The Mechanical Engineering Series features graduate texts and research monographs intended to address the need for information in contemporary areas of mechanical engineering.

The series is conceived as a comprehensive one that covers a broad range of concentrations important to mechanical engineering graduate education and research. We are fortunate to have a distinguished roster of consulting editors on the advisory board, each an expert in one of the areas of concentration. The names of the consulting editors are listed on the next page of this volume. The areas of concentration are applied mechanics, biomechanics, computational mechanics, dynamical systems and control, energetics, mechanics of materials, processing, production systems, thermal science, and tribology.

New York, New York

Frederick F. Ling

Mechanical Engineering Series

Frederick F. Ling
Series Editor

The Mechanical Engineering Series features graduate texts and research monographs to address the need for information in contemporary mechanical engineering, including areas of concentration of applied mechanics, biomechanics, computational mechanics, dynamical systems and control, energetics, mechanics of materials, processing, production systems, thermal science, and tribology.

Contents

Introduction

Today more than ever, modeling and simulation are central to a mechanical engineer's activity. Increasingly complex models are being used routinely on a daily basis. This revolution, which has just begun, is the result of the extraordinary progress in computer technology in terms of both hardware and software.

In order to represent a real problem, one does not use just a single model, but a series of models. Starting from a first model, called the reference model, practical or economic considerations, along with the wish to take advantage of certain particular situations, often lead to the introduction of additional simplifying hypotheses, called condensation hypotheses, which result in a new, more manageable model. This, for example, is the case of hypotheses which, starting from a continuous model of a medium subjected to a given environment, lead to a "finite element" model involving parameters such as the size and type of the elements, the number of iterations, the duration of the time increments....

Of course, it is imperative not to alter the reference model completely. Therefore, controlling the additional simplifying hypotheses is an obvious and major issue. This has been a constant preoccupation on the industrial level as well as in research. The new situation is that over the last twenty years truly quantitative tools for assessing the quality of a model compared to another reference model have appeared.

This work deals with the control of the hypotheses leading from a mechanical model, usually coming from continuum mechanics, to a numerical model, i.e. the mastery of the mechanical computation process itself. Particular attention is given to structural analysis which, in this context, is the most advanced domain. The term "structure" designates the material envelope, which can consist of metallic materials, composite materials, biomaterials ... in solid, fluid or gaseous environments. The models being studied are not necessarily linear and high degrees of nonlinearity may be present (plasticity, viscoplasticity, unilateral contact...). The objective of structural analysis is to simulate the behavior of a structure subject to various solicitations (prescribed displacements and forces) numerically; in particular, the aim is to evaluate the state of damage of the structure and compare it with one or several limit states. The final stage consists in optimizing the structural parameters. The practical problems concern the dimensioning, optimization, reliability and even the manufacturing process of the object being designed or built.

The basic problem consists in defining and evaluating a measure of the error due to the discretization performed, in this case, by the finite element method.

Two situations must be dealt with, depending on whether the error is evaluated before or after the finite element calculation has been performed.

Today, for the first situation corresponding to what one calls "a priori" errors, only coarse evaluations are available. The second situation is more favorable: the finite element solution constitutes an additional piece of information. It is in the corresponding field of "a posteriori" error evaluation that the first research works on linear problems were published about twenty years ago.

The numerous techniques proposed can be categorized into three approaches:

- the first approach relies on the concept of error in constitutive relation and on related field construction techniques [LADEVEZE, 1975];
- the second approach relies on the concept of error indicator associated with the satisfaction of the equilibrium equations [BABUSKA - RHEINBOLDT, 1978];
- the third approach is based on the unevenness of the finite element solution [ZIENKIEWICZ - ZHU, 1987].

In the present work, after having described the various approaches, we focus on the first family of estimators because, on the one hand, it has the strongest mechanical meaning and, on the other hand, contrary to the other two families, it can be extended without much difficulty to nonlinear evolution problems. This approach is based on a partition of the equations of the reference problem into:

- admissibility conditions (kinematic constraint equations, equilibrium equations, initial conditions);
- the constitutive relation.

Indeed, the constitutive relation has a special status: in practice, this is often the least reliable equation. Therefore, it is natural to set this equation apart and seek an approximate displacement-stress solution over the time interval being considered which verifies the most reliable group of equations (i.e. the admissibility conditions) exactly. This solution verifies a constitutive relation which, in general, differs from the constitutive relation of the material; thus, the quality of the approximation can be assessed by comparing this constitutive relation to that of the material. Energy norms or other norms which have a deep physical meaning are used to quantify this error.

An a priori obstacle is that it is difficult to construct admissible approximate solutions, i.e. solutions which verify the admissibility conditions exactly. Indeed, the usual approximations – particularly the approximations resulting from the application of the finite element method – fail to verify these conditions exactly because the calculated stresses are not in equilibrium with the applied forces. One circumvents this difficulty by a very general technique which enables one to construct an admissible approximate solution explicitly, therefore very inexpensively, starting from the approximate solution obtained by the finite element method. It should be noted that this construction technique takes advantage of the properties of the finite element solution.

Of course, we also present the other error estimators proposed in the literature and compare them to the constitutive relation error estimators.

Most error estimators do not provide information on local errors such as errors in the stresses. The construction of local error estimators is one of today's key issues on the research level. Very few works have been dedicated to this problem. Here, we present a recent theory which is an extension of the approach which led to our constitutive relation error estimators.

A significant part of this work is dedicated to the application of error estimators – whatever these estimators may be – to the control of the various parameters involved in a calculation, beginning with the parameters related to the mesh. Some examples illustrate the current state of the art.

Many developments presented here for the first time stem from recent research by the authors. Let us mention, for example, the extension of the concept of error in constitutive relation to nonlinear evolution problems and to dynamic problems, the adaptive improvements to nonlinear calculations in mechanics, the evaluation of local errors….

This work is addressed to all – students, researchers, engineers – who are interested in mechanics, from the construction of models to their simulation for industrial purposes.

The first chapter describes the reference problems, the approximate models obtained by the finite element method and the main sources of discretization errors.

Chapter 2 presents the bases of the constitutive relation error method for linear problems and outlines the techniques of construction of admissible fields. The concept of error in constitutive relation is also used to establish the theorems known as "energy theorems" which, in fact, result directly from the global formulation of the constitutive relation using overpotentials.

The other two major error estimation methods for linear problems proposed in the literature (error estimators based on the equilibrium deficiency and error indicators based on the smoothing of the finite element stresses) are presented in detail in Chapter 3.

For linear problems, simple examples of the use of the constitutive relation error measure and some elements of comparison of the global effectiveness of the various estimators are given in Chapter 4.

Chapter 5 is dedicated to the various techniques of finite element mesh adaptation. Particular emphasis is put on the "*h*" method, which is the most widely used today. Using the error estimates obtained in a preliminary calculation, it is possible to predict the element sizes necessary to achieve a predetermined level of quality. Examples of mesh adaptation in 2D and in 3D are given.

Chapter 6 (for nonlinear problems) and Chapter 7 (for vibration and transient dynamics problems) show how the constitutive relation error method enables one to derive consistent error estimates in these difficult situations.

Chapter 8 details the techniques used in the construction of admissible fields, whose central aspect is the construction of force densities on the interfaces between elements. The method is first introduced for the simpler case of 2D thermal problems, then detailed for 2D or 3D elasticity, incompressible elasticity and elastic plate problems. In this chapter, we also describe in detail an improved method of constructing the densities which increases the effectiveness of the error estimators in difficult situations (e.g. for elements with very high aspect ratios).

Chapter 9 presents recent works on the evaluation of local quantities (stresses, displacements...). Access to such estimates is crucial for industrial applications.

This book, as well as most of the corresponding work, was produced at LMT-Cachan (Ecole Normale Supérieure de Cachan/CNRS/Université Paris 6).

We thank J.-P. Combe, E. Florentin, L. Gallimardand N. Moes for their various remarks and corrections, as well as J.-P. Combe, P. Coorevits, J.-P. Dumeau, L. Gallimard, N. Moes and P. Rougeot for the preparation of some of the figures.

We wish to extend our very special thanks to our colleague J.-L. Chenot for his meticulous proofreading of this book, which we were able to improve thanks to his numerous suggestions.

April 13, 2001 Pierre Ladevèze and Jean-Pierre Pelle

Finally, we thank our colleague and friend T. Strouboulis warmly for the English translation of this book.

November 20, 2003 Pierre Ladevèze and Jean-Pierre Pelle

Chapter 1

The notion of quality of a finite element solution

1.1 Introduction

The process which leads one, starting from a physical problem, to carry out a finite element calculation is generally complex. It depends on the situations being considered and is the result of various hypotheses and simplifications made in order to represent the physical problem more or less accurately. Between the real problem and the finite element computational model, many modeling stages take place (Figure 1.1).

For example, in order to design a mechanical structure, one uses:

- a model of the geometry;
- a model of the loading case (or of the different loading cases) involved in the dimensioning;
- a model of the connections with the outside world;
- a model of the behavior of the material (or materials) the structure is made of.

Depending on the choices made in these different modeling stages, one ends up with various *mechanical models*. In most cases, the mechanical models constructed in this manner pertain to the mechanics of continuous media and lead to equations (differential equations, partial differential equations…) relating the unknowns (which in the most common cases are the displacement field $\underline{U}$ and the stress field $\underline{\underline{\sigma}}$) and the problem data (initial conditions, kinematic constraints, loads, coefficients characterizing the material behavior…).

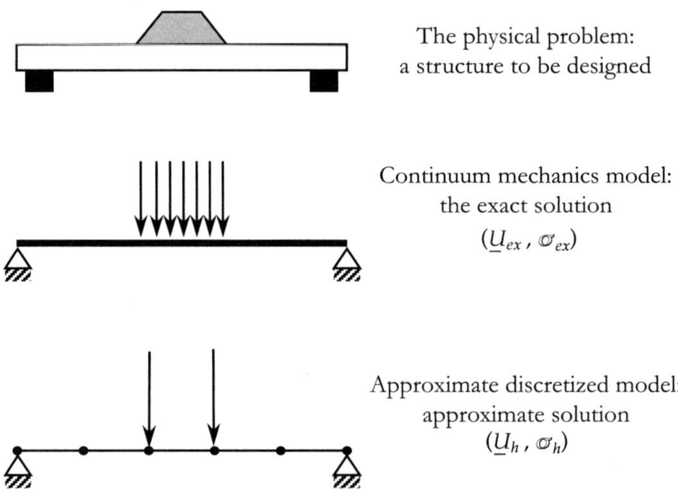

Figure 1.1. *From the physical problem to the numerical model.*

Since it is impossible, except in very simple situations, to determine the exact solution $(\underline{U}_{ex}, \underline{\underline{\sigma}}_{ex})$ of such a continuous mechanical model analytically, the user must be satisfied with approximations of $(\underline{U}_{ex}, \underline{\underline{\sigma}}_{ex})$. In practice, this is equivalent to replacing the continuous mechanical model by a simpler approximate model whose solution, considered to be an approximation of $(\underline{U}_{ex}, \underline{\underline{\sigma}}_{ex})$, can be determined. For example, the finite element method, which is currently the most widespread method for obtaining approximations, consists in replacing the continuous mechanical model by a discrete approximate model whose solution $(\underline{U}_h, \underline{\underline{\sigma}}_h)$ constitutes an approximation of $(\underline{U}_{ex}, \underline{\underline{\sigma}}_{ex})$.

Controlling the quality of the finite element calculation consists in considering the continuous mechanical model as a *reference model* and evaluating the quality of the calculated solution $(\underline{U}_h, \sigma_h)$ as an approximation of the exact solution $(\underline{U}_{ex}, \sigma_{ex})$ of that reference model.

1.2 The reference model

The medium being studied, which in our case can usually be described as a solid medium, occupies at the initial time $t = 0$ a domain Ω bounded by $\partial\Omega$.

In most of this work, we will be assuming small strains. Therefore, the various configurations occupied by the medium can be assumed to coincide with the initial configuration Ω. The evolution of the medium is studied over the time interval $[0, T]$.

We assume that the medium is subjected to a given environment which, at each instant *t*, can be typically represented (see Figure 1.2) by:

- a surface displacement field $\underline{U}_d$ on a part $\partial_1\Omega$ of the boundary $\partial\Omega$;
- a surface force density $\underline{F}_d$ on the part $\partial_2\Omega = \partial\Omega - \partial_1\Omega$;
- a volume force density $\underline{f}_d$ inside the domain Ω.

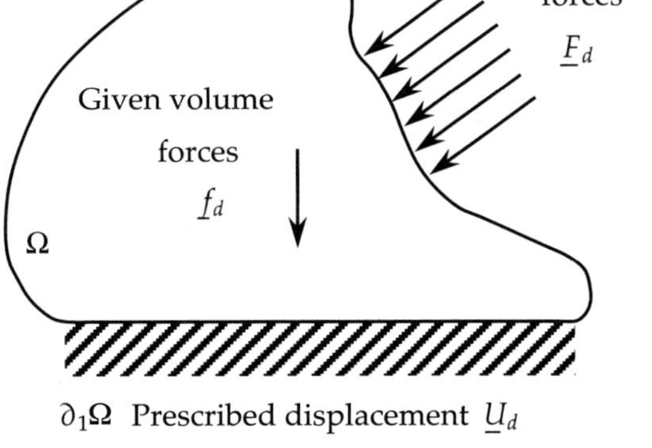

Figure 1.2. *The reference problem.*

At the initial time $t = 0$, we assume that the initial position $\underline{U}_0$ and the initial velocity $\underline{V}_0$ at any point M of Ω are also given

$$\forall M \in \Omega, \quad \underline{U}_{|t=0} = \underline{U}_0, \quad \text{and} \quad \left(\frac{d}{dt}\underline{U}\right)_{|t=0} = \dot{\underline{U}}_{|t=0} = \underline{V}_0$$

and that the reference system is Galilean.

The problem which describes the evolution of the medium in $[0, T]$ can be formulated as follows:

Find a displacement field $\underline{U}(M, t)$ *and a stress field* $\sigma(M, t)$ *defined on* $\Omega \times [0, T]$ *such that:*

- *Kinematic constraint equations and initial conditions*

$$\underline{U} \in \boldsymbol{\mathcal{U}}^{[0,T]}$$

$$\forall t \in [0, T], \quad \underline{U}_{|\partial_1\Omega} = \underline{U}_d \tag{1.1}$$

$$\forall M \in \Omega, \quad \underline{U}_{|t=0} = \underline{U}_0, \quad \dot{\underline{U}}_{|t=0} = \underline{V}_0 \tag{1.2}$$

- *Equilibrium equations*

$$\sigma \in \boldsymbol{\mathcal{S}}^{[0,T]}$$

$$\forall t \in [0, T], \quad \forall \underline{U}^* \in \boldsymbol{\mathcal{U}}_0$$

$$-\int_\Omega \mathrm{Tr}[\sigma \varepsilon(\underline{U}^*)]\, d\Omega + \int_\Omega \underline{f}_d \cdot \underline{U}^*\, d\Omega + \int_{\partial_2\Omega} \underline{F}_d \cdot \underline{U}^*\, dS = \int_\Omega \rho\, \ddot{\underline{U}} \cdot \underline{U}^*\, d\Omega \tag{1.3}$$

- *Constitutive relation*

$$\forall t \in [0, T], \quad \forall M \in \Omega$$

$$\sigma_{|t} = \mathbf{A}\left(\varepsilon(\dot{\underline{U}})_{|\tau}; \tau \in [0, t]\right) \tag{1.4}$$

In this formulation:

- the strain $\varepsilon(\underline{U})$ associated with $\underline{U}$ under the assumption of small perturbations is defined by[1]

$$\mathscr{E}(\underline{U}) = \frac{1}{2}\left[\frac{\partial \underline{U}}{\partial M} + \left(\frac{\partial \underline{U}}{\partial M}\right)^T\right]$$

or, in indicial notation[2],

$$[\mathscr{E}(\underline{U})]_{ij} = \frac{1}{2}(U_{i,j} + U_{j,i})$$

- ρ designates the mass density, which here is constant with respect to t;
- $\mathcal{U}^{[0,T]}$ designates the space in which the displacement field is sought;
- $\mathcal{S}^{[0,T]}$ designates the space in which the stress field is sought;
- $\mathcal{U}_0$ designates the space of the virtual fields chosen, which is of the form

$$\mathcal{U}_0 = \left\{\underline{U}^* \text{ “regular” and such that } \underline{U}^*{}_{|\partial_1\Omega} = 0\right\};$$

- **A** is an operator which depends on the material and characterizes its behavior; the value of the stress at time t is a function of the history of the strain rate until time t.

REMARKS

1. Relation (1.4) is a functional formulation of the constitutive behavior. For virtually all the constitutive relations commonly used, this formulation is equivalent to a formulation with internal variables using a free energy and a dissipation pseudopotential.

2. The regularity required depends on the problems being studied and is expressed in the choice of the spaces $\mathcal{U}^{[0,T]}$, $\mathcal{S}^{[0,T]}$ and $\mathcal{U}_0$.

Usually, this choice consists in imposing that the free energy and the kinetic energy of the fields being considered have finite values throughout the domain Ω at each time t in $[0, T]$. For more details, one can refer, for example, to

---

[1] The notation $\partial\underline{U}/\partial M$ designates the gradient of field $\underline{U}$. $\bullet^T$ designates the Euclidean transpose.

[2] The notation $\bullet_{,i}$ designates the partial derivative with respect to the *ith* coordinate.

[DUVAUT - LIONS, 1972], [BREZIS, 1973], [EKELAND - TEMAM, 1974], [NECAS - HLAVACEK, 1981], [DAUTRAY - LIONS, 1984], [POGU - TOURNEMINE, 1992]. Further on, we will assume that the mathematical framework chosen ensures the existence and uniqueness of the solution $(\underline{U}_{ex}, \sigma_{ex})$ of the reference problem. Let us note that a very general uniqueness condition can be found in [LADEVEZE, 1996].

3. The equilibrium equations (1.3) are written in global form; under the condition of regularity, they are equivalent to the local equations

$$\begin{aligned} &\forall t \in [0, T] \\ &\underline{\text{div}}\ \sigma + \underline{f}_d = \rho\, \ddot{\underline{U}} \quad \text{in } \Omega \\ &\sigma \underline{n} = \underline{F}_d \qquad\qquad \text{on } \partial_2\Omega \end{aligned} \tag{1.5}$$

where $\underline{n}$ denotes the unit outward normal to the boundary $\partial\Omega$.

4. We are using the language and notations of structural mechanics. However, the concepts and methods we will present are very general and can be applied, with some relatively simple adjustments, to numerous other physical situations.

Two simplified versions of the reference model [Equations (1.1) - (1.4)] will be used especially throughout this work.

The first, very simple version is that of linear statics in which the data $\underline{U}_d$, $\underline{f}_d$ and $\underline{F}_d$ and the unknown fields $\underline{U}$ and σ are independent of time, and the constitutive relation is a linear relation between the stress σ and the strain ε (linear elasticity). Of course, in this case, the initial conditions (1.2) are irrelevant and one is interested in the final configuration at time T. Thus, the reference problem becomes:

Find a displacement field $\underline{U}(M)$ *and a stress field* $\sigma(M)$ *defined on* Ω *such that:*

- *Kinematic constraint equations*

$$\underline{U} \in \boldsymbol{\mathcal{U}} \quad \text{and} \quad \underline{U}_{|\partial_1\Omega} = \underline{U}_d \tag{1.6}$$

- *Equilibrium equations*

$$\underline{\underline{\sigma}} \in \boldsymbol{S}, \quad \forall \underline{U}^* \in \boldsymbol{U}_0$$

$$\int_{\Omega} \mathrm{Tr}\left[\underline{\underline{\sigma}}\,\underline{\underline{\varepsilon}}(\underline{U}^*)\right] d\Omega = \int_{\Omega} \underline{f}_d \cdot \underline{U}^* \, d\Omega + \int_{\partial_2 \Omega} \underline{F}_d \cdot \underline{U}^* \, dS \tag{1.7}$$

- *Constitutive relation*

$$\underline{\underline{\sigma}} = \mathbf{K}\underline{\underline{\varepsilon}}(\underline{U}) \tag{1.8}$$

In this formulation, $\mathbf{K}$ designates the HOOKE operator of the material. For example, for an isotropic elastic material, one has

$$\mathbf{K}\underline{\underline{\varepsilon}} = \lambda \,\mathrm{Tr}(\underline{\underline{\varepsilon}})\mathbf{I} + 2\mu\underline{\underline{\varepsilon}}$$

where $\mathbf{I}$ is the identity operator and λ and μ are two LAME coefficients classically connected to the YOUNG modulus E and the POISSON ratio ν by the relations

$$\lambda = \frac{\nu E}{(1+\nu)(1-2\nu)} \qquad \text{and} \qquad \mu = \frac{E}{2(1+\nu)}$$

The spaces $\boldsymbol{U}$ and $\boldsymbol{S}$ are homologous to the spaces $\boldsymbol{U}^{[0,T]}$ and $\boldsymbol{S}^{[0,T]}$ for time-independent fields. For example, for linear static problems in 3D, one has

$$\boldsymbol{U} = \left[H^1(\Omega)\right]^3; \quad \boldsymbol{S} = \left[L^2(\Omega)\right]^6 \text{ and } \boldsymbol{U}_0 = \left\{\underline{U}^* \in \left[H^1(\Omega)\right]^3 / \underline{U}^*|_{\partial_1\Omega} = 0\right\}$$

where $L^2(\Omega)$ designates the space of the square-integrable functions in Ω, and $H^1(\Omega)$ the SOBOLEV space of the functions of $L^2(\Omega)$ with derivatives in the space $L^2(\Omega)$.

The second simplified version corresponds to the case where the acceleration terms can be considered negligible.

Under these conditions, one obtains the problem:

Find a displacement field $\underline{U}(M,t)$ *and a stress field* $\underline{\underline{\sigma}}(M,t)$ *defined on* $\Omega \times [0,T]$ *such that*

- *Kinematic constraint equations and initial conditions*

$$\underline{U} \in \boldsymbol{U}^{[0,T]}$$

$$\forall t \in [0, T], \quad \underline{U}_{|\partial_1\Omega} = \underline{U}_d \tag{1.9}$$

$$\forall M \in \Omega, \qquad \underline{U}_{|t=0} = \underline{U}_0, \quad \dot{\underline{U}}_{|t=0} = \underline{V}_0 \tag{1.10}$$

- *Equilibrium equations*

$$\sigma \in \boldsymbol{S}^{[0,T]}$$

$$\forall t \in [0, T], \quad \forall \underline{U}^* \in \boldsymbol{U}_0$$

$$\int_\Omega \mathrm{Tr}[\sigma \varepsilon(\underline{U}^*)]\, d\Omega = \int_\Omega \underline{f}_d \cdot \underline{U}^*\, d\Omega + \int_{\partial_2\Omega} \underline{F}_d \cdot \underline{U}^*\, dS \tag{1.11}$$

- *Constitutive relation*

$$\forall t \in [0, T], \quad \forall M \in \Omega$$

$$\sigma_{|t} = \mathbf{A}\left(\varepsilon(\dot{\underline{U}})_{|\tau}\,;\, \tau \in [0, t]\right) \tag{1.12}$$

This is called a "quasi-static" problem. Plasticity and viscoplasticity problems are often treated in this framework.

1.3 The approximate problem and discretization errors

Except in very exceptional cases, the solution $(\underline{U}_{ex}, \sigma_{ex})$ to the reference problem [Equations (1.1) - (1.4)], even in its simplest form [Equations (1.6) - (1.8)], cannot be obtained explicitly. In practice, by introducing simplifying hypotheses called *condensation hypotheses*, the reference problem is replaced by an approximate, simpler problem whose solution, which we will designate by $(\underline{U}_h, \sigma_h)$, can be obtained numerically. Of course, there are numerous methods for constructing approximate problems; these depend not only on the nature of the problem being considered, but also on the types of results being sought.

We are going to examine the cases of linear problems and of quasi-static nonlinear evolution problems, limiting ourselves to the most commonly used approximation methods. Here, for two characteristic cases, our objective is to emphasize the impact of a numerical approximation on the quality of the calculated approximate solution.

1.3.1 Linear problems

Today, the method used most often to obtain approximations of Equations [(1.6) - (1.8)] is the finite element displacement method.

The finite element displacement method

Let us formulate Problem [(1.6) - (1.8)] using the potential energy[3]

$$E_P(\underline{U}) = \frac{1}{2}\int_\Omega \mathrm{Tr}\left[\varepsilon(\underline{U})\mathbf{K}\varepsilon(\underline{U})\right] d\Omega - \left[\int_\Omega \underline{f}_d \cdot \underline{U}\, d\Omega + \int_{\partial_2\Omega} \underline{F}_d \cdot \underline{U}\, dS\right]$$

Indeed, the displacement $\underline{U}_{ex}$ is the solution of the minimization problem

$$E_P(\underline{U}_{ex}) = \underset{\underline{U}\ KA}{\mathrm{Min}}\ E_P(\underline{U}) \tag{1.13}$$

where $\underline{U}\ KA$ designates the Kinematically Admissible displacement fields, i.e. the fields which verify the kinematic constraint equations (1.6). Then, the solution stress field σ_{ex} is obtained through the constitutive relation

$$\sigma_{ex} = \mathbf{K}\varepsilon(\underline{U}_{ex})$$

The finite element displacement method consists in seeking the minimum of the potential energy only on a finite-dimension subspace of *KA* displacement fields rather than on the set of all the *KA* fields. Thus, the approximate field $\underline{U}_h$ is the solution of the problem

$$E_P(\underline{U}_h) = \underset{\underline{U}\ KA \text{ and of } FE \text{ type}}{\mathrm{Min}}\ E_P(\underline{U}) \tag{1.14}$$

or, in terms of extremum conditions:

Find a displacement field $\underline{U}_h$ *defined on* Ω *such that:*

$\underline{U}_h \in \boldsymbol{\mathcal{U}}_h$ *and* $\forall \underline{U}_h^* \in \boldsymbol{\mathcal{U}}_{h0}$

$$\int_\Omega \mathrm{Tr}\left[\varepsilon(\underline{U}_h)\mathbf{K}\varepsilon(\underline{U}_h^*)\right] d\Omega = \int_\Omega \underline{f}_d \cdot \underline{U}_h^*\, d\Omega + \int_{\partial_2\Omega} \underline{F}_d \cdot \underline{U}_h^*\, dS \tag{1.15}$$

where $\boldsymbol{\mathcal{U}}_h$ is the affine finite-dimension subspace of $\boldsymbol{\mathcal{U}}$ chosen and $\boldsymbol{\mathcal{U}}_{h0}$ is the associated vector subspace, which is a subspace of $\boldsymbol{\mathcal{U}}_0$.

---

[3] An introduction to the potential energy is proposed in Chapter 2.

The stress field σ_h is then obtained, element by element, through the constitutive relation

$$\sigma_h = \mathbf{K}\varepsilon(\underline{U}_h) \tag{1.16}$$

In practice, Problem (1.15) is expressed by the linear system

$$\mathbb{K}\mathbf{q} = \mathbf{F} \tag{1.17}$$

where:

- $\mathbf{q}$ is the vector of nodal displacements (degrees of freedom);
- $\mathbb{K}$ is the stiffness matrix;
- $\mathbf{F}$ is the vector of generalized loads.

For example, for $\underline{U}_d = 0$, one has

$$\mathbb{K} = \int_\Omega \mathbf{B}^T\mathbf{K}\mathbf{B}\, d\Omega\,; \qquad \mathbf{F} = \int_\Omega \mathbf{N}^T \underline{f}_d\, d\Omega + \int_{\partial_2\Omega} \mathbf{N}^T \underline{F}_d\, d\Gamma \tag{1.18}$$

with

$$\underline{U}_h(M) = \mathbf{N}(M)\mathbf{q} \quad \text{and} \quad \varepsilon(\underline{U}_h(M)) = \mathbf{B}(M)\mathbf{q} \tag{1.19}$$

where $\mathbf{N}$ is the matrix of shape functions.

For more details on the finite element displacement method, the reader may refer, for example, to [BATHE, 1982], [IMBERT, 1984], [HUGHES, 1987], [BATOZ - DHATT, 1990], [ZIENKIEWICZ - TAYLOR, 1988], and for the stochastic aspects to [KEIBER - HEIN, 1992].

Nonsatisfaction of the equilibrium equations

If one compares the reference problem and the approximate problem, one can observe that the approximate solution $(\underline{U}_h, \sigma_h)$ verifies, as does the exact solution $(\underline{U}_{ex}, \sigma_{ex})$, the kinematic constraint equations and the constitutive relation. However, the field σ_h does not verify the equilibrium equations:

> *In the displacement-type finite element method, the main approximation concerns the equilibrium equations.*

More specifically, the stress $\underline{\underline{\sigma}}_h$ presents three types of equilibrium deficiencies:

- the interior equilibrium equation is not verified

$$\underline{\text{div}}\,\underline{\underline{\sigma}}_h + \underline{f}_d \neq 0 \quad \text{in } \Omega \tag{1.20}$$

- the stress vector is not in equilibrium with the applied loads

$$\underline{\underline{\sigma}}_h \underline{n} \neq \underline{F}_d \quad \text{on } \partial_2\Omega \tag{1.21}$$

- the stress vector is discontinuous at the interface between two elements

$$\left[\underline{\underline{\sigma}}_h \underline{n}\right]_{|E_1} + \left[\underline{\underline{\sigma}}_h \underline{n}\right]_{|E_2} \neq 0 \quad \text{on } \Gamma_{E_1 E_2} \tag{1.22}$$

with the notations defined in Figure 1.3.

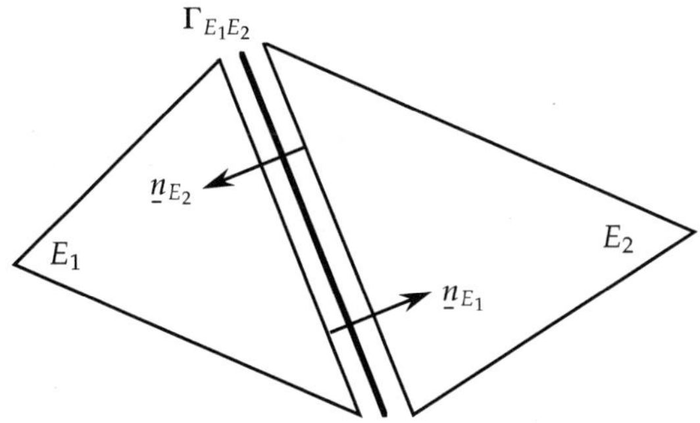

Figure 1.3. *Notations at the interface between two elements.*

Other sources of discretization errors

In fact, the above equilibrium deficiencies are not the only possible sources of errors. Depending on the continuum mechanics model being used, other sources of error exist or can exist:

- *Errors due to nonrespect of the geometry:* For example, the border of a circular plate is replaced by a polygon if one uses elements with straight edges, or by parabolic arcs if one uses isoparametric elements of degree 2.
- *Errors due to the approximation of the displacement boundary conditions:* If the prescribed displacements are nonzero, it is necessary, in order for the fields of

the finite element type to be *KA*, that the prescribed $\underline{U}_d$ be compatible with the type of element chosen; for example, on the boundary of an element in $\partial_1\Omega$, the given displacement $\underline{U}_d$ should be linear for first-order elements, or parabolic for second-order elements.

- *Errors due to the approximation of the applied loads:* In practice, the loads actually taken into account in finite element programs are approximations of the real loads; for example, for first-order elements, the loads $\underline{f}_d$ are assumed to be constant within each element.
- *Errors due to the numerical treatment of the approximate problem:* Numerical integration errors during the calculation of $\mathbb{K}$ and $\mathbf{F}$, errors during the resolution of the linear system (errors due to the ending of the iterations if an iterative method is being used, and, in all cases, roundoff errors).

In a preliminary stage, the first three types of errors can be disregarded, which is equivalent to assuming that the corresponding approximations have been made on the level of the continuum mechanics reference model. For more details on these types of errors, the reader could refer, for example, to [STRANG - FIX, 1976], [CIARLET, 1978]. The last type is of a different nature and cannot be avoided. However, considering the precision of modern computers, in numerous everyday situations and for most of the problems that we are considering here, these errors are completely negligible compared to the discretization errors due to the nonrespect of the equilibrium equations of the reference model. Nevertheless, methods of evaluation – or, at least, detection – of this type of error exist, for example, the method of [LA PORTE - VIGNE, 1974], which is stochastic. In practice, it is therefore necessary to perform several calculations in order to propagate the roundoff errors differently [DAUMAS - MULLER, 1997].

Remarks and comments

The finite element displacement method as described in the previous paragraphs is used in mechanics for the resolution of 2D and 3D linear elasticity problems. One should note that 2D and 3D thermal equilibrium problems also fall within the domain of application of these types of methods.

Nevertheless, there are also many other finite element methods: "equilibrium" methods based on the minimization of the complementary energy, originally developed by FRAEIJS DE VEUBEKE [FRAEIJS DE VEUBEKE, 1965], [FRAEIJS DE VEUBEKE - SANDER, 1968], [FRAEIJS DE VEUBEKE - HOGGE, 1970], as well as numerous variations of mixed methods based on more or less sophisticated mixed principles [WASHIZU, 1975], [ZIENKIEWICZ - TAYLOR, 1988], [VALID, 1995]. Let us make a special mention of the family of elements developed by JIROUSEK [JIROUSEK, 1985], [JIROUSEK - WROBLESKI, 1996].

For structures of the beam, plate and shell types, the numerous finite elements proposed in the literature can be interpreted in the framework of mixed formulations [VALID, 1995], [BATOZ - DHATT, 1990], [CRISFIELD, 1991].

Another large family of finite elements is that of boundary elements associated with the integral equation method, which is extremely effective for 3D homogeneous and isotropic media [BREBBIA - TELLES - WROBEL, 1984], but loses much of its interest outside of these assumptions, despite some exceptions, such as [BONNET, 1999].

1.3.2 Nonlinear problems

Here, we will consider the example of a nonlinear quasi-static problem [Equations (1.9) - (1.12)]. The classical treatment of this type of problem by the incremental method requires both a space discretization and a time discretization.

Discretization of the problem

In this section, to simplify the presentation, we will assume that the prescribed displacements are zero $\underline{U}_d = 0$.

Discretization in space

The displacement field is sought in the form

$$\underline{U}_h(M, t) = \mathbf{N}(M)\mathbf{q}(t) \tag{1.23}$$

where **N** is the matrix of (given) shape functions corresponding to a discretization of the finite element type and **q** is a vector of (unknown) functions of time which represent the nodal displacements at each instant.

Thus, one obtains the approximate problem:

Find $\underline{U}_h$ *of the finite element type and* σ_h *such that:*

$$\forall t \in [0, T], \quad \underline{U}_h \textit{ is KA and satisfies the initial conditions} \tag{1.24}$$

$$\forall t \in [0, T] \quad \textit{and} \quad \forall \underline{U}_h^* \in \boldsymbol{\mathcal{U}}_{h0}$$

$$\int_\Omega \mathrm{Tr}\left[\sigma_h \, \varepsilon(\underline{U}_h^*)\right] d\Omega = \int_\Omega \underline{f}_d \cdot \underline{U}_h^* \, d\Omega + \int_{\partial_2\Omega} \underline{F}_d \cdot \underline{U}_h^* \, dS \tag{1.25}$$

$$\forall t \in [0, T], \quad \forall M \in \Omega, \quad \sigma_{h|t} = \mathbf{A}\left(\varepsilon(\dot{\underline{U}}_h)_{|\tau}\, ; \tau \in [0, t]\right) \tag{1.26}$$

This problem is discrete in space, but continuous in time.

Discretization in time

In order to solve Equations [(1.24) - (1.26)] in an approximate way, one discretizes the interval $[0, T]$ by subdividing it into p subintervals $[t_n, t_{n+1}]$,

$$0 = t_0 < t_1 < t_2 < \quad \cdots \quad < t_p = T$$

The solution is determined step by step.

Assuming the approximate solution to be known until time t_n, the problem consists in determining the displacement and stress over the increment $[t_n, t_{n+1}]$. Since each interval has been chosen "small", one usually makes the hypothesis that the displacement varies linearly over $[t_n, t_{n+1}]$,

$$\underline{U}_h(M, t) = \left[\underline{U}_h(M, t_{n+1}) - \underline{U}_h(M, t_n)\right] \frac{t - t_n}{t_{n+1} - t_n} + \underline{U}_h(M, t_n)$$

i.e. in finite element notation

$$\mathbf{q}(t) = \left[\mathbf{q}(t_{n+1}) - \mathbf{q}(t_n)\right] \frac{t - t_n}{t_{n+1} - t_n} + \mathbf{q}(t_n) \tag{1.27}$$

The only unknown is then the displacement $\mathbf{q}(t_{n+1})$ at t_{n+1}.
To determine $\mathbf{q}(t_{n+1})$, one expresses Equilibrium (1.25) at t_{n+1}, which leads to

the problem:

$$\textit{Find } \mathbf{q}_{n+1} = \mathbf{q}(t_{n+1}) \textit{ and } \sigma_{h,\, n+1} = \sigma_h(t_{n+1}) \textit{ such that:}$$

$$\int_\Omega \mathbf{B}^T \sigma_{h,\, n+1}\, d\Omega = \mathbf{F}_{n+1} \quad \textit{and} \quad \sigma_{h,\, n+1} = \mathbf{A}\Big[\mathbf{B}\, \dot{\mathbf{q}}(\tau)\,|_{\tau \le t_{n+1}}\Big] \tag{1.28}$$

with

$$\mathbf{F}_{n+1} = \int_\Omega \mathbf{N}^T \underline{f}_d(t_{n+1})\, d\Omega + \int_{\partial_2\Omega} \mathbf{N}^T \underline{F}_d(t_{n+1})\, dS$$

Since the history of the strain rate until t_{n+1} depends on $\mathbf{q}_{n+1}$ alone, so does $\sigma_{h,\, n+1}$. Therefore, (1.28) is a nonlinear equation whose unknown is $\mathbf{q}_{n+1}$ and in which time is no longer present.

This equation is solved in an approximate way by an iterative method, usually a NEWTON-type method.

REMARK. One can also make more complex assumptions for the time variation of the displacement field over the increment: quadratic variation, cubic, etc. In such cases, one must also write the equilibrium at a certain number of intermediate times.

Origin of the discretization errors

The first consequence of the space discretization is, as in linear problems, that the calculated stress fields do not verify the equilibrium equations of the reference model, but, at best, verify only the weak equations in the finite element sense.

The time discretization introduces new approximations:

- the hypothesis on the time variation of the displacement field over the increment constitutes a first approximation which, if the time steps chosen are too large, can be a source of significant error;
- the solution of the nonlinear problem (1.28) over each increment by an iterative method also introduces approximations; when one ends the iterations, Equation (1.28) is not verified exactly;
- in the course of the above iterations, one must integrate the constitutive

relation, using a numerical scheme, for a given strain history; this numerical integration introduces approximations which can be significant if one uses an insufficiently accurate integration scheme;

- the stress field $\underline{\underline{\sigma}}_{h,\, n+1}$ sought at the end of an increment is known only through its values at a certain number of GAUSS points within each element.

1.3.3 Evaluation of the discretization errors

The evaluation of discretization errors has always been a major preoccupation of users of the finite element method. Here, we present the main lines of the various methods which have been proposed to attempt to estimate these discretization errors. The most popular methods will be detailed in the following chapters. It should be noted that most of the studies carried out on this matter concern linear problems.

1.3.4 A priori estimates

In many occurrences of linear problems in statics, the functional analysis and numerical analysis lead, under certain regularity assumptions, to *a priori estimates* of the form

$$\left\| \underline{U}_{ex} - \underline{U}_h \right\| \leq E(h, D, \underline{U}_{ex}) \tag{1.29}$$

where $\| \bullet \|$ is a norm of the displacement fields and E is a function of size h of the elements, of the problem's data D (applied forces, prescribed displacements...) and of the exact solution $\underline{U}_{ex}$. If the function E is such that

$$\lim_{h \to 0} E(h, D, \underline{U}_{ex}) = 0$$

this means that the finite element method used converges. If one can show hat there exists a real number $q > 0$ such that

$$E(h, D, \underline{U}_{ex}) \leq h^q \, \tilde{E}(D, \underline{U}_{ex}) \tag{1.30}$$

one also has an indication of the convergence rate as a function of the mesh size.

However, since these estimates involve the exact solution, they cannot be calculated explicitly and, thus, they provide no information on the magnitude of the discretization error.

REMARKS

1. The estimates (1.30) are asymptotic results which are proven only as h tends toward 0. They often yield extremely rough evaluations for finite values of h, particularly for coarse meshes.

2. The proof of the a priori estimates involves the regularity properties of the solution and of the meshes. For further details, the reader may refer, for example, to [AZZIZ - BABUSKA, 1972] and [CIARLET, 1978].

1.3.5 A posteriori estimates

Principle of the classical error estimators

The basic idea consists in using the approximate solution to evaluate the discretization error. Contrary to a priori estimates, these estimates can be used only once the approximate solution has been calculated, and are therefore called a posteriori estimates.

The great many suggestions which have been made and continue to be made in this area can be schematically classified into three categories:

- error measures based on the concept of error in constitutive relation and on specific techniques for constructing admissible fields [LADEVEZE, 1975...]; these are developed in this work.
- error estimators based on the equilibrium deficiencies of the finite element solution [BABUSKA - RHEINBOLDT, 1978...]; these estimators, which have been the subject of many numerical analysis works, are presented in Chapter 3.
- error indicators based on the unevenness of the finite element solution [ZIENKIEWICZ - ZHU, 1987...]; these indicators are also presented in Chapter 3.

All these approaches provide the user with an estimate $\mathbf{e}$ of the global discretization error, whose accuracy depends on the type of method, and an estimate $\mathbf{e}_E$ of the contribution of each element E of the mesh being used to the global error. Usually, these two estimates are related by

$$\mathbf{e}^2 = \sum_E \mathbf{e}_E^2 \tag{1.31}$$

Classically, one associates these "absolute" errors with "relative" errors $\varepsilon = \mathbf{e}/\mathbf{d}$ and $\varepsilon_E = \mathbf{e}_E/\mathbf{d}$, where $\mathbf{d}$ is a quantity chosen for the purpose of normalizing these various estimates.

Classification of a posteriori error estimators

It is interesting to analyze the reasoning which leads to a posteriori estimators of the error. The first step consists in associating with the finite element solution $(\underline{U}_h, \sigma_h)$ a new pair $(\hat{\underline{U}}_h, \hat{\sigma}_h)$ which is considered to be a better representation of the exact solution.

A first series of error estimators is obtained by requiring that $(\hat{\underline{U}}_h, \hat{\sigma}_h)$ verify two of the three equations of the reference problem. Thus, the associated estimator is a measure of the residual of the equation which is not verified a priori. The estimators based on the error in constitutive relation belong to this category; in this case, it is the constitutive relation which is not verified a priori. The estimators based on the equilibrium deficiencies of the finite element solution are in the same group; in this case, the constitutive relation and the kinematic constraints (connections and compatibility conditions) are verified exactly.

Another family of estimators can be obtained by requiring $(\hat{\underline{U}}_h, \hat{\sigma}_h)$ to verify only one of the equations of the reference problem exactly. Then, the estimator measures the quality of satisfaction of the other two equations. The proposal from [DEBONGNIE - ZHONG - BECKERS, 1995], [DEBONGNIE - BECKERS, 1999] belongs to this category; the equation which is verified exactly is the equilibrium equation and the residuals concern the constitutive relation and the compatibility equations.

The family of estimators based on the unevenness of the solution constitutes a group in which none of the equations of the reference problem is verified exactly.

1.3.6 Qualities of an estimator

The various methods developed for constructing a posteriori estimators of the error differ, of course, in the techniques used, but also in the quality of the results obtained. Many criteria can be used to compare these different methods:

- Intrinsic quality criteria which enable the measured error to be compared with the "true" error actually made.
- Reliability criteria: Is the estimated error an upper bound of the "true" error?
- Sensitivity to particular mesh configurations: Coarse meshes, elements with high aspect ratios, distorted elements....
- Sensitivity to particular types of behavior: Material anisotropy....
- Economic criteria: Ease of implementation, cost of utilization....
- Extent of the domain of application.

We will now present some intrinsic quality criteria which are generally accepted today. Other aspects will be addressed later in this work for the various types of estimators which we will introduce.

Exhaustive studies of the quality and behavior of error estimators can be found in [ODEN - DEMKOWICZ - RACHOWICZ - WESTERMANN, 1989], [BECKERS - ZHONG, 1991], [STROUBOULIS - HAQUE, 1992a, 1992b], [BABUSKA - STROUBOULIS - UPADHYAY - GANGARAJ, 1994], [BABUSKA - STROUBOULIS - UPADHYAY - GANGARAJ - COPPS, 1994].

Effectivity index

The effectivity index is defined as

$$\gamma = \frac{\mathbf{e}}{\mathbf{e}_{ex}} = \frac{\text{estimated measure of the global error}}{\text{measure of the true global error}} \tag{1.32}$$

In practice, for a good estimator, the global effectivity index should be close to one. The most reliable method for evaluating γ consists in using test problems whose exact solutions are known analytically. Another widely used technique consists in using as the "exact" solution a finite element solution obtained with a very fine mesh. In order to obtain reliable results, it is essential that this mesh be much finer than that for which the error is to be estimated, and some precautions must be taken when transferring fields from one mesh to the other.

Similarly, the local effectivity index in an element (or in a group of elements) is defined as

$$\gamma_E = \frac{\mathbf{e}_E}{\mathbf{e}_{ex}|_E} = \frac{\text{estimated measure of the error on E}}{\text{measure of the true error on E}} \tag{1.33}$$

Asymptotic behavior

For some error estimators, there are two positive constants C_{min} and C_{max} independent of the sizes of the elements such that

$$C_{min} \le \gamma \le C_{max} \tag{1.34}$$

These inequalities show that the "true" error and the error estimate being used tend toward zero simultaneously.

Moreover, if an inequality such as (1.30) is available, one has

$$\mathbf{e} \le C\, h^q \tag{1.35}$$

This inequality, which provides information on the convergence rate of the error estimator as a function of the mesh size, is very useful in adapting the meshes.

REMARKS

1. Of course, it is preferable that constants C_{min} and C_{max} be close to 1. From a reliability point of view, it is obviously advantageous to use estimators such that

$$C_{min} = 1 \tag{1.36}$$

2. An even more desirable property would be to have

$$\lim_{h \to 0} \gamma = 1 \tag{1.37}$$

Unfortunately, it seems that none of today's proposed estimators verifies this condition, except in very specific situations.

1.4 Extensions and general references

This work focuses mainly on problems encountered in the mechanics of solids and structures. However, the methods presented can be applied to many other situations. For formulations in fluid mechanics, one can consult [TEMAM, 1977], [POGU - TOURNEMINE, 1992] and, more generally for the mechanics of continuous media, [MANDEL, 1966], [TRUSDELL, 1971], [GERMAIN, 1973], [SALENÇON, 1988], [DESTUYNDER, 1991], [VALID, 1995], [ROUGEE, 1997]. Fluid-structure interaction problems and their numerical treatment are described in particular in [MORAND - OHAYON, 1992] and [LEWIS - SCHREFLER, 1998]. For acoustic problems, one can refer to [BRUNEAU, 1999], and for dynamic problems to [GIBERT, 1988] and [GERADIN - RIXEN, 1991].

Chapter 2

The constitutive relation error method for linear problems

2.1 Introduction

The reference problem considered here is the linear problem described by Equations [(1.6) - (1.8)]. It is discretized by the finite element displacement method. In order to evaluate the quality of a finite element calculation, one must:

- define the notion of error;
- define measures of this error.

The definition of the error is closely connected to the definition of the notion of approximate solution. Classically, in the framework of the finite element displacement method, the displacement field plays a special role and the error is defined as the difference between the exact displacement field and the finite element displacement field

$$\underline{e}_h = \underline{U}_{ex} - \underline{U}_h \qquad (2.1)$$

From this definition, one can obviously deduce an error in the stresses

$$\sigma_{ex} - \sigma_h = \mathbf{K}\varepsilon(\underline{e}_h) \tag{2.2}$$

The major drawback of this approach from a mechanical point of view is that one considers as an approximation of the exact pair $(\underline{U}_{ex}, \sigma_{ex})$ a pair $(\underline{U}_h, \sigma_h)$ whose stress part does not verify the equilibrium equations – the very equations which constitute one of the foundations of the formulation. This observation is at the root of the introduction of the notion of error in constitutive relation presented for the first time in [LADEVEZE, 1975]. The works related to this early period have been collected in [LADEVEZE, 1995]. This approach is based on a partitioning of the equations of the problem to be solved into two groups:

- the admissibility conditions: kinematic constraint equations, equilibrium equations and initial conditions;
- the constitutive relation.

Indeed, the constitutive relation has a special status; in practice, it is often the least reliable equation. Therefore, it is natural to set this equation apart and to ensure that an approximate displacement-stress solution over the time interval of interest verifies the group of the most reliable equations (i.e. the admissibility conditions) exactly. This approximate solution verifies a certain constitutive relation which, in general, is different from the constitutive relation of the material. Thus, the quality of the approximation can be evaluated by comparing this constitutive relation with that of the material. In order to quantify this error, energy-type norms or other norms which have a strong physical meaning are used.

The difficult part, which is a priori an obstacle, is to develop admissible approximate solutions, i.e. solutions which satisfy the conditions of admissibility exactly. Indeed, the usual approximations, including those obtained by the finite element method, do not satisfy these conditions because the stress σ_h is not in equilibrium with the given loads. For the finite element method, the completely new technique proposed in [LADEVEZE, 1975, 1977] and subsequently expanded enables one to circumvent this difficulty. For linear problems, one can refer particularly to [LADEVEZE - LEGUILLON, 1983], [LADEVEZE - PELLE - ROUGEOT, 1991], [LADEVEZE - ROUGEOT, 1997]. This technique is very general; it enables the explicit (therefore inexpensive) construction of an admissible solution starting from the solution calculated by the finite element method and taking advantage of its properties. In particular, this method does not depend on a finite element computation by the dual approach developed by FRAEIJS DE VEUBEKE [FRAEIJS DE VEUBEKE, 1965]. In [DEBONGNIE - BECKERS, 1999], the approach described here is called the *non-conventional dual approach.*

The global error in constitutive relation is related to the classical error between the exact solution and the finite element solution by the well-known PRAGER - SYNGE theorem [PRAGER - SYNGE, 1947]. Its potential in the domain of error evaluation has been known for a long time [TOTTENHAM, 1970], [AUBIN - BURCHARD, 1970]. Of course, the concept of error in constitutive relation is much more general and mechanically very different. Therefore, it should not be confused with the approach underlying the PRAGER - SYNGE hypercircle theorem; for example, the concept of error in constitutive relation is definitely not limited to linear problems.

Several review articles have presented the state of progress of the general concept [LADEVEZE, 1990, 1992], [PELLE, 1995]. This chapter outlines the approach which leads to estimators of the error in constitutive relation. The more technical aspects will be detailed in Chapter 8.

2.2 Error in constitutive relation

2.2.1 Admissible approximate solution

In accordance with what was said in the Introduction, an approximate solution of the reference problem is, by definition, an admissible displacement-stress pair $(\hat{\underline{U}}, \hat{\sigma})$, i.e. a pair such that:

- $\hat{\underline{U}}$ is kinematically admissible

$$\hat{\underline{U}} \in \boldsymbol{\mathcal{U}} \quad \text{and} \quad \hat{\underline{U}}_{|\partial_1\Omega} = \underline{U}_d \tag{2.3}$$

- $\hat{\sigma}$ is statically admissible

$$\hat{\sigma} \in \boldsymbol{S} \quad \text{and} \quad \forall \underline{U}^* \in \boldsymbol{\mathcal{U}}_0$$

$$\int_{\Omega} \mathrm{Tr}\left[\hat{\sigma}\varepsilon(\underline{U}^*)\right] d\Omega = \int_{\Omega} \underline{f}_d \cdot \underline{U}^* \, d\Omega + \int_{\partial_2\Omega} \underline{F}_d \cdot \underline{U}^* \, dS \tag{2.4}$$

Under these conditions, the approximate character of an admissible displacement-stress pair $(\hat{\underline{U}}, \hat{\sigma})$ resides in the nonsatisfaction of the constitutive relation.

2.2.2 Error in constitutive relation in the linear case

Let $(\hat{\underline{U}}, \hat{\sigma})$ be an admissible pair. If it verifies the constitutive relation $\hat{\sigma} - \mathbf{K}\varepsilon(\hat{\underline{U}}) = 0$, then this pair is the exact solution of the reference problem

$$\hat{\underline{U}} = \underline{U}_{ex} \quad \text{and} \quad \hat{\sigma} = \sigma_{ex}$$

If the pair does not verify the constitutive relation $\hat{\sigma} - \mathbf{K}\varepsilon(\hat{\underline{U}}) \neq 0$, then $(\hat{\underline{U}}, \hat{\sigma})$ is an approximation of $(\underline{U}_{ex}, \sigma_{ex})$, and the manner in which $(\hat{\underline{U}}, \hat{\sigma})$ verifies the constitutive relation makes it possible to assess its quality. One is thus led to introduce the following definition:

Definition

One calls error in constitutive relation associated with the admissible pair $(\hat{\underline{U}}, \hat{\sigma})$ *the quantity defined at all points of* Ω *by*

$$e_{RdC} = \hat{\sigma} - \mathbf{K}\varepsilon(\hat{\underline{U}}) \tag{2.5}$$

Thus, the error in constitutive relation e_{RdC} is a quantity of the stress type.

This approach has certain advantages. The first advantage is that it directs the doubt to the constitutive relation, i.e. the equation of the reference problem which is the least reliable. The second advantage is that, in contrast with classical errors of the $\underline{e}_h$ type, the expression of e_{RdC} does not contain the exact solution of the reference problem. Therefore, in order to calculate the error in constitutive relation, it is not necessary to know the exact solution of the problem: the knowledge of the approximate solution is sufficient! Finally, in the context of linear problems, the error measures show up very naturally.

However, the implementation of this approach requires the knowledge of an admissible displacement-stress pair. Therefore, in order to evaluate the error in constitutive relation, it is necessary to reconstruct an admissible pair by post-processing the solution of the approximate model, which is generally not an admissible pair; we will see that this step poses certain technical difficulties.

Measures associated with the error

Classically, the error in constitutive relation is measured using the energy norm on the structure being considered. We will therefore define the absolute global error as

$$\mathbf{e}_{RdC}(\underline{\hat{U}}, \hat{\sigma}) = \left\| \hat{\sigma} - \mathbf{K}\varepsilon(\underline{\hat{U}}) \right\|_{\sigma,\Omega} \tag{2.6}$$

with

$$\| \cdot \|_{\sigma,\Omega} = \left[\int_{\Omega} \mathrm{Tr}(\cdot \mathbf{K}^{-1} \cdot)\, d\Omega \right]^{1/2} \tag{2.7}$$

When there is no risk of confusion, we will simply write $\mathbf{e}_{RdC}$ instead of $\mathbf{e}_{RdC}(\underline{\hat{U}}, \hat{\sigma})$.

To the above global error, one associates the relative error

$$\varepsilon = \frac{\left\| \hat{\sigma} - \mathbf{K}\varepsilon(\underline{\hat{U}}) \right\|_{\sigma,\Omega}}{\left\| \hat{\sigma} + \mathbf{K}\varepsilon(\underline{\hat{U}}) \right\|_{\sigma,\Omega}} \tag{2.8}$$

Thus, $\boldsymbol{\varepsilon}$ is a global accuracy parameter which enables one to estimate the global quality of the approximation $(\underline{\hat{U}}, \hat{\sigma})$.

It is also easy to define the contribution of a part E of $\boldsymbol{\Omega}$ (which, in practice, can be any element of the mesh) to the global accuracy by

$$\varepsilon_E = \frac{\left\| \hat{\sigma} - \mathbf{K}\varepsilon(\underline{\hat{U}}) \right\|_{\sigma,E}}{\left\| \hat{\sigma} + \mathbf{K}\varepsilon(\underline{\hat{U}}) \right\|_{\sigma,\Omega}} \tag{2.9}$$

with

$$\| \cdot \|_{\sigma,E} = \left[\int_{E} \mathrm{Tr}(\cdot \mathbf{K}^{-1} \cdot)\, dE \right]^{1/2} \tag{2.10}$$

These local contributions enable one to localize the discretization errors. From (2.8) and (2.9), it follows that

$$\varepsilon^2 = \sum_E \varepsilon_E^2 \tag{2.11}$$

REMARKS

1. Other measures of the error in constitutive relation can also be used. For example, one can use the following measure:

$$\varepsilon = \frac{\underset{M\in\Omega}{\mathrm{Sup}} \left| \mathrm{Tr}[(\hat{\sigma} - \mathbf{K}\varepsilon(\hat{\underline{U}}))\mathbf{K}^{-1}(\hat{\sigma} - \mathbf{K}\varepsilon(\hat{\underline{U}}))\right|}{[\mathrm{mes}(\Omega)]^{-1/2} \left\| \hat{\sigma} + \mathbf{K}\varepsilon(\hat{\underline{U}}) \right\|_{\sigma,\Omega}}$$

which is very close to an L^{∞} type norm [LADEVEZE - LEGUILLON, 1983].

2. Similarly, using the local contributions ε_E, one can define the quantities

$$\bar{\varepsilon}_E = \left[\frac{\mathrm{mes}(\Omega)}{\mathrm{mes}(E)}\right]^{1/2} \cdot \varepsilon_E = \frac{\dfrac{1}{[\mathrm{mes}(E)]^{1/2}} \left\| \hat{\sigma} - \mathbf{K}\varepsilon(\hat{\underline{U}}) \right\|_{\sigma,E}}{\dfrac{1}{[\mathrm{mes}(\Omega)]^{1/2}} \left\| \hat{\sigma} + \mathbf{K}\varepsilon(\hat{\underline{U}}) \right\|_{\sigma,\Omega}} \tag{2.12}$$

2.3 Properties of the error in constitutive relation

The objective of this section is to show that the error in constitutive relation – more precisely, the energy norm of this error – is a very convenient tool for introducing the classical potential energy and complementary energy theorems.

2.3.1 New formulation for an elasticity problem

The reference problem [(1.6) - (1.8)] can be formulated in the following manner:

Find $(\underline{U}, \sigma)$ *such that:*

- $\underline{U}$ *is KA (Kinematically Admissible);*
- σ *is SA (Statically Admissible);*

- *$\underline{U}$ and σ verify the constitutive relation:*

  at every point M *of* Ω, $\sigma = \mathbf{K}\varepsilon(\underline{U})$.

Let

$$\Psi(\underline{U}, \sigma) = \frac{1}{2} \| \sigma - \mathbf{K}\varepsilon(\underline{U}) \|^2_{\sigma,\Omega}$$
$$= \frac{1}{2}\int_\Omega \text{Tr}(\, [\sigma - \mathbf{K}\varepsilon(\underline{U})]\mathbf{K}^{-1} [\sigma - \mathbf{K}\varepsilon(\underline{U})]\,)\; d\Omega \qquad (2.13)$$

The constitutive relation is verified everywhere if and only if $\Psi(\underline{U}, \sigma) = 0$. Therefore, the exact solution $(\underline{U}_{ex}, \sigma_{ex})$ of the reference problem verifies

$$\Psi(\underline{U}_{ex}, \sigma_{ex}) = 0$$

Since one always has $\Psi(\underline{U}, \sigma) \geq 0$, one can deduce that $(\underline{U}_{ex}, \sigma_{ex})$ is the solution of the minimization problem

$$\underset{\hat{\underline{U}}\, KA \text{ and } \hat{\sigma}\, SA}{\text{Min}} \quad \Psi(\hat{\underline{U}}, \hat{\sigma}) \qquad (2.14)$$

Under the hypotheses of Chapter 1, which guarantee the existence of a unique solution of the reference problem, Formulation (2.14) is equivalent to the classical formulation [(1.6) - (1.8)].

REMARK. In some situations, particularly for problems described as "ill-posed", Formulation [(1.6) - (1.8)] may have no solution; this is the case, for example, in identification problems in which both the forces and the displacements are prescribed simultaneously on a part (with nonzero measure) of $\partial\Omega$.

In this case, Problem (2.14) still has solutions, but if $(\tilde{\underline{U}}, \tilde{\sigma})$ is one of these solutions one has $\Psi(\tilde{\underline{U}}, \tilde{\sigma}) > 0$.

For these identification problems and, more generally, for inverse problems, one can refer to [BUI, 1992].

2.3.2 Potential energy and complementary energy

Let us consider an admissible pair $(\hat{\underline{U}}, \hat{\sigma})$. One has the following decoupling property:

Property

$$\Psi(\hat{\underline{U}}, \hat{\sigma}) = \mathrm{E}_p(\hat{\underline{U}}) + \mathrm{E}_c(\hat{\sigma}) \tag{2.15}$$

with

$$\mathrm{E}_p(\hat{\underline{U}}) = \frac{1}{2}\int_\Omega \mathrm{Tr}\left[\varepsilon(\hat{\underline{U}})\mathbf{K}\varepsilon(\hat{\underline{U}})\right] d\Omega - \left[\int_\Omega \underline{f}_d \cdot \hat{\underline{U}}\, d\Omega + \int_{\partial_2\Omega} \underline{F}_d \cdot \hat{\underline{U}}\, dS\right]$$

and

$$\mathrm{E}_c(\hat{\sigma}) = \frac{1}{2}\int_\Omega \mathrm{Tr}\left[\hat{\sigma}\,\mathbf{K}^{-1}\hat{\sigma}\right] d\Omega - \int_{\partial_1\Omega} \hat{\sigma}\,\underline{n} \cdot \underline{U}_d\, dS$$

Proof

By expanding $\Psi(\hat{\underline{U}}, \hat{\sigma})$, one obtains

$$\Psi(\hat{\underline{U}}, \hat{\sigma}) = \frac{1}{2}\int_\Omega \mathrm{Tr}(\hat{\sigma}\mathbf{K}^{-1}\hat{\sigma})\, d\Omega + \frac{1}{2}\int_\Omega \mathrm{Tr}(\varepsilon(\hat{\underline{U}})\mathbf{K}\varepsilon(\hat{\underline{U}}))\, d\Omega - \int_\Omega \mathrm{Tr}(\hat{\sigma}\varepsilon(\hat{\underline{U}}))\, d\Omega$$

Since $(\hat{\underline{U}}, \hat{\sigma})$ is admissible, one has

$$\begin{aligned}\int_\Omega \mathrm{Tr}(\hat{\sigma}\varepsilon(\hat{\underline{U}}))\, d\Omega &= \int_\Omega \left[\mathrm{div}(\hat{\sigma}\hat{\underline{U}}) - \underline{\mathrm{div}}\,\hat{\sigma} \cdot \hat{\underline{U}}\right] d\Omega \\ &= \int_{\partial\Omega} \hat{\sigma}\underline{n} \cdot \hat{\underline{U}}\, d\Omega - \int_\Omega \underline{\mathrm{div}}\,\hat{\sigma} \cdot \hat{\underline{U}}\, d\Omega \\ &= \int_\Omega \underline{f}_d \cdot \hat{\underline{U}}\, d\Omega + \int_{\partial_1\Omega} \hat{\sigma}\underline{n} \cdot \underline{U}_d\, d\Omega + \int_{\partial_2\Omega} \underline{F}_d \cdot \hat{\underline{U}}\, d\Omega\end{aligned}$$

The combination of these results yields Property (2.15).

This decoupling property enables one to split the minimization problem (2.14) into two minimization problems, one in terms of displacements and the other in terms of stresses:

- The potential energy theorem:

  The displacement field $\underline{U}_{ex}$ is the solution of the problem

$$\underset{\hat{\underline{U}}\ CA}{\mathrm{Min}}\ \mathrm{E}_p(\hat{\underline{U}}) \tag{2.16}$$

- The complementary energy theorem:

The stress field $\underline{\underline{\sigma}}_{ex}$ *is the solution of the problem*

$$\underset{\hat{\underline{\underline{\sigma}}}\ SA}{\text{Min}}\ \mathrm{E}_c(\hat{\underline{\underline{\sigma}}}) \tag{2.17}$$

2.3.3 Extensions

Many constitutive relations can be defined starting from a pair of convex dual functions φ and φ^* such that

$$\varphi(\underline{\underline{\varepsilon}}) + \varphi^*(\underline{\underline{\sigma}}) - \mathrm{Tr}[\underline{\underline{\sigma}}\,\underline{\underline{\varepsilon}}] \geq 0$$

Besides, at a point M, the equality $\varphi(\underline{\underline{\varepsilon}}) + \varphi^*(\underline{\underline{\sigma}}) - \mathrm{Tr}[\underline{\underline{\sigma}}\,\underline{\underline{\varepsilon}}] = 0$ is equivalent to the satisfaction of the constitutive relation at that point. The global error in constitutive relation is then defined by

$$\Psi(\underline{U}, \underline{\underline{\sigma}}) = \int_\Omega \varphi[\underline{\underline{\varepsilon}}(\underline{U})]\, d\Omega + \int_\Omega \varphi^*(\underline{\underline{\sigma}})\, d\Omega - \int_\Omega \mathrm{Tr}[\underline{\underline{\sigma}}\,\underline{\underline{\varepsilon}}(\underline{U})]\, d\Omega$$

The dual convex functions defined by

$$\Phi(\underline{\underline{\varepsilon}}) = \int_\Omega \varphi(\underline{\underline{\varepsilon}})\, d\Omega \quad \text{and} \quad \Phi^*(\underline{\underline{\sigma}}) = \int_\Omega \varphi^*(\underline{\underline{\sigma}})\, d\Omega$$

are the overpotentials of the constitutive relation.

The minimization of the error in constitutive relation leads to two theorems similar to the energy theorems:

- *The displacement field* $\underline{U}_{ex}$ *is the solution of the problem*

$$\underset{\hat{\underline{U}}\ CA}{\text{Min}}\ \mathrm{E}_p(\hat{\underline{U}})$$

with

$$\mathrm{E}_p(\hat{\underline{U}}) = \Phi[\underline{\underline{\varepsilon}}(\hat{\underline{U}})] - \left[\int_\Omega \underline{f}_d \cdot \hat{\underline{U}}\, d\Omega + \int_{\partial_2\Omega} \underline{F}_d \cdot \hat{\underline{U}}\, dS\right]$$

- *The stress field* $\underline{\underline{\sigma}}_{ex}$ *is the solution of the problem*

$$\underset{\hat{\underline{\underline{\sigma}}}\ SA}{\text{Min}}\ \mathrm{E}_c(\hat{\underline{\underline{\sigma}}})$$

with

$$\mathrm{E}_c(\hat{\boldsymbol{\sigma}}) = \Phi^*(\hat{\boldsymbol{\sigma}}) - \int_{\partial_1\Omega} \hat{\boldsymbol{\sigma}}\, \underline{n} \cdot \underline{U}_d \, dS$$

REMARKS

1. In elasticity, one has

$$\Phi(\boldsymbol{\varepsilon}) = \frac{1}{2}\int_\Omega \mathrm{Tr}[\boldsymbol{\varepsilon}\mathbf{K}\boldsymbol{\varepsilon}]\, d\Omega \quad \text{and} \quad \Phi^*(\boldsymbol{\sigma}) = \frac{1}{2}\int_\Omega \mathrm{Tr}[\boldsymbol{\sigma}\,\mathbf{K}^{-1}\boldsymbol{\sigma}]\, d\Omega$$

2. The denomination "energy theorems" is not really appropriate; these theorems are related to the global formulation of the constitutive relation, not to the notion of energy.

3. This formalism owes much to the works of J.-J. MOREAU [MOREAU, 1966, 1974] and to the further developments by NAYROLES [NAYROLES, 1973].

2.4 Utilization in finite element calculations

2.4.1 Construction of an admissible pair

The calculation of the error in constitutive relation cannot be made directly on the pair $(\underline{U}_h, \boldsymbol{\sigma}_h)$ because this pair is not admissible.

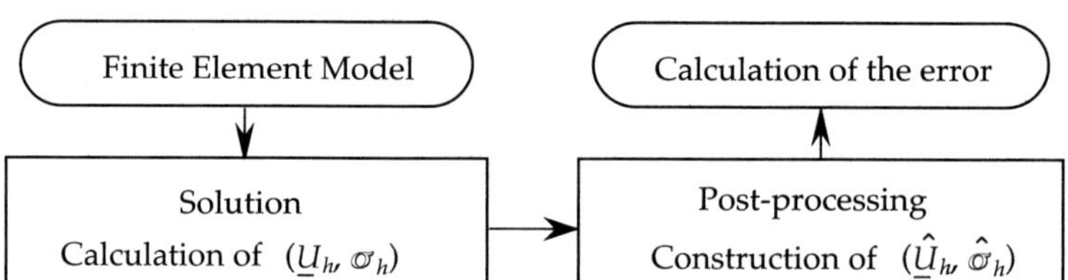

Figure 2.1. *Principle of utilization.*

Our method consists in post-processing the finite element solution in order to construct an admissible displacement-stress pair $(\hat{\underline{U}}_h, \hat{\boldsymbol{\sigma}}_h)$, in the sense of the relations

(2.3) and (2.4), starting from (Figure 2.1):

- the data of the reference problem;

- the finite element solution $(\underline{U}_h, \sigma_h)$.

From a technical standpoint, this is a crucial aspect of the method. Indeed, on the one hand, the quality of the constructed pair $(\hat{\underline{U}}_h, \hat{\sigma}_h)$ determines the quality of the error measure and, on the other hand, the numerical cost of this construction should be small compared to the cost of the finite element analysis.

2.4.2 Displacement field

In the context of finite element displacement methods of the conforming type, the displacement field $\underline{U}_h$ is admissible. Therefore, for the sake of simplicity, one generally chooses

$$\hat{\underline{U}}_h = \underline{U}_h \tag{2.18}$$

REMARKS

1. Of course, $\underline{U}_h$ is admissible only with the provision that the displacement boundary conditions on $\partial_1\Omega$ be compatible with the interpolation chosen for the discretization.

2. An example of the construction of a field $\hat{\underline{U}}_h$ different from $\underline{U}_h$ can be found, in the framework of thermal problems, in [LADEVEZE, 1977], [LADEVEZE - LEGUILLON, 1983].

3. For calculations in incompressible elasticity, the choice $\hat{\underline{U}}_h = \underline{U}_h$ is unsuitable. A specific construction of fields $\hat{\underline{U}}_h$ which verify the incompressibility condition $\mathrm{div}(\hat{\underline{U}}_h) = 0$ was developed in [GASTINE - LADEVEZE - MARIN - PELLE, 1992]. We will return to this point in Chapter 8.

2.4.3 Stress field

Conversely, the stress field σ_h calculated by finite elements is not statically admissible. It is therefore necessary to reconstruct a field $\hat{\sigma}_h$ which verifies the equilibrium equations exactly. A priori, it is possible to construct such a field using a finite element method of the equilibrium type [FRAEIJS DE VEUBEKE,

1965], [STEIN - AHMAD, 1977]. In spite of their remarkable effectiveness, these methods have been used very rarely because of the difficulty of their implementation. In addition, the use of this dual method along with the finite element displacement method is not realistic today. Let us also note that the codes which are presently used in industry do not include equilibrium elements.

The method presented below was introduced in [LADEVEZE, 1975, 1977]. It relies on strong extension conditions. Today, we use a variant of this method, based on weak extension conditions, which is more effective, but also more costly (Chapter 8). It is described in [LADEVEZE, 1994] and developed in [LADEVEZE - ROUGEOT, 1997].

Strong extension condition

The basic idea consists in seeking a field $\hat{\sigma}_h$ as a prolongation (or an extension) of the finite element solution in the following sense:

For any basis function φ_i of the finite element type and for any element E of the mesh

$$\int_E (\hat{\sigma}_h - \sigma_h)\, \underline{\text{grad}}\, \varphi_i \, dE = 0 \tag{2.19}$$

Here, we merely outline the technique of construction of the stress field. This construction will be reintroduced in detail in Chapter 8.

REMARK. The extension condition (2.19) has a strong mechanical content. It requires, of the stress field $\hat{\sigma}_h$ that one is seeking, that it yield the same work in each element and for each finite element displacement field as the finite element stress field σ_h. In particular, for 3-node triangles, this condition imposes equality of the mean stresses in each element.

Notations

On the boundary ∂E of each element E, one defines a function η_E, constant on each side of E and with the value $+1$ or -1, such that on the side $\Gamma_{E_1E_2}$ common to two adjacent elements E_1 and E_2 one has $\eta_{E_1} + \eta_{E_2} = 0$; we will designate by $\underline{n}_E$ the outward unit normal on ∂E.

Let i be a node of the mesh. We will denote by $E_1, E_2, \ldots, E_N$ the N elements connected to node i and by $\Gamma_E^1, \Gamma_E^2, \ldots, \Gamma_E^R$ the R sides of an element E which are connected to node i.

Principles of the construction

The construction of an equilibrated stress field $\hat{\sigma}_h$ is carried out in two stages. First, one constructs force densities $\hat{\underline{F}}$ on the element sides. The purpose of these densities is to represent the stress vectors on the element sides

$$\hat{\sigma}_{h|E}\, \underline{n}_E = \eta_E\, \hat{\underline{F}} \quad \text{on } \partial E \tag{2.20}$$

In this manner, the continuity of the stress vector in the $\underline{n}_E$ direction is automatically verified across an interface.

Naturally, on the sides Γ which are part of $\partial_2\Omega$, one imposes

$$\eta_E\, \hat{\underline{F}} = \underline{F}_d \tag{2.21}$$

Finally, these force densities are constructed in such a way that the densities $\eta_E\, \hat{\underline{F}}$ are in equilibrium with the given interior loads $\underline{f}_d$ on each element E of the mesh:

For any field of rigid body displacement $\underline{U}_S$,

$$\int_E \underline{f}_d \cdot \underline{U}_S\, dE + \int_{\partial E} \eta_E\, \hat{\underline{F}} \cdot \underline{U}_S\, dS = 0 \tag{2.22}$$

In the second stage, it suffices to construct a simple solution of the local problem on each element E,

$$\begin{aligned} &\underline{\text{div}}\, \hat{\sigma}_h + \underline{f}_d = 0 && \text{in } E \\ &\hat{\sigma}_h\, \underline{n}_E = \eta_E\, \hat{\underline{F}} && \text{on } \partial E \end{aligned} \tag{2.23}$$

Construction of the densities

Let us consider a node i of an element E. By applying the extension condition (2.19) to the shape function φ_i of node i and using the admissibility conditions (2.4) verified by $\hat{\underline{\underline{\sigma}}}_h$, one obtains

$$\int_{\partial E} \eta_E \hat{\underline{F}} \varphi_i \, d\Gamma = -\int_E \underline{f}_d \varphi_i \, dE + \int_E \underline{\underline{\sigma}}_h \, \underline{\text{grad}} \, \varphi_i \, dE \tag{2.24}$$

i.e., observing that φ_i is zero on the sides which do not contain i,

$$\sum_{r=1}^{R} \left[\int_{\Gamma_E^r} \eta_E \hat{\underline{F}} \varphi_i \, d\Gamma \right] = \underline{Q}_E(i) \tag{2.25}$$

where $\underline{Q}_E(i)$ denotes the right-hand side of (2.24), which is a known vector which depends on the data and on the finite element solution.

Determination of projections of the densities

By applying Equations (2.25) to all the elements connected to node i, one finds that the projections of the densities $\eta_E \hat{\underline{F}}$,

$$\underline{b}_n^r(i) = \int_{\Gamma_{E_n}^r} \eta_E \, \hat{\underline{F}} \, \varphi_i \, d\Gamma \tag{2.26}$$

are solutions of the linear system

$$\sum_{r=1}^{R_n} \underline{b}_n^r(i) = \underline{Q}_{E_n}(i) \quad \text{for } n \in \{1, 2, \dots, N\} \tag{2.27}$$

To make things clear, let us assume that i is an interior node of Ω.

The number of equations of System (2.27) is less than or equal to the number of unknowns. Moreover, these equations are not independent. Indeed, let us consider one of the unknowns $\underline{b}_n^r(i)$ which appear in the nth equation. There is only one element E_m, distinct from E_n, which has $\Gamma_{E_n}^r$ as a side. It follows that this unknown $\underline{b}_n^r(i)$ also appears in the mth equation (and only in that equation other than the nth equation). But because of the property $\eta_{E_n} + \eta_{E_m} = 0$ it appears in that equation with the opposite sign. Consequently, the sum of the rows of the matrix of System (2.27) is zero. Hence, System (2.27) has solutions only if the following compatibility condition is verified:

$$\sum_{n=1}^{N} \underline{Q}_{E_n}(i) = 0 \tag{2.28}$$

By summing the $Q_{E_n}(i)$, one gets

$$\sum_{n=1}^{N} Q_{E_n}(i) = \sum_{n=1}^{N} \left[-\int_{E_n} \underline{f}_d \varphi_i \, dE + \int_{E_n} \sigma_h \, \underline{\text{grad}} \, \varphi_i \, dE \right]$$
$$= -\int_{\Omega} \underline{f}_d \varphi_i \, dE + \int_{\Omega} \sigma_h \, \underline{\text{grad}} \, \varphi_i \, dE \tag{2.29}$$

and, due to the equilibrium in the finite element sense (1.15),

$$-\int_{\Omega} \sigma_h \, \underline{\text{grad}} \, \varphi_i \, dE + \int_{\Omega} \underline{f}_d \, \varphi_i \, dE = 0 \tag{2.30}$$

the compatibility condition (2.28) is always verified and System (2.27) always has an infinite number of solutions.

REMARKS

1. Different techniques for choosing a particular solution of (2.27) will be studied in Chapter 8.

2. We will show in Chapter 8 that situations with nodes located on the boundary $\partial\Omega$ can be treated in an similar way.

Determination of the densities

Therefore, we can assume that a solution of (2.27) is known for each node of the mesh being considered. Thus, if Γ is a side of the mesh, for each node $i_1, i_2, \ldots, i_M$ of Γ (where M depends on the degree and type of the elements), one knows a vector

$$\int_{\Gamma} \hat{\underline{F}} \, \varphi_{i_m} \, d\Gamma \quad \text{for } m \in \{1, 2, \ldots, M\} \tag{2.31}$$

By choosing for $\hat{\underline{F}}$ an interpolation of the same type as that used for the displacement field on Γ, one can determine $\hat{\underline{F}}$ uniquely on Γ.

Equilibrium with the interior loads

Let us consider Equations (2.24) for a particular element E and for all the shape functions φ_i of this element. Any rigid body displacement field $\underline{U}_S$ on E

can be written as a linear combination of the shape functions of the element

$$\underline{U}_S = \sum_i \underline{V}_i \, \varphi_i \tag{2.32}$$

By carrying out the same linear combination on Equations (2.24), one obtains the equilibrium conditions (2.22). This important property is a direct consequence of the extension condition.

Construction of an admissible field

The objective is to find a solution of (2.23), the local problem on the element E.

Analytical construction

When the given interior loads $\underline{f}_d$ are polynomials, the basic idea is to look for a solution $\hat{\sigma}_{h|E}$ which is also a polynomial.

However, in the context of elasticity problems, for such a solution to exist the densities $\hat{\underline{F}}$ must verify at the vertices of the element compatibility conditions which result from the symmetry of the operator $\hat{\sigma}_{h|E}$. For example, for triangular elements in 2D, $\hat{\underline{F}}$ must verify at each vertex the conditions

$$\eta_E \, \hat{\underline{F}}_{|\Gamma} \cdot \tilde{\underline{n}}_E = \eta_E \, \hat{\underline{F}}_{|\tilde{\Gamma}} \cdot \underline{n}_E$$

with the notations described in Figure 2.2.

However, the densities $\hat{\underline{F}}$ constructed by the above method usually do not verify these compatibility conditions.

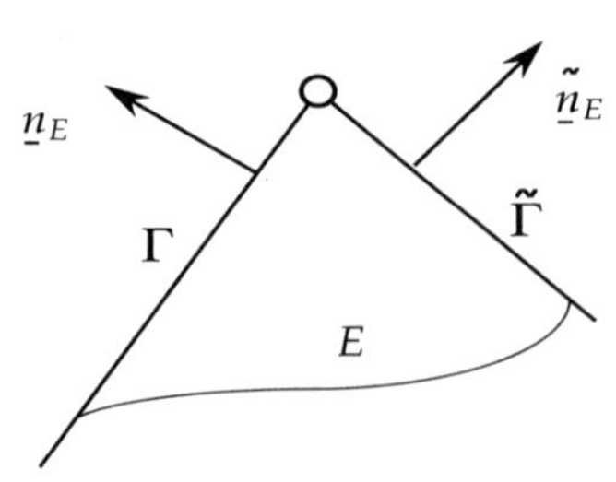

Figure 2.2. *Notations.*

This leads one into decomposing each triangle into three subtriangles (Figure 2.3) and seeking $\hat{\sigma}_{h|E}$ as a polynomial in each of the subtriangles.

Similar decomposition techniques are used for quadrilaterals, tetrahedra and bricks (see Chapter 8).

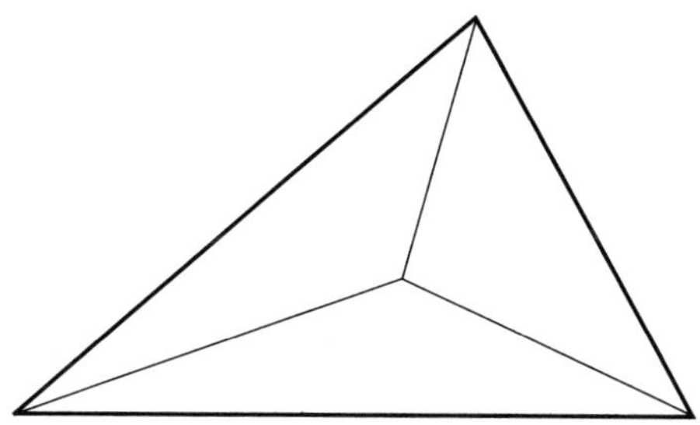

Figure 2.3. *Decomposition of an element.*

Approximate numerical construction

Once a set of interface force densities $\hat{\underline{F}}$ has been chosen, the best field $\hat{\sigma}_{h|E}$ is the solution of the problem

$$\underset{\sigma_{|E}}{\text{Min}} \left\| \sigma_{|E} - \sigma_h \right\|_{\sigma,E} \tag{2.33}$$

for $\sigma_{|E}$ such that

$$\begin{aligned} &\underline{\text{div}}\, \sigma_{|E} + \underline{f}_d = 0 \quad && \text{in } E \\ &\sigma_{|E}\, \underline{n}_E = \eta_E \hat{\underline{F}} && \text{on } \partial\text{E} \end{aligned} \tag{2.34}$$

One has

$$\left\| \sigma_{|E} - \sigma_h \right\|^2_{\sigma,E} = \left\| \sigma_{|E} \right\|^2_{\sigma,E} + \left\| \sigma_h \right\|^2_{\sigma,E} + 2\int_\Omega \text{Tr}\left[\sigma_{|E}\mathbf{K}^{-1}\sigma_h\right] d\Omega$$

and

$$\begin{aligned} \int_E \text{Tr}\left[\sigma_{|E}\mathbf{K}^{-1}\sigma_h\right] dE &= \int_E \text{Tr}\left[\sigma_{|E}\varepsilon(\underline{U}_h)\right] dE \\ &= \int_E \underline{f}_d \cdot \underline{U}_h \, d\Omega + \int_{\partial E} \eta_E \hat{\underline{F}} \cdot \underline{U}_h \, d\Gamma \end{aligned}$$

It follows that (2.33) is equivalent to the minimization problem

$$\underset{\sigma_{|E}}{\mathrm{Min}} \frac{1}{2}\int_E \mathrm{Tr}\left[\sigma_{|E}\,\mathbf{K}^{-1}\sigma_{|E}\right] dE \tag{2.35}$$

where $\sigma_{|E}$ still verifies (2.34).

By duality, this is the same as seeking a displacement field $\underline{V}_E$ defined on E such that

$$\underline{V}_E \in \boldsymbol{\mathcal{U}}(E) \quad \text{and} \quad \forall \underline{V}^* \in \boldsymbol{\mathcal{U}}(E)$$

$$\int_E \mathrm{Tr}(\, \varepsilon(\underline{V}_E)\mathbf{K}\varepsilon(\underline{V}^*)\,)\, dE = \int_E \underline{f}_d \cdot \underline{V}^* dE + \int_{\partial E} \eta_E \hat{\underline{F}} \cdot \underline{V}^* d\Gamma \tag{2.36}$$

where $\boldsymbol{\mathcal{U}}(E)$ denotes the space of the restrictions to E of the fields in $\boldsymbol{\mathcal{U}}$.

Then, $\sigma_{|E}$ is given by

$$\sigma_{|E} = \mathbf{K}\varepsilon(\underline{V}_E) \tag{2.37}$$

One can therefore obtain an approximation of $\sigma_{|E}$ by solving Problem (2.36) by a standard finite element method in E.

In practice, in order to obtain a good approximation, it is sufficient to consider a discretization of E by a single element with an interpolation of degree $p+k$, where p is the degree of interpolation used in the finite element analysis and k is a positive integer.

Numerical experiments showed that with $k \geq 3$ one obtains a good quality approximate field.

REMARKS

1. Of course, the analytical construction yields strictly admissible fields only if $\underline{f}_d$ is a polynomial with a degree compatible with the polynomial form chosen for the field $\hat{\sigma}_{h|E}$. For example, for 3-node triangles, the load $\underline{f}_d$ must be constant within each element. In practice, this limitation is hardly restrictive. Nevertheless, if the load $\underline{f}_d$ is a higher-degree polynomial, one can construct a strictly admissible field $\hat{\sigma}_{h|E}$ by seeking this field in a space of higher-order polynomials. Thus, it is theoretically possible to get a very good approximation of any type of load $\underline{f}_d$ which is not a polynomial.

2. The technique for the approximate construction of $\hat{\sigma}_{h|E}$ works in the same manner regardless of the type of load f_d. But, of course, the field thus constructed is only a good approximation of an admissible field.

3. This approximate construction technique can also be used with no modification for curved isoparametric elements [COOREVITS - DUMEAU - PELLE, 1999].

4. The use of elements of degree $p+k$, where p is the degree of the finite element interpolation and k is a positive integer for the approximate construction of the stress $\hat{\sigma}_{h|E}$, was proposed by BABUSKA and STROUBOULIS. In their terminology, they talk about "equilibration" of degree k. In particular, they showed that the error indicator associated with $k=1$ is extremely effective.

2.4.4 Relationship with the errors in the solution

In the context of linear problems, it is easy to connect the error in constitutive relation with the classically used errors in the solution:

- error in the solution for the displacement

  $\underline{e}_h = \underline{U}_{ex} - \underline{U}_h$

- error in the solution for the stress

  $\mathbf{K}\varepsilon(\underline{e}_h) = \sigma_{ex} - \sigma_h$

Let $\hat{\sigma}_h$ be a statically admissible stress field. The relationship between the error in the solution and the error in constitutive relation $\hat{\sigma}_h - \mathbf{K}\varepsilon(\underline{U}_h)$ is given by the popular hypercircle theorem [PRAGER - SYNGE, 1947].

Theorem

$$\left[\mathbf{e}_{RdC}(\underline{U}_h, \hat{\sigma}_h)\right]^2 = \| \sigma_{ex} - \hat{\sigma}_h \|^2_{\sigma,\Omega} + \| \underline{U}_{ex} - \underline{U}_h \|^2_{u,\Omega} \tag{2.38}$$

where

$$\| \cdot \|^2_{u,\Omega} = \int_\Omega \mathrm{Tr}\left[\varepsilon(\cdot)\mathbf{K}\varepsilon(\cdot)\right] d\Omega$$

Proof

One has

$$\| \hat{\sigma}_h - \mathbf{K}\varepsilon(\underline{U}_h) \|^2_{\sigma,\Omega} = \| (\hat{\sigma}_h - \sigma_{ex}) + (\sigma_{ex} - \mathbf{K}\varepsilon(\underline{U}_h)) \|^2_{\sigma,\Omega}$$

$$= \| \hat{\sigma}_h - \sigma_{ex} \|^2_{\sigma,\Omega} + \| \underline{e}_h \|^2_{u,\Omega} + 2\int_\Omega \mathrm{Tr}\left[(\hat{\sigma}_h - \sigma_{ex})\varepsilon(\underline{e}_h)\right] d\Omega$$

The displacement field $\underline{e}_h$, which is the difference between two *KA* fields, is a field *KA* to zero $\underline{e}_h \in \boldsymbol{\mathcal{U}}_0$. Similarly, the stress field $\hat{\sigma}_h - \sigma_{ex}$, which is the difference between two *SA* fields, is *SA* to zero. It follows that

$$\int_\Omega \mathrm{Tr}\left[(\hat{\sigma}_h - \sigma_{ex})\varepsilon(\underline{e}_h)\right] d\Omega = 0$$

which proves Theorem (2.38).

REMARKS

1. The hypercircle theorem is nothing other than a form of the PYTHAGORAS theorem in the space of the stresses $\boldsymbol{\mathcal{S}}$ (Figure 2.4) endowed with the inner product

$$(\sigma_1, \sigma_2) \mapsto \int_\Omega \mathrm{Tr}\left[\sigma_1 \mathbf{K}^{-1} \sigma_2\right] d\Omega$$

2. If one introduces as the approximate stress the average of the finite element stress and the admissible stress

$$\sigma^* = \frac{1}{2}(\hat{\sigma}_h + \sigma_h) \qquad (2.39)$$

one gets

$$\mathbf{e}_{RdC} = \| e_{RdC} \|_{\sigma,\Omega} = 2 \| \sigma_{ex} - \sigma^* \|_{\sigma,\Omega} \quad \text{and} \quad \varepsilon = \frac{\| \sigma_{ex} - \sigma^* \|_{\sigma,\Omega}}{\| \sigma^* \|_{\sigma,\Omega}} \qquad (2.40)$$

Thus, the measure of the error in constitutive relation enables one to calculate the error between σ_{ex} and σ^* *exactly*, in the sense of the energy norm, without knowing the exact solution σ_{ex}.

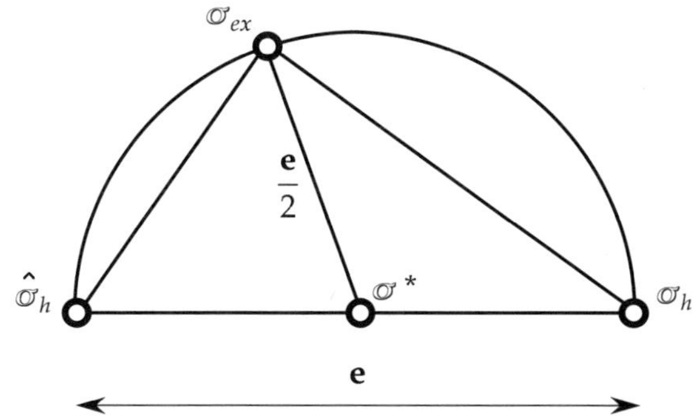

Figure 2.4. *The* PRAGER - SYNGE *hypercircle.*

3. An alternative and more mathematical presentation of the concept of error in constitutive relation starting from a mixed formulation is given in [DESTUYNDER - METIVET, 1996], [DESTUYNDER - COLLOT - SALAUN, 1997].

4. Several error estimators use the method of construction of admissible stress fields described in Section 2.4.3. This is the case, for example, of the variant proposed for thermal problems in [BABUSKA - STROUBOULIS - UPADHYAY - GANGARAJ - COPPS, 1994] and of the estimators given in [STEIN - OHNIMUS, 1999], [STEIN - BARTHOLD - OHNIMUS - SCHMIDT, 1998]. This is also the case of the estimators developed for electromagnetism in [DEMKOWICZ - KIM, 2000] and [REMACLE - DULAR - GENON - LEGROS, 1996].

5. A dose of equilibrium, i.e. of "mechanics", is introduced in most of today's error estimators – a tendency which seems to be increasing – in order to improve their effectiveness. We will clarify this point in Chapter 3.

2.4.5 Asymptotic behavior

The hypercircle theorem leads in particular to the following inequalities:

$$\| \sigma_{ex} - \sigma_h \|_{\sigma,\Omega} \leq \mathbf{e}_{RdC} \quad \text{and} \quad \| \underline{U}_{ex} - \underline{U}_h \|_{u,\Omega} \leq \mathbf{e}_{RdC} \tag{2.41}$$

Therefore, the measure of the error in constitutive relation is a reliable error measure; its effectivity index is always greater than 1, i.e. $\gamma \geq 1$.

Besides, we will show in Chapter 8 that there exists a constant C independent of the size of the elements such that

$$\mathbf{e}_{RdC} \leq C \| \underline{U}_{ex} - \underline{U}_h \|_{u,\Omega}$$

In summary, one has

$$\| \underline{U}_{ex} - \underline{U}_h \|_{u,\Omega} \leq \mathbf{e}_{RdC} \leq C \| \underline{U}_{ex} - \underline{U}_h \|_{u,\Omega} \tag{2.42}$$

Chapter 3

Other methods for linear problems

3.1 Introduction

In this chapter, we give an overview of the two other main methods which have been proposed for estimating discretization errors for linear problems. [LADEVEZE - ODEN, 1998] gave a precise idea of the state of the art.

Section 3.2 is devoted to estimators based on the residuals of the equilibrium equations first proposed by BABUSKA and RHEINBOLDT [BABUSKA - RHEINBOLDT, 1978]. Several variants of these (of which one of the most recent can be found in [DIEZ - EGOZENZ - HUERTA, 1998]) have since been developed by many authors: explicit estimators and implicit estimators by element or by groups of elements.

In particular, it is worth noting that the "Er*p*B*p*+*k*" family of estimators in the context of thermal problems developed by BABUSKA and STROUBOULIS [BABUSKA - STROUBOULIS - UPADHYAY - GANGARAJ - COPPS,1994] [BABUSKA - STROUBOULIS - GANGARAJ, 1997] is the variant of the error in constitutive relation described in Section 2.4.3, built starting from the strong extension condition and using the solution of a local problem on a single element of degree $p+k$. This variant is also used in the iterative bounding technique proposed in [BABUSKA - STROUBOULIS - GANGARAJ, 1999], [STROUBOULIS - BABUSKA - GANGARAJ, 2000], a very general approach which is outlined in Section 3.4.

In Section 3.3, we present the principle of the construction of estimators using a smoothed stress introduced by ZIENKIEWICZ and ZHU [ZIENKIEWICZ - ZHU, 1987, 1992].

Different variants proposed subsequently in order to obtain better estimators are also presented. These improvements consist essentially in enforcing in the smoothing technique the use of superconvergence points (when they exist), of more or less weakened conditions of equilibrium, or of other conditions similar to the extension conditions.

For more details, one can refer to [BABUSKA - SROUBOULIS - UPADHYAY - GANGARAJ, 1994, 1995] and also to [AINSWORTH - ODEN, 1997]. In particular, these references describe the various suggestions of more or less partial inclusion of equilibrium conditions.

The estimators which are presented here in elasticity can obviously be extended to problems of fluid mechanics, to which many works have been devoted [JOHNSON, 1990], [ODEN - WU - AINSWORTH, 1994], [HACKBRICH - WITTRUM, 1994], [AINSWORTH, 1999], [ACHCHAB - AGOUZAL - BARANGER - MAITRE, 2000].

3.2 Methods based on equilibrium residuals

3.2.1 General principle

In the finite element displacement method, as pointed out in Chapter 1, the main approximation concerns the equilibrium equations. These equilibrium deficiencies, also called residuals, are explained below along with their connection with the error in the solution.

Equilibrium residuals

For each element E of the mesh, one defines the interior residual $\underline{r}_E$ as

$$\underline{r}_E = \underline{\mathrm{div}}\, \sigma_{h|E} + \underline{f}_d \tag{3.1}$$

The residuals $\underline{r}_E$ measure the non-satisfaction of the interior equilibrium equation.

Likewise, on each side Γ of an element not included in $\partial_1\Omega$, one defines the residual $\underline{t}_\Gamma$ as

$$\underline{t}_\Gamma = \sigma_{h|E}\, \underline{n}_E + \underline{\Phi}_E \tag{3.2}$$

with:

- if Γ is not a side of $\partial_2\Omega$: $\underline{\Phi}_E = \sigma_{h|\tilde{E}}\, \underline{n}_{\tilde{E}}$,

where $\tilde{E}$ is the element adjacent to element E with respect to side Γ;

- if Γ is a side of $\partial_2\Omega$: $\underline{\Phi}_E = -\underline{F}_d$;

The residuals $\underline{t}_\Gamma$ measure the discontinuity of the stress vector in the direction $\underline{n}_E$ at the interface between two elements, and the nonsatisfaction of the equilibrium conditions $\partial_2\Omega$.

Relation with the error in the solution

Property

The error in the solution $\underline{e}_h = \underline{U}_{ex} - \underline{U}_h$ is the solution of the problem

$$\underline{e}_h \in \boldsymbol{\mathcal{U}}_0 \quad \textit{and} \quad \forall \underline{U}^* \in \boldsymbol{\mathcal{U}}_0,$$

$$\int_\Omega \mathrm{Tr}\left[\varepsilon(\underline{e}_h)\mathbf{K}\varepsilon(\underline{U}^*)\right] d\Omega = \sum_E \int_E \underline{r}_E \cdot \underline{U}^* dE - \sum_\Gamma \int_\Gamma \underline{t}_\Gamma \cdot \underline{U}^* d\Gamma \tag{3.3}$$

Thus, the error $\underline{e}_h$ is the solution of an elasticity problem whose applied loads (in the interior and on the interfaces of the mesh being used) are the residuals of the equilibrium equations.

Proof

One has

$$\int_\Omega \mathrm{Tr}\left[\varepsilon(\underline{e}_h)\mathbf{K}\varepsilon(\underline{U}^*)\right] d\Omega = \int_\Omega \mathrm{Tr}\left[\varepsilon(\underline{U}_{ex})\mathbf{K}\varepsilon(\underline{U}^*)\right] d\Omega - \int_\Omega \mathrm{Tr}\left[\varepsilon(\underline{U}_h)\mathbf{K}\varepsilon(\underline{U}^*)\right] d\Omega$$

$$\int_\Omega \mathrm{Tr}\left[\varepsilon(\underline{U}_{ex})\mathbf{K}\varepsilon(\underline{U}^*)\right] d\Omega = \int_\Omega \underline{f}_d \cdot \underline{U}^* d\Omega + \int_{\partial_2\Omega} \underline{F}_d \cdot \underline{U}^* d\Gamma$$

One also has

$$\int_\Omega \mathrm{Tr}\left[\varepsilon(\underline{U}_h)\mathbf{K}\varepsilon(\underline{U}^*)\right] d\Omega = \sum_E \int_E \left[\mathrm{div}(\sigma_{h|E}\,\underline{U}^*) - \underline{\mathrm{div}}\,\sigma_{h|E} \cdot \underline{U}^*\right] dE$$

$$= \sum_E \left[\sum_{\Gamma\subset\partial E} \int_\Gamma \sigma_{h|E}\,\underline{n}_E \cdot \underline{U}^* d\Gamma - \int_E \underline{\mathrm{div}}\,\sigma_{h|E} \cdot \underline{U}^* dE \right]$$

The combination of these results, noting that each side Γ interior to Ω occurs exactly twice in the summation, yields the announced result.

Corollary

For any $\underline{U}_h^* \in \boldsymbol{\mathcal{U}}_{h0}$, *one has*

$$\sum_E \int_E \underline{r}_E \cdot \underline{U}_h^* \; dE - \sum_\Gamma \int_\Gamma \underline{t}_\Gamma \cdot \underline{U}_h^* \; d\Gamma = 0 \tag{3.4}$$

Proof

Using (1.7) for $\underline{U}^* = \underline{U}_h^*$, one obtains

$$\int_\Omega \mathrm{Tr}\left[\sigma_{ex}\varepsilon(\underline{U}_h^*)\right] d\Omega = \int_\Omega \underline{f}_d \cdot \underline{U}_h^* \; d\Omega + \int_{\partial_2\Omega} \underline{F}_d \cdot \underline{U}_h^* \; d\Gamma$$

Besides, the finite element field verifies (1.15),

$$\int_\Omega \mathrm{Tr}\left[\varepsilon(\underline{U}_h)\mathbf{K}\varepsilon(\underline{U}_h^*)\right] d\Omega = \int_\Omega \underline{f}_d \cdot \underline{U}_h^* \; d\Omega + \int_{\partial_2\Omega} \underline{F}_d \cdot \underline{U}_h^* \; d\Gamma$$

Subtracting these equations member by member yields

$$\int_\Omega \mathrm{Tr}\left[\varepsilon(\underline{e}_h)\mathbf{K}\varepsilon(\underline{U}_h^*)\right] d\Omega = 0 \tag{3.5}$$

and Property (3.4) is obtained using (3.3) for $\underline{U}^* = \underline{U}_h^*$.

All a posteriori error estimators based on the residuals are constructed using (3.3). In broad terms, one can distinguish:

• explicit estimators, which use (3.3) directly to obtain a global estimate of certain norms of the error $\underline{e}_h$;

• implicit estimators, which use approximations of the solution of (3.3) obtained by solving local problems element by element or by groups of elements.

3.2.2 Explicit estimators

A typical result

The following result enables one to evaluate the energy norm of $\underline{e}_h$:

Theorem

There exists a constant C independent of the sizes of the elements such that

$$\| \underline{e}_h \|_{u,\Omega} \leq C \left[\sum_E h_E^2 \| \underline{r}_E \|_{L^2(E)}^2 + \sum_{\Gamma \not\subset \partial_1 \Omega} \ell_\Gamma \| \underline{t}_\Gamma \|_{L^2(\Gamma)}^2 \right]^{1/2} \tag{3.6}$$

where h_E *denotes the size of element* E *and* ℓ_Γ *the size of side* Γ *defined, respectively, by*

$$h_E = \operatorname*{Sup}_{M \in E \text{ and } N \in E} \left\| \overrightarrow{MN} \right\| \quad \textit{and} \quad \ell_\Gamma = \operatorname*{Sup}_{M \in \Gamma \text{ and } N \in \Gamma} \left\| \overrightarrow{MN} \right\|$$

Proof

Let Π_h be a linear operator from $\boldsymbol{U}_0$ onto $\boldsymbol{U}_{h0}$: $\underline{U}^* \mapsto \Pi_h \underline{U}^*$.
Using (3.3) and (3.4) for $\underline{U}_h^* = \Pi_h \underline{U}^*$, one gets

$$\int_\Omega \operatorname{Tr}\left[\varepsilon(\underline{e}_h) \mathbf{K} \varepsilon(\underline{U}^*) \right] d\Omega = \sum_E \int_E \underline{r}_E \cdot (\underline{U}^* - \Pi_h \underline{U}^*) \, dE$$
$$- \sum_{\Gamma \not\subset \partial_1 \Omega} \int_\Gamma \underline{t}_\Gamma \cdot (\underline{U}^* - \Pi_h \underline{U}^*) \, d\Gamma \tag{3.7}$$

Let: $A = \left| \int_\Omega \operatorname{Tr}\left[\varepsilon(\underline{e}_h) \mathbf{K} \varepsilon(\underline{U}^*) \right] d\Omega \right|$.

Applying the CAUCHY - SCHWARZ inequality, one deduces

$$A \le \sum_{E} \| \underline{r}_E \|_{L^2(E)} \| \underline{U}^* - \Pi_h \underline{U}^* \|_{L^2(E)} + \sum_{\Gamma \not\subset \partial_1 \Omega} \| \underline{t}_\Gamma \|_{L^2(\Gamma)} \| \underline{U}^* - \Pi_h \underline{U}^* \|_{L^2(\Gamma)} \tag{3.8}$$

By using for the operator Π_h a local interpolation operator [CLEMENT, 1975], [VERFÜRTH, 1996], one can show that there exists a constant $\tilde{C}$ independent of the sizes of the elements such that

$$\| \underline{U}^* - \Pi_h \underline{U}^* \|_{L^2(E)} \le \tilde{C}\, h_E \| \underline{U}^* \|_{u,\tilde{E}}$$

and

$$\| \underline{U}^* - \Pi_h \underline{U}^* \|_{L^2(\Gamma)} \le \tilde{C}\, \ell_\Gamma^{1/2} \| \underline{U}^* \|_{u,\tilde{\Gamma}}$$

where $\tilde{E}$ is the set of the elements whose intersections with E are nonempty and $\tilde{\Gamma}$ is the set of the elements whose intersections with Γ are nonempty (Figure 3.1).

Introducing these inequalities into (3.8), one gets

$$A \le \tilde{C} \sum_{E} h_E \| \underline{r}_E \|_{L^2(E)} \| \underline{U}^* \|_{u,\tilde{E}} + \tilde{C} \sum_{\Gamma \not\subset \partial_1 \Omega} \ell_\Gamma^{1/2} \| \underline{t}_\Gamma \|_{L^2(\Gamma)} \| \underline{U}^* \|_{u,\tilde{\Gamma}}$$

Applying the discrete CAUCHY - SCHWARZ inequality, one gets

$$A \le \tilde{C} \left[\sum_{E} h_E^2 \| \underline{r}_E \|_{L^2(E)}^2 + \sum_{\Gamma \not\subset \partial_1 \Omega} \ell_\Gamma \| \underline{t}_\Gamma \|_{L^2(\Gamma)}^2 \right]^{1/2} * \left[\sum_{E} \| \underline{U}^* \|_{u,\tilde{E}}^2 + \sum_{\Gamma \not\subset \partial_1 \Omega} \| \underline{U}^* \|_{u,\tilde{\Gamma}}^2 \right]^{1/2}$$

Given that

$$\sum_{E} \| \underline{U}^* \|_{u,\tilde{E}}^2 \le \hat{C} \| \underline{U}^* \|_{u,\Omega}^2 \quad \text{and} \quad \sum_{\Gamma \not\subset \partial_1 \Omega} \| \underline{U}^* \|_{u,\tilde{\Gamma}}^2 \le \hat{C} \| \underline{U}^* \|_{u,\Omega}^2 \tag{3.9}$$

where $\hat{C}$ is a constant independent of the sizes, but which depends on the topology of the mesh and on the shapes of the elements, one gets ($C = 2\tilde{C}\hat{C}$)

$$A \leq C\| \underline{U}^*\|_{u,\Omega} \left[\sum_E h_E^2 \| \underline{r}_E \|_{L^2(E)}^2 + \sum_{\Gamma \not\subset \partial_1\Omega} \ell_\Gamma \| \underline{t}_\Gamma \|_{L^2(\Gamma)}^2 \right]^{1/2}$$

which, for $\underline{U}^* = \underline{e}_h$, yields (3.6).

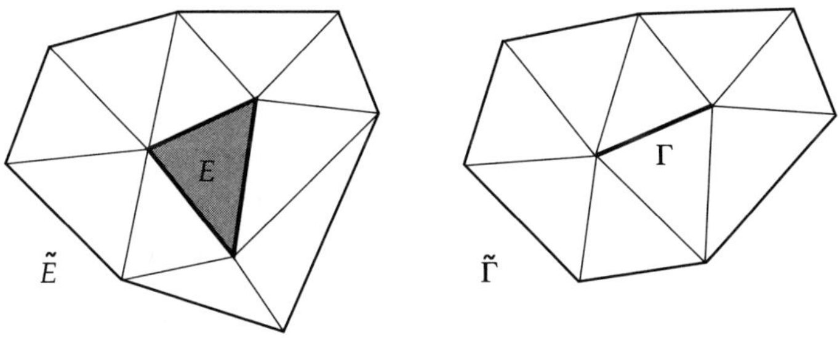

Figure 3.1. *Definition of* $\tilde{E}$ *and of* $\tilde{\Gamma}$.

Associated error estimator

In the estimate defined in (3.6), the right-hand side, except for the constant C, can be calculated explicitly from the data and the finite element solution. One can therefore use this expression as an estimator of the error. More precisely, on each element E, one defines the local estimator θ_E by

$$\theta_E^2 = h_E^2 \| \underline{r}_E \|_{L^2(E)}^2 + \sum_{\Gamma \subset \partial E} \beta_\Gamma \ell_\Gamma \| \underline{t}_\Gamma \|_{L^2(\Gamma)}^2 \tag{3.10}$$

with

$$\beta_\Gamma = \frac{1}{2} \quad \text{if } \Gamma \not\subset \partial\Omega, \qquad \beta_\Gamma = 1 \quad \text{if } \Gamma \subset \partial_2\Omega, \qquad \beta_\Gamma = 0 \quad \text{if } \Gamma \subset \partial_1\Omega$$

and the global estimator θ by

$$\theta^2 = \sum_E \theta_E^2 \tag{3.11}$$

Thus, one has

$$\| \underline{e}_h \|_{u,\Omega} \leq C\theta \tag{3.12}$$

Similarly, with the provision that the data $\underline{f}_d$ and $\underline{F}_d$ be polynomials of at most the same degree as the finite elements being used, one can show [VERFÜRTH, 1996], [AINSWORTH - ODEN, 1997] that there exists a constant C' independent of the sizes of the elements such that

$$\theta \leq C' \| \underline{e}_h \|_{u,\Omega} \tag{3.13}$$

which, along with (3.12), shows that the estimator θ has the same asymptotic behavior as the error in the solution.

REMARKS

1. If the loads $\underline{f}_d$ and $\underline{F}_d$ are not polynomials, one replaces them by polynomial approximations $\underline{f}_{dh}$ and $\underline{F}_{dh}$. One can then show that

$$\theta \leq C' \left[\| \underline{e}_h \|^2_{u,\Omega} + \sum_E h_E^2 \| \underline{f}_d - \underline{f}_{dh} \|^2_{L^2(E)} + \sum_{\Gamma \subset \partial_2 \Omega} \ell_\Gamma \| \underline{F}_d - \underline{F}_{dh} \|^2_{L^2(\Gamma)} \right]^{1/2}$$

2. It is possible, using similar techniques, to construct estimators of $\underline{e}_h$ for norms other than the energy norm.

3. One can also construct explicit estimators of the form

$$\theta_E^2 = a_E \| \underline{r}_E \|^2_{L^2(E)} + \sum_{\Gamma \subset \partial E} b_\Gamma \| \underline{t}_\Gamma \|^2_{L^2(\Gamma)}$$

where a_E and b_Γ are positive constants. Obviously, the difficulty is to choose these constants wisely in order to obtain a estimator of good quality.

3.2.3 Implicit estimators

Another means of utilizing Problem (3.3) is to seek local approximations of $\underline{e}_h$ as solutions of local problems defined on small groups of elements, then to use the results of these ancillary local calculations to construct error estimators. To distinguish them from the previous estimators which are calculated explicitly as functions of the finite element solution, these estimators are termed *implicit*.

Estimators by subdomains

For any node i of the mesh, let us denote by Ω_i the set of the elements connected to i (Figure 3.2) and by $\partial_0\Omega_i$ the part of $\partial\Omega_i$ not included in $\partial_2\Omega$.

Let us define an approximation $\underline{V}_i$ of $\underline{e}_h$ on Ω_i as the solution of the local problem:

Find the displacement field $\underline{V}_i$ *which is zero on* $\partial_0\Omega_i$ *and such that:*

$\forall \underline{U}^*$ *zero on* $\partial_0\Omega_i$,

$$\int_{\Omega_i} \mathrm{Tr}\left[\varepsilon(\underline{V}_i)\mathbf{K}\varepsilon(\underline{U}^*)\right] d\Omega = \sum_{E \subset \Omega_i} \int_E \underline{r}_E \cdot \underline{U}^* dE - \sum_{\Gamma \subset \Omega_i} \int_\Gamma \underline{t}_\Gamma \cdot \underline{U}^* d\Gamma \tag{3.14}$$

The idea is then to use as the global estimator the quantity

$$\theta^2 = \sum_i \int_{\Omega_i} \mathrm{Tr}\left[\varepsilon(\underline{V}_i)\mathbf{K}\varepsilon(\underline{V}_i)\right] d\Omega \tag{3.15}$$

This estimator can also be written in the form

$$\theta^2 = \sum_E \theta_E^2$$

where θ_E is the local estimator defined by

$$\theta_E^2 = \sum_{i \in E} \int_E \mathrm{Tr}\left[\varepsilon(\underline{V}_i)\mathbf{K}\varepsilon(\underline{V}_i)\right] dE \tag{3.16}$$

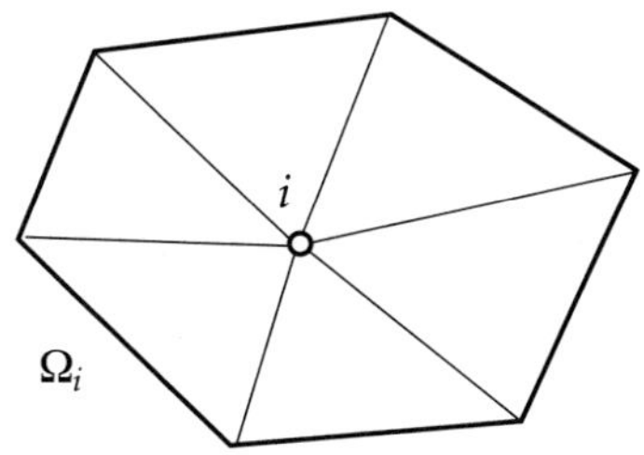

Figure 3.2. *Elements connected to node i.*

For example, one can find in [AINSWORTH - ODEN, 1997] the proof of the existence of constants C and C' such that

$$C\theta \leq \| \underline{e}_h \|_{u,\Omega} \leq C'\theta \tag{3.17}$$

These constants C and C' are independent of the sizes of the elements.

Nevertheless, the exact determination of the field $\underline{V}_i$ solution of (3.14) is impossible; this is again a matter of solving an elasticity problem! In practice, one discretizes Problem (3.14) by a finite element-type method which yields an approximation $\underline{V}_{ih}$ of $\underline{V}_i$. According to Property (3.5), the discretization must be more refined than that which was used for the initial analysis. For example, for triangular meshes in 2D, one can seek $\underline{V}_{ih}$ in the space $\boldsymbol{\mathcal{V}}_{ih}$ constructed as follows:

$\underline{V} \in \boldsymbol{\mathcal{V}}_{ih}$ *if,* $\forall E \subset \Omega_i$, *the restriction of* $\underline{V}$ *to* E *is of the form*

$$\underline{V}_{|E} = \lambda_1\lambda_2\lambda_3\, \underline{A}_E + \sum_{\substack{\Gamma \text{ edge of } E \\ \Gamma \not\subset \partial_0\Omega_i}} \lambda_i\lambda_j \underline{B}_\Gamma$$

where λ_1, λ_2, λ_3 are the barycentric coordinates of element E, i and j are the end nodes of Γ, and $\underline{A}_E$ and $\underline{B}_\Gamma$ are constant vectors.

Then, one can calculate a global estimator $\tilde{\theta}$ (and the associated local estimators $\tilde{\theta}_E$) by replacing in (3.15) and (3.16) the field $\underline{V}_i$ by the field $\underline{V}_{ih}$ which has been actually calculated.

These types of techniques can be easily extended to quadrilateral elements and to 3D elements (tetrahedra and bricks).

If the space $\boldsymbol{\mathcal{V}}_{ih}$ is properly chosen, the estimator $\tilde{\theta}$ also verifies inequalities similar to (3.17).

REMARKS

1. Other types of subdomains can also be used. For example, one can find in [VERFÜRTH, 1996] an example whose subdomains are the parts $\tilde{E}$ defined in Figure 3.1.

2. The method described above determines approximations of $\underline{e}_h$ by solving local problems with DIRICHLET-type boundary conditions. This leads to an underestimation of the true error [DIEZ - EGOZENZ - HUERTA, 1998]. In [STROUBOULIS - HAQUE, 1992a, 1992b], one can find an example using NEUMANN-type boundary conditions.

Estimators by elements

These types of estimators are constructed by seeking locally in each element E an approximation $\underline{W}_E$ of $\underline{e}_h$ as the solution of a NEUMANN-type problem in E. Let

$$\partial_1 E = \partial E \cap \partial_1 \Omega \quad \text{and} \quad \partial_2 E = \partial E \cap \partial_2 \Omega$$

One seeks $\underline{W}_E$ as the solution of the problem:

Find a displacement field $\underline{W}_E$ *which is zero on* $\partial_1 E$*, such that:*

$\forall \underline{U}^*$ *zero on* $\partial_1 E$,

$$\int_E \mathrm{Tr}\left[\varepsilon(\underline{W}_E)\mathbf{K}\varepsilon(\underline{U}^*)\right] dE = \int_E \underline{r}_E \cdot \underline{U}^* dE + \int_{\partial E - \partial_1 E} \underline{R} \cdot \underline{U}^* d\Gamma \tag{3.18}$$

where $\underline{R}$ is a surface force density which is prescribed on the sides of E which are not part of $\partial_1 \Omega$.

At this point, one must still choose the density $\underline{R}$, then discretize Problem (3.18) to obtain an approximation of $\underline{W}_E$.

In particular, if one chooses the densities $\underline{R}$ arbitrarily, Problem (3.18) may very well have no solution for the elements E which do not have a side in $\partial_1 \Omega$. Indeed, in this case, (3.18) has a solution if and only if the global equilibrium conditions on E are satisfied

$\forall \underline{U}_S$ *rigid body displacement field on* E,

$$\int_E \underline{r}_E \cdot \underline{U}_S \, dE + \int_{\partial E} \underline{R} \cdot \underline{U}_S \, d\Gamma = 0 \tag{3.19}$$

Two main methods have been proposed to eliminate these difficulties:

- the first method consists in discretizing (3.18) on a subspace $\boldsymbol{W}_{Eh}$ on which the approximate problem is regular, then choosing $\underline{R}$ in the best possible way;

• the second method consists in constructing $\underline{R}$ in such a way that the equilibrium conditions (3.19) on E are satisfied, then discretizing Problem (3.18).

Regularization of the problem

To make things clear, let us consider triangular elements and choose the subspace $\boldsymbol{W}_{Eh}$ defined by

$$\underline{W} \in \boldsymbol{W}_{Eh} \quad \textit{if and only if} \quad \underline{W} = \lambda_1\lambda_2\lambda_3\underline{A}_E + \sum_{\Gamma \subset \partial E - \partial_1 E} \lambda_i\lambda_j\underline{B}_\Gamma$$

where λ_1, λ_2, λ_3 are the barycentric coordinates of element E, i and j are the end nodes of Γ, and $\underline{A}_E$ and $\underline{B}_\Gamma$ are constant vectors.

Then, the problem:

Find a displacement field $\underline{W} \in \boldsymbol{W}_{Eh}$ *such that:*

$\forall \underline{U}^* \in \boldsymbol{W}_{Eh}$,

$$\int_E \mathrm{Tr}\left[\varepsilon(\underline{W}_{Eh})\mathbf{K}\varepsilon(\underline{U}^*)\right] dE = \int_E \underline{r}_E \cdot \underline{U}^*\, dE + \int_{\partial E - \partial_1 E} \underline{R} \cdot \underline{U}^*\, d\Gamma \tag{3.20}$$

is regular in each element, regardless of the choice of $\underline{R}$.

Usually, one chooses for the density $\underline{R}$,

$$\text{if} \quad \Gamma \subset \partial_2 E : \ \underline{R} = -\underline{t}_\Gamma \quad \text{and if} \quad \Gamma \subset \partial E - \partial_1 E \cup \partial_2 E : \ \underline{R} = -\frac{1}{2}\underline{t}_\Gamma \tag{3.21}$$

Thus, one obtains the local estimator

$$\theta_E^2 = \int_E \mathrm{Tr}\left[\varepsilon(\underline{W}_{Eh})\mathbf{K}\varepsilon(\underline{W}_{Eh})\right] dE \tag{3.22}$$

and the global estimator

$$\theta^2 = \sum_E \theta_E^2 \tag{3.23}$$

One can show [AINSWORTH - ODEN, 1997] that this estimator verifies inequalities which are similar to (3.17).

These techniques can also be extended to quadrilateral elements and to 3D elements (tetrahedra and bricks).

REMARK. On a side $\Gamma \subset \partial E - \partial_1 E \cup \partial_2 E$, the choice in (3.21) is equivalent to replacing the unknown vector $\sigma_{ex|E}\, \underline{n}_E$ in $\mathbf{K}\varepsilon(\underline{e}_h)\, \underline{n}_E = \sigma_{ex|E}\, \underline{n}_E - \sigma_{h|E}\, \underline{n}_E$ by the average of the finite element fields on both sides of Γ,

$$(\sigma_{h|E}\, \underline{n}_E + \sigma_{h|\tilde{E}}\, \underline{n}_E)/2$$

where $\tilde{E}$ is the element adjacent to E with respect to edge Γ.

Equilibrated densities

The objective is to construct the densities $\underline{R}$ in such a way that they verify the global equilibrium (3.19) on E. A simple and efficient method consists in using the densities $\eta_E\, \hat{\underline{F}}$ introduced in Section 2.4.3.

Then, one gets the following result:

Theorem

In each element E, the densities defined by

$$\underline{R} = \eta_E\, \hat{\underline{F}} - \sigma_{h|E}\, \underline{n}_E \tag{3.24}$$

are in equilibrium with the volume loads in the sense of Problem (3.18).

Proof

For any rigid body field $\underline{U}_S$, one has

$$\int_E \underline{f}_d \cdot \underline{U}_S\, dE + \int_{\partial E} \eta_E\, \hat{\underline{F}} \cdot \underline{U}_S\, d\Gamma = 0$$

Thus, one gets

$$\int_E \underline{f}_d \cdot \underline{U}_S\, dE + \int_{\partial E} (\underline{R} + \sigma_{h|E}\, \underline{n}_E) \cdot \underline{U}_S\, d\Gamma = 0$$

Noting that

$$\int_{\partial E} \sigma_{h|E}\, \underline{n}_E \cdot \underline{U}_S\, dS = \int_E \operatorname{div}\left[\sigma_{h|E}\, \underline{U}_S\right] dE = \int_E \underline{\operatorname{div}}\, \sigma_{h|E} \cdot \underline{U}_S\, dE$$

one finally has

$$\int_E \left[\underline{f}_d + \underline{\operatorname{div}}\, \sigma_{h|E}\right] \cdot \underline{U}_S\, dE + \int_{\partial E} \underline{R} \cdot \underline{U}_S\, d\Gamma = 0$$

which is the equilibrium condition (3.19).

REMARKS

1. For thermal problems, KELLY [KELLY, 1984] proposed a method of construction of equilibrated densities for one-dimensional elements and for quadrangular elements.

2. The method (called "flux splitting") introduced for thermal-type problems in [AINSWORTH - ODEN, 1993, 1997] consists, in the case of elasticity, in seeking $\underline{R}$ in the form

$$\underline{R}_{|\Gamma} = \eta_E \underline{A}_\Gamma - \frac{1}{2}\underline{t}_\Gamma$$

where $\underline{A}_\Gamma$ is an affine vector field on Γ and is zero if $\Gamma \subset \partial_2\Omega$.

Therefore, this method is a particular version of our method for constructing densities $\hat{\underline{F}}$ which was outlined in Chapter 2 and which we detail in Chapter 8. Its authors viewed that technique as an extension of the technique developed, again for thermal problems discretized with first-degree elements, in [BANK - WEISER, 1985].

To solve Problem (3.18), which is now well-posed, one can use all the techniques described in Chapter 2, which involve either a local approximate resolution by the dual method (seeking a simple piecewise polynomial solution by subelements) or an approximate solution by the primal method. For example, BABUSKA and STROUBOULIS recommend the use of a single finite element of degree $p+k$, where k is a small strictly positive integer.

3.2.4 Comments

This family of error estimators, first introduced by BABUSKA and RHEINBOLDT in 1978, is based on a strong mechanical property:

The nonsatisfaction of the equilibrium equations by the finite element solution.

Its implementation is relatively straightforward. However, it runs into a major difficulty:

How can one define a simple and mechanically meaningful norm for the residuals of the equilibrium equations, which are split into volume terms and boundary terms on the sides of the elements?

In mathematical terms, this is an H^{-1} norm; in practice, one uses an expression of the (3.6) type, which involves constants whose values are open to discussion. In addition, the effectivity index is, in many cases, relatively mediocre.

For nonlinear problems, there is an additional difficulty: the error $\underline{e}_h$ is no longer always indicative of the distance between the finite element solution and the exact reference solution. In particular, an upper bound such as (3.6) may no longer exist.

3.3 Methods based on a smoothing of the stresses

3.3.1 General principle

The principle of these methods is extremely simple. Starting from the finite element solution σ_h, which has discontinuities across the interfaces between elements, one reconstructs by a smoothing technique a stress field $\tilde{\sigma}_h$ which is continuous in Ω.

Then, the error $\sigma_{ex} - \sigma_h$ is estimated by replacing the unknown exact solution by the smoothed stress

$$\sigma_{ex} - \sigma_h \approx \tilde{\sigma}_h - \sigma_h$$

Thus, one ends up with the error indicator

$$\theta = \left[\int_\Omega Tr\left[(\tilde{\sigma}_h - \sigma_h)\mathbf{K}^{-1}(\tilde{\sigma}_h - \sigma_h) \right] d\Omega \right]^{1/2} \tag{3.25}$$

and, of course, the local contributions

$$\theta_E = \left[\int_E \mathrm{Tr}\left[(\tilde{\sigma}_h - \sigma_h)\mathbf{K}^{-1}(\tilde{\sigma}_h - \sigma_h) \right] dE \right]^{1/2}$$

The differences among the various indicators proposed in the literature reside in the smoothing method used to construct $\tilde{\sigma}_h$.

3.3.2 The ZZ1 method

This method [ZIENKIEWICZ - ZHU, 1987] consists in seeking $\tilde{\mathbb{\sigma}}_h$ as a linear combination of the shape functions used in the calculation of $\mathbb{\sigma}_h$,

$$\tilde{\mathbb{\sigma}}_h = \sum_{i=1}^{N} \mathbb{A}_i \varphi_i \tag{3.26}$$

where the $\mathbb{A}_i$ are constant symmetric operators.

Several techniques have been proposed for determining the operators $\mathbb{A}_i$.

Global projection

The distance between $\tilde{\mathbb{\sigma}}_h$ and $\mathbb{\sigma}_h$ is minimized using a least-squares technique

$$\underset{\{\mathbb{A}_i\}}{\text{Min}} \int_\Omega \text{Tr}\left[(\tilde{\mathbb{\sigma}}_h - \mathbb{\sigma}_h)\mathbf{K}^{-1}(\tilde{\mathbb{\sigma}}_h - \mathbb{\sigma}_h)\right] d\Omega \tag{3.27}$$

which leads to the resolution of the global linear system

$$\forall i = 1, 2, \dots, N: \quad \int_\Omega \mathbf{K}^{-1}(\tilde{\mathbb{\sigma}}_h - \mathbb{\sigma}_h)\varphi_i \, d\Omega = 0 \tag{3.28}$$

To cut costs, (3.28) is solved by an iterative method.

REMARK. If the HOOKE operator is constant, System (3.28) becomes

$$\forall i = 1, 2, \dots, N: \quad \int_\Omega (\tilde{\mathbb{\sigma}}_h - \mathbb{\sigma}_h)\varphi_i \, d\Omega = 0$$

i.e.

$$\sum_k \left[\int_\Omega \varphi_i \varphi_k \, d\Omega\right] \mathbb{A}_k = \int_\Omega \mathbb{\sigma}_h \varphi_i \, d\Omega$$

Thus, for each component of the $\mathbb{A}_i$ one must solve a linear system whose matrix is the standard mass matrix with a density equal to 1,

$$\mathbb{M} = \left[\int_\Omega \varphi_i \varphi_k \, d\Omega\right]$$

An easy way of simplifying this resolution consists in replacing $\mathbb{M}$ by a lumped, i.e. diagonal, mass matrix.

Local averages

In order to avoid having to solve the global problem (3.28), more local methods of calculation of the $\mathbb{A}_i$ have been used. For example, a very simple method consists in defining the stress $\mathbb{A}_i$ as a weighted average of the values of the stress σ_h at the GAUSS points in the vicinity of node I,

$$\mathbb{A}_i = \left[\sum_{k=1}^{K} \alpha_k \, \sigma_h(P_k) \right] \left[\sum_{k=1}^{K} \alpha_k \right]^{-1} \tag{3.29}$$

where P_k are the GAUSS points and α_k are the weighting coefficients which depend on the geometry of the mesh.

REMARKS

1. The implementation of this estimator is simple. Unfortunately, it is not rooted in a strong mechanical concept. Clearly, it cannot be adequate for coarse meshes; indeed, its value is zero if the mesh consists of only one element. Moreover, in the case of interpolations of even degree, it has been shown in [STROUBOULIS - HAQUE, 1992a, 1992b] that it can greatly underestimate the true error. Today, this estimator is becoming obsolete.

2. A priori, any smoothing technique can be used. For example, one can find a use of HINTON and CAMPBELL's smoothing techniques [HINTON - CAMPBELL, 1974] in [ZIENKIEWICZ - ZHU, 1992].

3.3.3 Recovery by local smoothing (ZZ2)

This method, initially developed by ZIENKIEWICZ and ZHU [ZIENKIEWICZ - ZHU, 1992], is known under the name *Superconvergent Patch Recovery* for reasons that will be explained below. It leads to a ZZ2 estimator which does not present the shortcomings of the ZZ1 estimator, but which is also more costly to calculate.

Principle of the method

- One chooses a family of domains Ω_R (a connected union of elements of the mesh) covering Ω.
- In each element E of the mesh, one chooses a (finite) set of sampling points G_E. Let

$$G_R = \bigcup_{E \subset \Omega_R} G_E$$

- On each Ω_R, one seeks a polynomial stress field in the form

$$\sigma_R(M) = \sum_{q=1}^{Q} \mathbb{B}_q \, p_q(M) \quad (M \in \Omega_R) \tag{3.30}$$

where $\mathbb{B}_q$ are symmetric constant operators yet to be determined and $\{p_1, p_2, \ldots, p_Q\}$ is a given basis of polynomials.

This field σ_R is determined by a least-squares method by minimizing

$$\underset{\{\mathbb{B}_q\}}{\text{Min}} \left[\sum_{P \in G_R} \text{Tr}\left[\left(\sigma_R(P) - \sigma_h(P) \right)^2 \right] \right] \tag{3.31}$$

- For the nodes i belonging to a subset N_R chosen among the nodes belonging to Ω_R, let

$$\mathbb{A}_i^R = \sigma_R(i)$$

- Then, for a particular node i, the operator $\mathbb{A}_i$ is the average of the various operators $\mathbb{A}_i^R$ calculated by the above method

$$\mathbb{A}_i = \frac{1}{m} \sum_R \mathbb{A}_i^R$$

where m is the number of domains Ω_R such that $i \in N_R$.

Implementation

Usually, the domains Ω_R are the domains Ω_i already introduced above (Figure 3.2), except possibly for the nodes located on the boundary, for which it is often preferable to choose Ω_R larger than Ω_i. On a domain Ω_R, the basis of polynomials $\{p_1, p_2, \ldots, p_Q\}$ is the basis associated with the type of finite element being used. For example $\{1, x, y\}$ for 3-node triangles, $\{1, x, y, x^2, xy, y^2\}$ for 6-node triangles, $\{1, x, y, xy\}$ for 4-node quadrilaterals..., and similar choices in three dimensions.

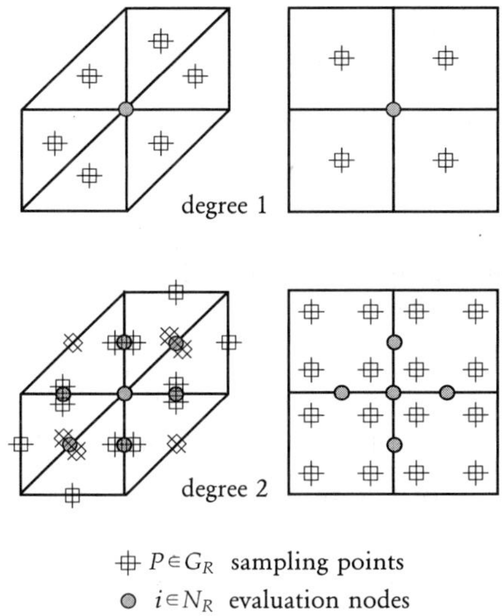

Figure 3.3. *Examples of domains* Ω_R.

The choice of the sampling points is more difficult. In the 1D case, the sampling points are usually the GAUSS points of the elements being used, which are superconvergence points for the derivative quantities of the stress type [LIEUSAINT - ZLAMAL, 1979], [WHALBIN, 1995]. In 2D or 3D, superconvergence properties are more uncommon; in fact, they require very specific meshes which are generally incompatible with the objective of mesh adaptation.

In these circumstances, the choice of the sampling points exploits either an extrapolation of the superconvergence properties, or the results of numerical experimentations [ZIENKIEWICZ - ZHU, 1992], [ZHONG, 1991], [DUFEU, 1997].

Figure 3.3 gives some examples of choices along with the corresponding evaluation nodes for standard 2D elements. Similar techniques are used in 3D.

3.3.4 Some variants

The ZZ2 estimator works remarkably well on the global level. Different variants and improvements have been proposed; one can especially mention the partial accounting for the equilibrium conditions and the substitution of averages over subdomains for values at the superconvergence points.

The WIBERG - ABDULWAHAB variant

This version consists in the partial introduction of equilibrium conditions in the construction of the field σ_R defined in (3.30). More precisely, one replaces the functional to be minimized (3.31) by the new functional

$$\underset{\{\mathbb{B}_q\}}{\text{Min}} \left[\sum_{P \in G_R} \text{Tr}\left[\left(\sigma_R(P) - \sigma_h(P) \right)^2 \right] + \beta \int_{\Omega_R} (\text{div } \sigma_R(P) + \underline{f}_d)^2 \, d\Omega \right] \tag{3.32}$$

where β is a weighting parameter allowing the importance of the role played by the equilibrium conditions to be adjusted. One can also interpret (3.32) by considering that in order to determine σ_R one introduces the satisfaction of the interior equilibrium of Ω_R as a constraint; thus, β is interpreted as a penalty factor. In the applications given in [WIBERG -ABDULWAHAB, 1992], the parameter β is most often taken equal to 1. Similar variants can be found in [WIBERG - ABDULWAHAB - ZIUKAS, 1994] and in [BLACKER - BELYTSCHKO, 1994]. These variants enable the equilibrium conditions on $\partial_2\Omega$ to be enforced partially.

The BOROOMAND - ZIENKIEWICZ and BECKERS - DUFEU variant

This more recent variant consists in requiring, for the stress σ_R to verify on Ω_R, the condition

$$\int_{\Omega_R} (\sigma_R - \sigma_h) \, \underline{\text{grad}} \, p_q \, d\Omega = 0 \quad \text{for } q = 1, 2, \ldots, Q \tag{3.33}$$

which is similar to the extension condition (2.19).

Since System (3.33) is generally overdetermined, it is verified only in a least-squares sense.

Thus, the stress σ_R is the solution of the minimization problem

$$\underset{\{\mathbb{B}_q\}}{\text{Min}} \left[\int_{\Omega_R} (\sigma_R - \sigma_h)\, \underline{\text{grad}}\, p_q \, d\Omega \right]^2 \tag{3.34}$$

This version was developed independently by ZIENKIEWICZ under the name REP (Recovery by Equilibrium in Patches) and by BECKERS under the name IPR (Integrated Patch Recovery).

Similar applications and variants can be found in [BOROOMAND - ZIENKIEWICZ, 1997] and [BECKERS - DUFEU, 1996] for 1D and 2D, and in [DUFEU, 1997] for the 3D elasticity case.

3.3.5 Comments

According to the works of BABUSKA and STROUBOULIS [BABUSKA - STROUBOULIS - UPADHYAY - GANGARAJ, 1994], [BABUSKA - STROUBOULIS - GANGARAJ, 1997], WIBERG's variant does not always result in an improvement. In fact, these authors showed that the original ZZ2 estimator is often more effective.

The variant which exploits the extension conditions (3.33) appears to give the same quality of results as the ZZ2 estimator. It is not easy to explain why, because in our experience in Cachan the quantity

$$\int_{\Omega_R} \sigma_h \, \underline{\text{grad}}\, p_q \, d\Omega$$

is definitely not superconvergent. In other terms, if the conditions (3.33) were satisfied exactly, the associated error estimator should be clearly inferior to ZZ2. What makes the associated estimators work well is simply that the choices made in practice lead to only approximate satisfaction of the conditions (3.33).

3.4 Iterative bounding techniques

These techniques introduced in [BABUSKA - STROUBOULIS - GANGARAJ, 1999], [GANGARAJ, 1999], [STROUBOULIS - BABUSKA - GANGARAJ, 2000] lead to bounding intervals of the error getting narrower and narrower as the number of iterations increases.

At each iteration, one must solve one standard global finite element problem plus two local problems on each element.

In our presentation, we will assume that these local problems can be solved exactly. In practice, one could be satisfied with an approximate solution using a displacement method with a single element of degree $p+k$ with, for example, $k=3$ as in Section 2.4.3.

3.4.1 The basic identity

Our starting point is Problem (3.3), which is verified by the error $\underline{e}_h = \underline{U}_{ex} - \underline{U}_h$,

$$\underline{e}_h \in \boldsymbol{\mathcal{U}}_0 \quad \text{and} \quad \forall \underline{U}^* \in \boldsymbol{\mathcal{U}}_0,$$

$$\int_\Omega \mathrm{Tr}\left[\varepsilon(\underline{e}_h)\mathbf{K}\varepsilon(\underline{U}^*)\right] d\Omega = \sum_E \int_E \underline{r}_E \cdot \underline{U}^* dE - \sum_\Gamma \int_\Gamma \underline{t}_\Gamma \cdot \underline{U}^* d\Gamma \tag{3.35}$$

For $\Gamma \subset \partial\mathrm{E}$, let us introduce, as in (3.24), the densities

$$\underline{R}_\Gamma^E = \eta_E \hat{\underline{F}} - \sigma_{h|E}\underline{n}_E \tag{3.36}$$

For each element E, these densities are in equilibrium inside E with the residual $\underline{r}_E$, and on each side Γ one has:

- if Γ is a side in the interior of Ω,

$$\underline{R}_\Gamma^E + \underline{R}_\Gamma^{\tilde{E}} = -\,\underline{t}_\Gamma;$$

- if $\Gamma \subset \partial_2\Omega$,

$$\underline{R}_\Gamma^E = -\underline{t}_\Gamma.$$

Consequently, the right-hand side of (3.35) can be rewritten as

$$\sum_E \left[\int_E \underline{r}_E \cdot \underline{U}^* dE + \int_{\partial E} \underline{R}_\Gamma^E \cdot \underline{U}^* d\Gamma\right] \tag{3.37}$$

Let us now introduce the following two local problems defined on each element E:

- *Problem* $\boldsymbol{\mathcal{P}}_{\mathrm{E}}$:

  Find $\hat{\underline{e}}_{Eh} \in \boldsymbol{\mathcal{U}}_{\mathrm{E}}$ *such that:*

  $\forall \underline{U}^* \in \boldsymbol{\mathcal{U}}_{\mathrm{E}},$

$$\int_E \mathrm{Tr}[\varepsilon(\hat{\underline{e}}_{Eh})\mathbf{K}\varepsilon(\underline{U}^*)]\,dE = \int_E \underline{r}_E \cdot \underline{U}^*\,dE + \int_{\partial E} \underline{R}_\Gamma^E \cdot \underline{U}^*\,d\Gamma \tag{3.38}$$

- *Problem* $\boldsymbol{\mathcal{P}}_{\mathrm{E}}^0$:

  Find $\tilde{\underline{e}}_{Eh} \in \boldsymbol{\mathcal{U}}_{0\mathrm{E}}$ *such that:*

  $\forall \underline{U}^* \in \boldsymbol{\mathcal{U}}_{0\mathrm{E}},$

$$\int_E \mathrm{Tr}[\varepsilon(\tilde{\underline{e}}_{Eh})\mathbf{K}\varepsilon(\underline{U}^*)]\,dE = \int_E \underline{r}_E \cdot \underline{U}^*\,dE + \int_{\partial E} \underline{R}_\Gamma^E \cdot \underline{U}^*\,d\Gamma \tag{3.39}$$

where $\boldsymbol{\mathcal{U}}_{\mathrm{E}}$ denotes the set of the displacement fields which are regular on E and $\boldsymbol{\mathcal{U}}_{0\mathrm{E}}$ the subspace of the displacement fields which are zero on ∂E.

REMARK. The existence of solutions to problem $\boldsymbol{\mathcal{P}}_{\mathrm{E}}$ follows from the equilibrium properties verified by the data $\underline{r}_E$ and $\underline{R}_\Gamma^E$ on element E. To ensure their uniqueness, one can, for example, require that the resultant and the moment of field $\hat{\underline{e}}_{Eh}$ be zero on E.

One gets the following properties:

Properties

$$\| \underline{e}_h \|_{u,\Omega}^2 = \| \underline{\delta}_h \|_{u,\Omega}^2 + \sum_E \left[\| \hat{\underline{e}}_{Eh} \|_{u,E}^2 - \| \hat{\underline{e}}_{Eh} - \tilde{\underline{e}}_{Eh} \|_{u,E}^2 \right] \tag{3.40}$$

$$\| \underline{e}_h \|_{u,\Omega}^2 = \| \underline{\delta}_h \|_{u,\Omega}^2 + \sum_E \| \tilde{\underline{e}}_{Eh} \|_{u,E}^2 \tag{3.41}$$

where $\underline{\delta}_h$ *is the displacement field solution of the problem:*
Find $\underline{\delta}_h \in \boldsymbol{\mathcal{U}}_0$ *such that:*

$$\forall \underline{U}^* \in \boldsymbol{\mathcal{U}}_0, \quad \int_\Omega \mathrm{Tr}\left[\varepsilon(\underline{\delta}_h)\mathbf{K}\varepsilon(\underline{U}^*)\right] d\Omega = \sum_E \int_E \mathrm{Tr}\left[\varepsilon(\hat{\underline{e}}_{Eh} - \tilde{\underline{e}}_{Eh})\mathbf{K}\varepsilon(\underline{U}^*)\right] dE \tag{3.42}$$

Proof

Let

$$A = \| \underline{e}_h \|^2_{u,\Omega} + \sum_E \left[\left\| \underline{\hat{e}}_{Eh} - \underline{\tilde{e}}_{Eh} \right\|^2_{u,E} - \left\| \underline{\hat{e}}_{Eh} \right\|^2_{u,E} \right]$$

Noting that

$$\| \bullet \|^2_{u,\Omega} = \sum_E \| \bullet \|^2_{u,E}$$

one gets

$$A = \sum_E \left[\| \underline{e}_h \|^2_{u,E} + \left\| \underline{\tilde{e}}_{Eh} \right\|^2_{u,E} - 2 \int_E \mathrm{Tr}\left[\varepsilon(\underline{\hat{e}}_{Eh}) \mathbf{K} \varepsilon(\underline{\tilde{e}}_{Eh}) \right] dE \right]$$

Let us now consider the field $\underline{\tilde{e}}_h$ defined by

$$\forall E, \quad \underline{\tilde{e}}_{h|E} = \underline{\tilde{e}}_{Eh}$$

Using (3.35) for $\underline{U}^* = \underline{\tilde{e}}_{Eh}$ on E and $\underline{U}^* = 0$ otherwise, then (3.38) for $\underline{U}^* = \underline{\tilde{e}}_{Eh}$, one gets

$$\int_E \mathrm{Tr}\left[\varepsilon(\underline{\hat{e}}_{Eh}) \mathbf{K} \varepsilon(\underline{\tilde{e}}_{Eh}) \right] dE = \int_E \mathrm{Tr}\left[\varepsilon(\underline{e}_h) \mathbf{K} \varepsilon(\underline{\tilde{e}}_{Eh}) \right] dE$$

Therefore

$$A = \sum_E \left[\left\| \underline{e}_h - \underline{\tilde{e}}_{Eh} \right\|^2_{u,E} \right] = \left\| \underline{e}_h - \underline{\tilde{e}}_h \right\|^2_{u,\Omega}$$

Besides, noting that, for $\underline{U}^* \in \boldsymbol{\mathcal{U}}_0$,

$$\sum_E \int_E \mathrm{Tr}\left[\varepsilon(\underline{\hat{e}}_{Eh}) \mathbf{K} \varepsilon(\underline{U}^*) \right] dE = \sum_E \left[\int_E \underline{r}_E \cdot \underline{U}^* dE + \int_{\partial E} \underline{R}^E_\Gamma \cdot \underline{U}^* d\Gamma \right]$$

and using (3.35) and (3.39), one obtains

$$\underline{\delta}_h = \underline{e}_h - \underline{\tilde{e}}_h$$

Therefore, one has $A = \| \underline{\delta}_h \|^2_{u,\Omega}$, which proves Property (3.40).

To prove Property (3.41), it suffices to observe that

$$\| \hat{\underline{e}}_{Eh} \|^2_{u,E} - \| \hat{\underline{e}}_{Eh} - \tilde{\underline{e}}_{Eh} \|^2_{u,E} = 2\int_E \mathrm{Tr}\big[\varepsilon(\hat{\underline{e}}_{Eh})\,\mathbf{K}\,\varepsilon(\tilde{\underline{e}}_{Eh})\big]\,dE - \| \tilde{\underline{e}}_{Eh} \|^2_{u,E}$$

$$= \int_E \mathrm{Tr}\big[\varepsilon(\tilde{\underline{e}}_{Eh})\mathbf{K}\varepsilon(2\hat{\underline{e}}_{Eh} - \tilde{\underline{e}}_{Eh})\big]\,dE = \| \tilde{\underline{e}}_{Eh} \|^2_{u,E}$$

3.4.2 The iterative bounding technique

Initial upper and lower bounds

Property

One has the following upper and lower bounds

$$0 \le \| \underline{\delta}_h \|^2_{u,\Omega} \le \sum_E \| \hat{\underline{e}}_{Eh} - \tilde{\underline{e}}_{Eh} \|^2_{u,E} \tag{3.43}$$

Proof

It is obvious that $\| \underline{\delta}_h \|^2_{u,\Omega} \ge 0$. Besides, from (3.42), one gets

$$\left| \int_\Omega \mathrm{Tr}\big[\varepsilon(\underline{\delta}_h)\mathbf{K}\varepsilon(\underline{U}^*)\big]\,d\Omega \right| \le \sum_E \| \hat{\underline{e}}_{Eh} - \tilde{\underline{e}}_{Eh} \|_{u,E} \| \underline{U}^* \|_{u,E}$$

then

$$\left| \int_\Omega \mathrm{Tr}\big[\varepsilon(\underline{\delta}_h)\mathbf{K}\varepsilon(\underline{U}^*)\big]\,d\Omega \right| \le \left[\sum_E \| \hat{\underline{e}}_{Eh} - \tilde{\underline{e}}_{Eh} \|^2_{u,E} \right]^{1/2} \left[\sum_E \| \underline{U}^* \|^2_{u,E} \right]^{1/2}$$

which, by taking $\underline{U}^* = \underline{\delta}_h$, yields (3.43).

One deduces the following upper and lower bounds for $\| \underline{e}_h \|_{u,\Omega}$:

Bounding property

$$\left[\sum_E \| \tilde{\underline{e}}_{Eh} \|^2_{u,E} \right]^{1/2} \le \| \underline{e}_h \|_{u,\Omega} \le \left[\sum_E \| \hat{\underline{e}}_{Eh} \|^2_{u,E} \right]^{1/2} \tag{3.44}$$

Proof

The upper bound in (3.44) follows immediately from (3.40) and (3.43), and the lower bound results from (3.41).

Iteration of the process

The basic objective is to improve the upper and lower bounds (3.43) of $\| \underline{\delta}_h \|^2_{u,\Omega}$. In order to do that, one performs a new standard finite element analysis of Problem (3.42) which yields an approximation $\underline{\delta}_{hh}$ of $\underline{\delta}_h$.

The corresponding error is

$$\underline{e}^{\delta}_{hh} = \underline{\delta}_h - \underline{\delta}_{hh}$$

which is the solution of the problem

$$\underline{e}^{\delta}_{hh} \in \boldsymbol{\mathcal{U}}_0 \quad and \quad \forall \underline{U}^* \in \boldsymbol{\mathcal{U}}_0,$$

$$\int_\Omega \mathrm{Tr}\left[\varepsilon(\underline{e}^{\delta}_{hh})\mathbf{K}\varepsilon(\underline{U}^*)\right] d\Omega = \sum_E \int_E \underline{r}^{\delta}_E \cdot \underline{U}^* dE - \sum_\Gamma \int_\Gamma \underline{t}^{\delta}_\Gamma \cdot \underline{U}^* d\Gamma$$

One can now bound $\| \underline{e}^{\delta}_{hh} \|_{u,\Omega}$ using the same technique as above

$$\left[\sum_E \| \underline{\tilde{e}}^{\delta}_{Ehh} \|^2_{u,E}\right]^{1/2} \leq \| \underline{e}^{\delta}_{hh} \|_{u,\Omega} \leq \left[\sum_E \| \underline{\hat{e}}^{\delta}_{Ehh} \|^2_{u,E}\right]^{1/2} \tag{3.45}$$

where fields $\underline{\hat{e}}^{\delta}_{Ehh}$ and $\underline{\tilde{e}}^{\delta}_{Ehh}$ are solutions of local problems similar to (3.38) and (3.39), but based on the residuals $\underline{r}^{\delta}_E$ and $\underline{t}^{\delta}_\Gamma$.

Observing that

$$\| \underline{\delta}_h \|^2_{u,\Omega} = \| \underline{\delta}_{hh} \|^2_{u,\Omega} + \| \underline{e}^{\delta}_{hh} \|^2_{u,\Omega}$$

one can derive from (3.45) new upper and lower bounds of $\| \underline{\delta}_h \|_{u,\Omega}$ and, thus, new upper and lower bounds of the error $\| \underline{e}_h \|_{u,\Omega}$.

The results given in [STROUBOULIS - BABUSKA - GANGARAJ, 2000] clearly show that the upper and lower bound pairs improve during the iterations.

3.4.3 Comments

The principle of the strategy described above is very general. The quality of the upper and lower bounds depends on the numerical effort invested.

Carrying out the procedure requires the resolution of local problems similar to $\boldsymbol{\mathcal{P}}_E$ and $\boldsymbol{\mathcal{P}}^0_E$ at each iteration.

In practice, one could be satisfied with the approximate solution obtained with a single element of order $p+k$ (with, for example, $p=3$). The resulting errors are very small.

Of course, it is also possible to take these errors into account in the upper and lower bounds proposed.

Chapter 4

Principles of the comparison of the various estimators in the linear case

4.1 Introduction

The objective of this chapter is to give some basic principles of comparison among the various types of estimators introduced in Chapters 2 and 3 in the context of linear problems.

To begin with, very simple examples of the implementation of errors in constitutive relation will be given for 2D and 3D problems.

Then, using the notion of global effectivity index, the global quality of the error estimators will be examined on the basis of results obtained at LMT-Cachan and additional results from the literature. The most comprehensive comparative analysis of the various estimators available was conducted by BABUSKA and STROUBOULIS [STROUBOULIS - HAQUE, 1992a, 1992b], [BABUSKA – STROUBOULIS - UPADHYAY - GANGARAJ, 1994, 1995], [BABUSKA - STROUBOULIS - UPADHYAY - GANGARAJ - COPPS, 1994]. That study, which was performed following a clearly defined protocol, used the theory of homogenization of continuous periodic media [SANCHEZ - PALENCIA, 1980], [DUVAUT, 1976]. Other studies have also been published, e.g. [ODEN - DEMKOWICZ - RACHOWICZ - WESTERMANN, 1989], [BECKERS - ZHONG, 1991], [LADEVEZE - ROUGEOT, 1997].

4.2 Errors in constitutive relation

4.2.1 A 2D example

Let us consider (Figure 4.1) a short beam in bending with applied loads such that the exact solution for the stress is known. In order to achieve this, one assumes that $\underline{f}_d = 0$ and at the extremities one prescribes the following load densities:

on $y = 0$ and on $y = L$: $\underline{F}_d = 0$

on $x = 0$: $\underline{F}_d = \left(\frac{2y}{L} - 1\right)\underline{x} - y\left(1 - \frac{y}{L}\right)\underline{y}$

on $x = L$: $\underline{F}_d = y\left(1 - \frac{y}{L}\right)\underline{y}$

The exact stress field is then given by

$$\sigma_{ex} = \sigma_0 \begin{bmatrix} \left(\frac{2y}{L} - 1\right)\left(\frac{x}{L} - 1\right) & \frac{y}{L}\left(1 - \frac{y}{L}\right) \\ \frac{y}{L}\left(1 - \frac{y}{L}\right) & 0 \end{bmatrix}_{(\underline{x}, \underline{y})} \tag{4.1}$$

Thus, it is possible to calculate the exact error between the true solution and the finite element solution. These calculations were carried out on uniform meshes composed of 3- or 6-node triangles and 4- or 8-node quadrilaterals. The results obtained are plotted in Figure 4.2 against the computation time. (All calculations were made on the same machine, and in every case we used the same program to solve the linear system by the CHOLESKI method.)

This diagram reflects numerically the hierarchy among the four types of elements tested which is well-known to users.

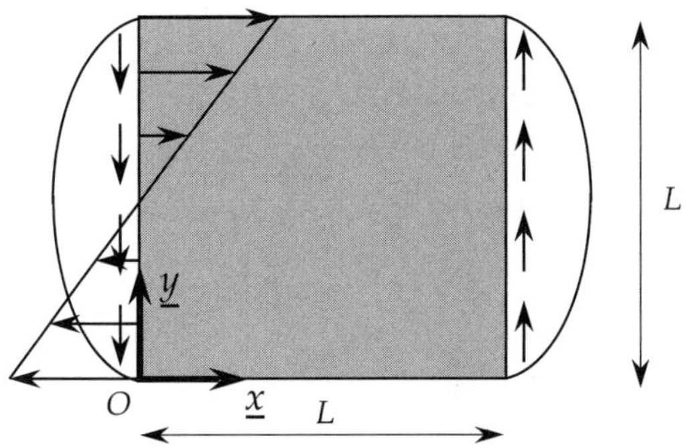

Figure 4.1. *Problem of mechanics - known exact solution.*

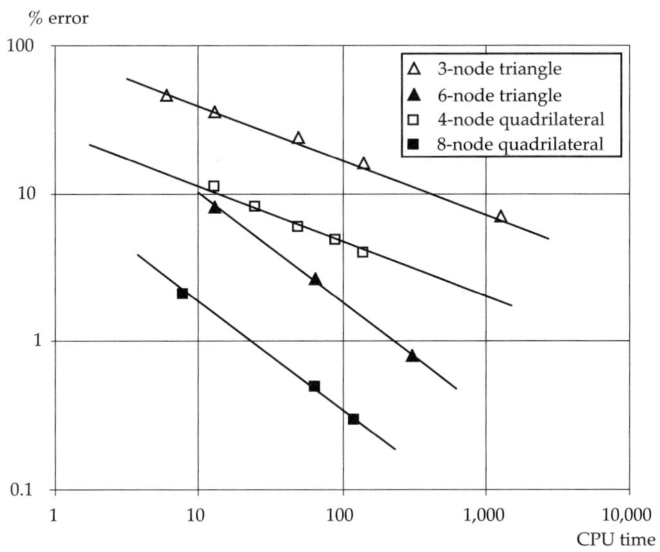

Figure 4.2. *Comparison of element types.*

4.2.2 Comparison of the different methods of construction

The objective of this example is to compare the different methods of construction of admissible stress fields: analytical construction as well as approximate construction techniques using a finite element of degree "$p+k$". This discussion is taken from [COOREVITS - DUMEAU - PELLE, 1999].

For these comparison tests, we will use the short beam of Figure 4.1 again. The solution is given in 2D by (4.1), and in 3D by

$$\boldsymbol{\sigma}_{ex} = \sigma_0 \begin{bmatrix} \left(\frac{2y}{L}-1\right)\left(\frac{x}{L}-1\right) & \frac{y}{L}\left(1-\frac{y}{L}\right) & 0 \\ \frac{y}{L}\left(1-\frac{y}{L}\right) & 0 & 0 \\ 0 & 0 & 0 \end{bmatrix}_{(\underline{x}, \underline{y}, \underline{z})}$$

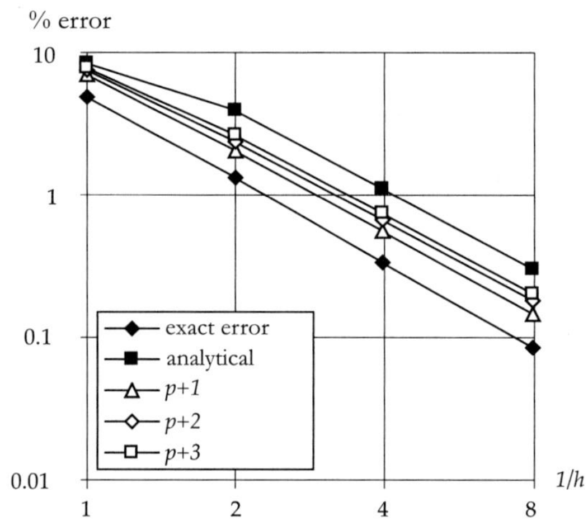

Figure 4.3. *Comparison of error estimates in 2D.*

In the 2D case, discretization was performed using 3-node triangles ($p = 1$). Figure 4.3 shows the error estimates obtained by analytical construction of $\hat{\boldsymbol{\sigma}}$ and by the approximate constructions (using in each element a $p + k$ method with $k = 1, 2$ or 3).

For the 3D calculations, discretization was performed using 4-node tetrahedra. The estimates obtained by analytical construction and by the approximate constructions (using in each element a $p + k$ method with $k = 1$ or 2) are compared in Figure 4.4. As observed in the works of BABUSKA and STROUBOULIS, the quality of the estimator obtained with $k = 1$ is very good. In 2D, we also note that for $k = 3$ we obtained an estimate very close to that obtained by analytical construction.

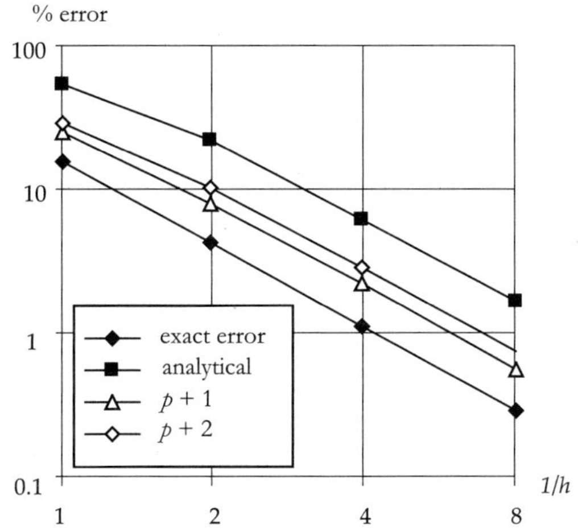

Figure 4.4. *Comparison of error estimates in 3D.*

4.3 Global effectivity index

It has been proven (Chapter 2) that for errors in constitutive relation the global effectivity index is always greater than 1; usually, in numerical experiments, it ranges between 1 and 2.5. However, if the mesh contains a large number of elements with very high aspect ratios, its value can be much larger. Let us now examine this effectivity index for various estimators.

4.3.1 Comparison of estimators for a thermal problem

This comparison is taken from [BABUSKA - STROUBOULIS - UPADHYAY - GANGARAJ, 1994]. The problem being considered is a 2D thermal problem whose analytical solution is known according to

$$u_{ex}(x, y) = \alpha(x^2 - y^2) + \beta xy$$

The mesh used was uniform; it is shown in Figure 4.5.

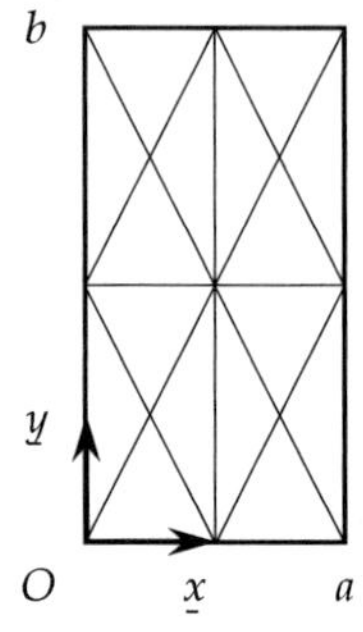

Figure 4.5. *Uniform mesh.*

Figure 4.6 shows, as functions of the ratio b/a, the effectivity indexes of the following estimators:

- constitutive relation error estimator $\mathbf{e}_{RdC}$;
- explicit estimator according to (3.10) $\mathbf{e}_{explicit}$;
- implicit estimator according to (3.23) $\mathbf{e}_{implicit}$;
- subdomain estimator according to (3.15) $\mathbf{e}_{subdomain}$.

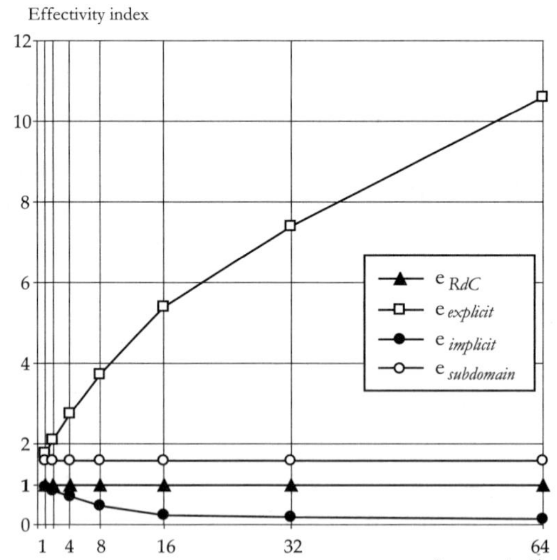

Figure 4.6. *Effectivity indexes of several estimators for the thermal problem.*

For thermal problems, in contrast to the estimators constructed using the residuals of the equilibrium equations, the constitutive relation estimator behaves very well.

4.3.2 Comparison for an elasticity problem

The objective of this example, taken from [LADEVEZE - ROUGEOT, 1997], is to study the influence of the aspect ratios of the elements on the quality of the constitutive relation error estimator and of the ZZ2 estimator (Section 3.3.3).

This is a problem in a rectangular domain of dimensions a and b whose exact solution is known. The finite element calculation was performed with 6-node triangles. The corresponding mesh is shown in Figure 4.7.

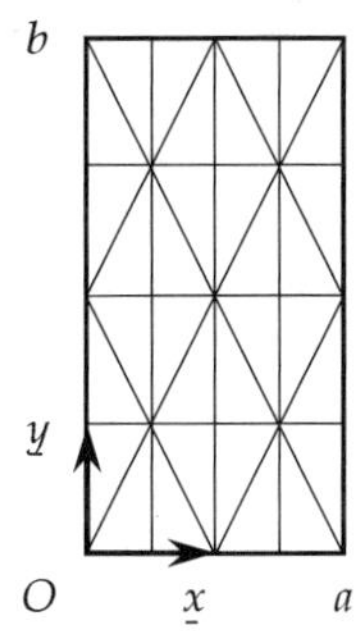

Figure 4.7. *Uniform mesh.*

The boundary conditions were chosen such that the exact solution takes one of the following four forms:

$$\sigma_{ex}^{1} = \alpha \begin{bmatrix} -2xy & y^2 - x^2 \\ y^2 - x^2 & 2xy \end{bmatrix} \qquad \sigma_{ex}^{2} = \alpha \begin{bmatrix} 2xy & -x^2 - y^2 \\ -x^2 - y^2 & 2xy \end{bmatrix}$$

$$\sigma_{ex}^{3} = \alpha \begin{bmatrix} -y^2 & 0 \\ 0 & x^2 \end{bmatrix} \qquad \sigma_{ex}^{4} = \alpha \begin{bmatrix} y^2 - x^2 & 2xy \\ 2xy & x^2 - y^2 \end{bmatrix}$$

In each case, the calculation was performed for ratio b/a varying between 2 and 64. In each case, we tested the constitutive relation estimator in its standard version, the improved estimator (see Section 8.6) and the ZZ2 estimator. The results are given in Figures 4.8 to 4.11.

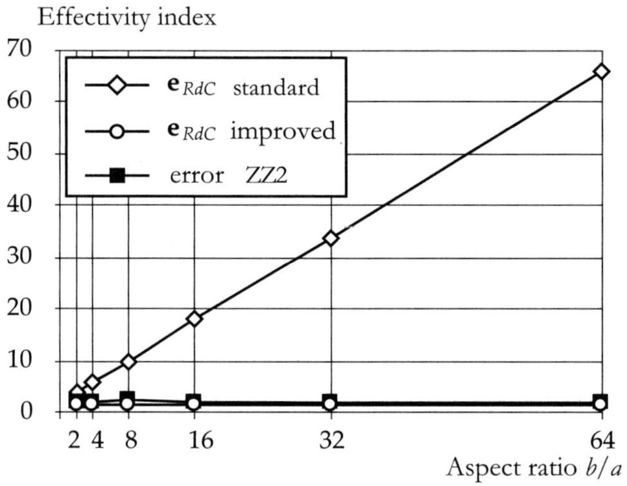

Figure 4.8. *Effectivity indexes - exact solution* σ^1_{ex}.

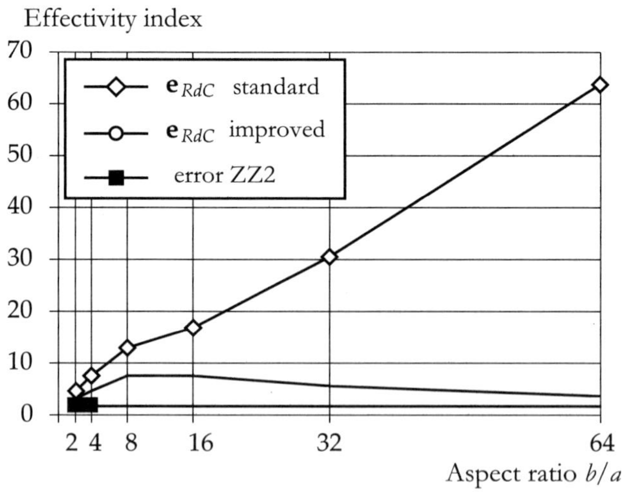

Figure 4.9. *Effectivity indexes - exact solution* σ^2_{ex}.

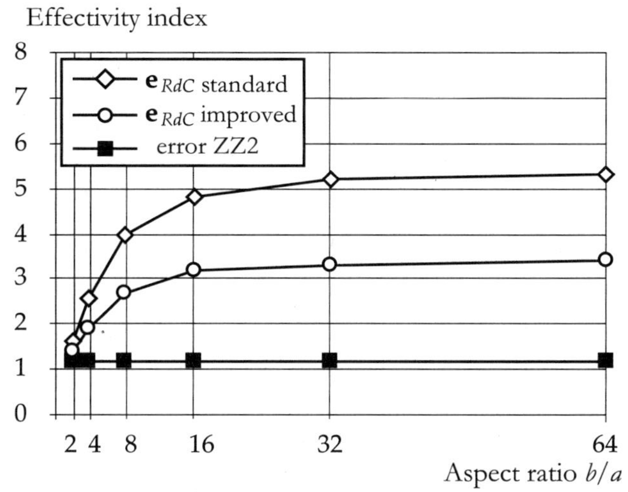

Figure 4.10. *Effectivity indexes - exact solution* σ_{ex}^{3}.

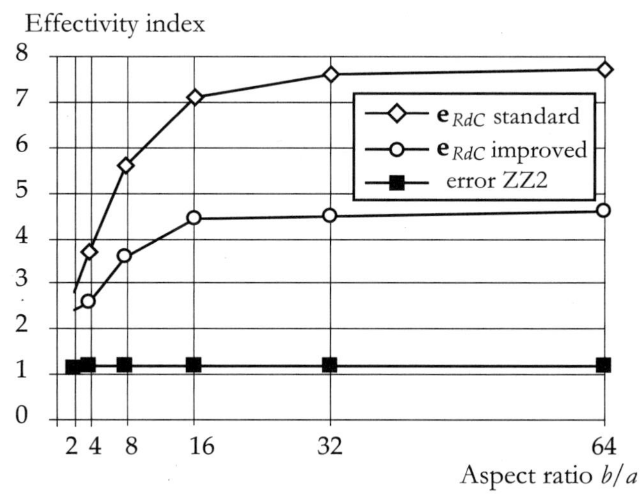

Figure 4.11. *Effectivity indexes - exact solution* σ_{ex}^{4}.

The classical error in constitutive relation is very sensitive to the aspect ratios of the elements. The improved construction (see Section 8.6) led to a much better quality estimator. In all the situations considered, the ZZ2 gave excellent global estimates.

REMARKS

1. The effectiveness of the ZZ2 estimator can be defeated in the case of very coarse meshes on which the smoothing technique gives bad results. We will see in Chapter 9 that on the local level, contrary to the error in constitutive relation, it gives unsatisfactory results; its local effectivity index is often quite poor.

2. The improved version of the constitutive relation error estimator is more effective than the standard version. For elements with low aspect ratios, however, the two versions give very similar results.

4.3.3 Analysis of the quality of the improved estimator

The improved version of the constitutive relation error estimator involves, compared with the standard version, two modifications:

- an optimization of the construction of the force densities $\hat{\underline{F}}$ at the interfaces (see Section 8.6);
- the resolution of the local problem within the element by the "$p+k$" technique with $k=3$ (see Section 2.4.3).

Let us study the improvements due to each of these two points. In order to do that, we consider again the example of Section 4.3.2 with the prestress σ_0^1.

The calculation of the effectivity index was carried out for:

- the standard construction of $\hat{\underline{F}}$ and the analytical construction of $\hat{\sigma}$;
- the standard construction of $\hat{\underline{F}}$ and the exact solution of (2.35);
- the standard construction of $\hat{\underline{F}}$ and the approximate solution of (2.35) by the $p+3$ technique;
- the optimized construction of $\hat{\underline{F}}$ with the exact solution of Problem (2.35);
- the optimized construction of $\hat{\underline{F}}$ with an approximate solution of (2.35) by the $p+3$ technique.

The results obtained are given in Table 4.1 for the case of elements with a standard aspect ratio $b/a = 2$, and in Table 4.2 for the case of elements with a very high aspect ratio $b/a = 128$.

It is interesting to note that the main contribution to the improvement came from the use of an approximate solution of Problem (2.35) by the $p + k$ technique, whereas the analytical construction can lead to admissible fields of mediocre quality. Using this approximate technique, the results of the two variants of the construction of the densities become much closer. Nevertheless, the optimization of the densities brings an additional improvement; this optimization reduces the artificial singularities which are inherent in the resolution of the elasticity problem (2.35) on the element with given inter-forces.

$b/a = 2$	Analytical construction	Exact solution	$p + k$ solution
$\hat{\underline{F}}$ standard	4.49	3.51	3.35
$\hat{\underline{F}}$ improved		2.89	2.75

Table 4.1. *Global effectivity indexes.*

$b/a = 128$	Analytical construction	Exact solution	$p + k$ solution
$\hat{\underline{F}}$ standard	267	2.33	2.17
$\hat{\underline{F}}$ improved		1.92	1.78

Table 4.2. *Global effectivity indexes (elements with very high aspect ratios).*

4.4 Common versions of the constitutive relation estimators

In practice, since the beginning of the 1990s, we have been using the constitutive relation error estimator built on the strong extension condition and on the technique of approximate resolution of the local problem within an element which, starting from the densities and the data, leads to the stress field. Generally, this approximate solution uses a single element of degree $p+3$, where p is the degree of interpolation of the initial finite element analysis. This idea was proposed by BABUSKA and STROUBOULIS. Consequently, the constitutive relation error estimator which is used in practice coincides exactly with the "ER*p*B3" estimator developed by these authors. Thc numerous tests they published corroborate the results regarding the quality of this estimator given here. Of course, in zones of high element aspect ratios or high gradients, the quality may be affected. In these zones alone, we use the version of the error in constitutive relation built on the weak extension condition which is presented in Section 8.6 (still with a single element of degree $p+3$ to solve the local problem).

Chapter 5

Mesh adaptation for linear problems

5.1 Introduction

It is a well-known fact that the quality of a linear finite element calculation depends on the choice of the mesh and on the type of interpolation used in the elements. To improve this quality, one must modify the parameters of the discretization, i.e. refine the mesh, increase the degree of the interpolation used in the elements, or use both techniques simultaneously. Indeed, a priori error evaluations show that the discretization error diminishes if the sizes of the elements are reduced or if the degree of interpolation is increased.

Nevertheless, these very simple rules cannot be applied indiscriminately. For example, if the calculation performed on an initial mesh is considered to be insufficiently accurate, it is, in theory, sufficient to refine the mesh until the quality obtained is considered acceptable.

While brute-force mesh refinement may be conceivable for linear problems in 2D, this is completely unrealistic in 3D. To be convinced, one only needs to consider a cube uniformly meshed with 20-node bricks, each edge being subdivided into N elements. Table 5.1 shows, for some values of N, the number of elements, the number of degrees of freedom and the approximate size (in machine words) required to store the stiffness matrix in banded form.

N	*No. of elements*	*No. of d.o.f.*	*Stiffness matrix size*
20	8,000	107,163	$5.5 * 10^8$
50	125,000	1,568,403	$4.8 * 10^{10}$
100	1,000,000	12,271,803	$1.5 * 10^{12}$

Table 5.1. *Examples of problem sizes.*

Thus, it is clear that today's workstations would become saturated very quickly, even for a simple linear resolution and all the more for a nonlinear or dynamic calculation.

Likewise, the degree of interpolation cannot be increased easily either, if only because the great majority of industrial codes do not provide elements of degree greater than 2.

Thus, in order to improve the quality of a calculation through the modification of the discretization parameters, it is necessary to introduce efficient procedures enabling the level of quality desired by the user to be attained while minimizing, to the greatest extent possible, the calculation costs.

In broad terms, these adaptive procedures (Figure 5.1) consist in performing an initial calculation on a mesh (which may be relatively coarse) with low-degree (usually linear) interpolations, and evaluating the discretization errors made. Next, using these results, the goal is to determine the parameters (element sizes and degree of interpolation) which yield a given accuracy while minimizing costs. Finally, a new discretization is defined according to these parameters and a new finite element analysis is carried out.

The advantage of these adaptive procedures is to reduce to a minimum, for a given level of quality, the cost of finite element analyses. In the context of linear 3D calculations, they enable good-quality calculations to be performed on ordinary workstations. In the nonlinear domains (plasticity, viscoplasticity, large strains...) and in dynamics, they make complex simulations which would otherwise be completely inaccessibly manageable.

In the first part, we present the techniques of mesh modification which were proposed under the names of *r*-, *h*- and *p*-versions.

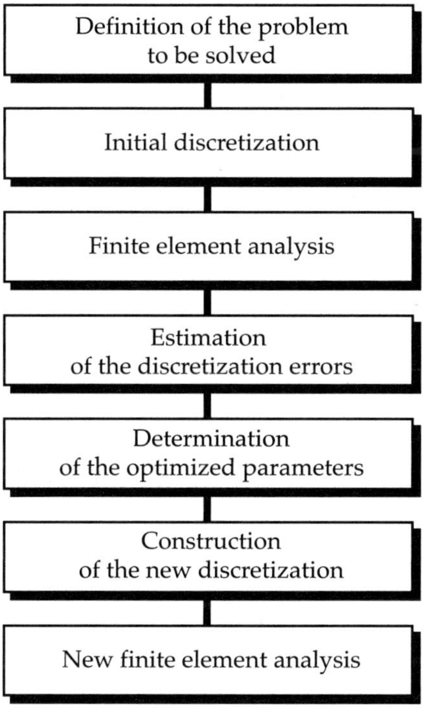

Figure 5.1. *Adaptive procedure.*

In the framework of the *h*-version, which is used most today, we define the concept of optimum mesh for a given error level, and we detail the techniques for calculating the optimum sizes. An efficient implementation of these adaptive techniques requires the use of robust automatic mesh generators capable of producing the prescribed mesh sizes accurately. The techniques available today are presented very briefly. For more details, the reader could refer to [GEORGE, 1991] and [BOROUCHAKI - GEORGE, 1997]. Special attention is given to the specific treatment of high-gradient zones, which is indispensable for obtaining properly optimized meshes.

In conclusion, we show how, with a combination of reliable error estimators, robust techniques for predicting the sizes, and automatic mesh generators, one can envisage in situations which are not exceedingly complex the complete automation of finite element analyses.

5.2 Mesh adaptation techniques

For many years, various mesh adaptation techniques have been developed in order to improve the quality of solutions, but it was primarily the emergence of efficient methods of error control which boosted the development and the use of mesh adaptivity techniques. Today, one generally distinguishes three adaptivity methods known by the names of *r*-, *h*- and *p*-versions.

5.2.1 The *r*-version

This method, which came out in the 1970s [CARROL - BARKER, 1973], [TURCKE - NEICE, 1974], consists in retaining the topology of the mesh (number of nodes and connectivity) as well as the type of interpolation, and seeking to reposition the nodes in order to minimize the error between the exact solution and the finite element solution (Figure 5.2). The application of this method leads to the resolution on the structure of an optimization problem whose optimization variables are then nodal coordinates. Such a procedure is numerically very costly and of limited effectiveness.

Different variants have been proposed to reduce the cost, such as the repositioning of the nodes on the isostrain curves [TURCKE - NEICE, 1974] or the construction of specific iterative algorithms [DIAZ - KIKUCHI - TAYLOR, 1983], [KIKUCHI, 1986]. Nevertheless, these methods have not been used very much in structural calculations since the emergence of automatic mesh generators. Still, combined with a *p*-version or an *h*-version, they can be useful in the application of ALE methods [PIJAUDIER-CABOT - BODE - HUERTA, 1995].

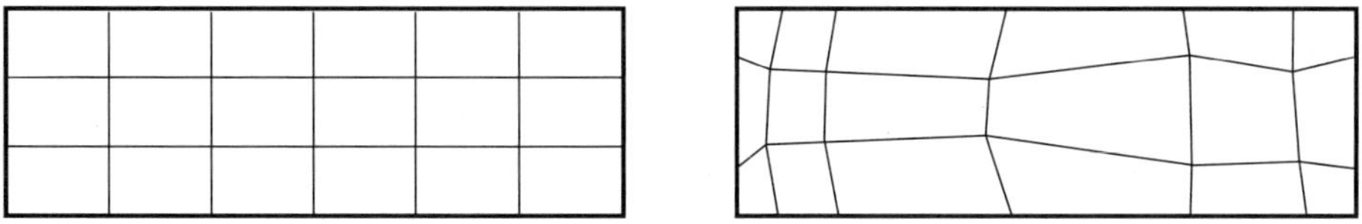

Figure 5.2. *The r-version.*

5.2.2 The *h*-version

This method consists in keeping the type of interpolation while partially or completely modifying the mesh (number and positions of the nodes, connectivity…). In practice, it has been used in two different forms:

- the local h-refinement/derefinement method;
- the global h-remeshing method.

Local h-refinement/derefinement

In this case, one keeps the framework of the initial mesh, but the elements can be either subdivided or grouped (Figure 5.3).

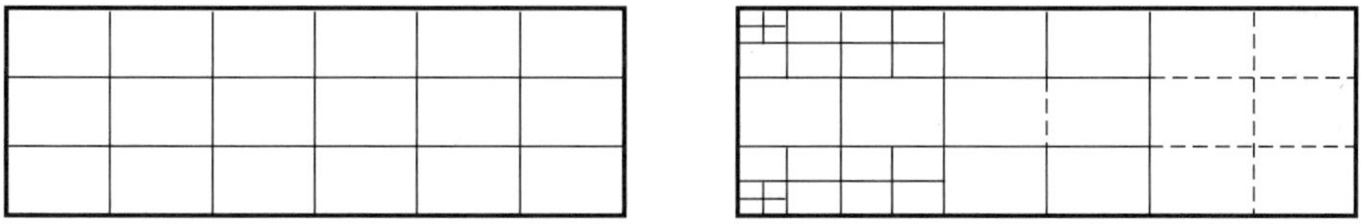

Figure 5.3. *h-refinement/derefinement.*

In practice, the local h-refinement method is the only method which can actually be used when the initial meshes contain more than just 2D rectangles or 3D straight bricks.

Limiting oneself to local refinements, this method is easy to implement and has been used extensively in connection with the following error indicators: [BABUSKA - SZABO, 1982], [KELLY - GAGO - ZIENKIEWICZ - BABUSKA, 1983], [RIVARA, 1984], [BAUDRON - TROMPETTE, 1986], [AUBRY - TIE, 1991], [ONATE -

BUGEDA, 1993], [DIEZ - HUERTA, 1999].

Nevertheless, some practical difficulties must be dealt with. Indeed, in the course of performing a local refinement one creates “hanging” nodes which require special treatment to enforce the continuity of the displacement field. One technique consists of making the displacements of these nodes dependent on the displacements of the neighboring nodes in such a way that continuity at the interface is ensured (Figure 5.4).

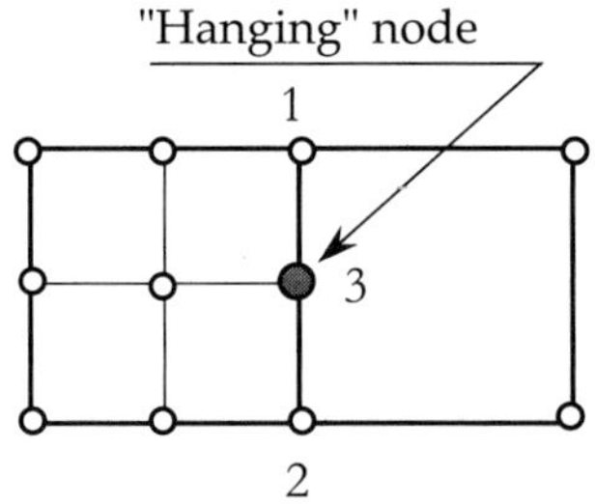

Figure 5.4. *Hanging node.*

For example, in the case of 4-node quadrangles, one sets

$$\underline{U}_3 = \frac{1}{2}(\underline{U}_1 + \underline{U}_2)$$

However, if the mesh contains many such nodes, this results in a significant increase of the solution cost (resolution of a linear system with many linear relations among the degrees of freedom). Another technique consists in introducing special “transition” finite elements to take these “hanging” nodes into account.

The most elegant treatment of the “hanging” nodes, proposed by ZIENKIEWICZ [ZIENKIEWICZ - CRAIG, 1986], consists in using hierarchical shape functions in the elements. The principle of hierarchical bases is described in Figure 5.5 for the 1D case and in Figure 5.6 for 2D triangles with linear interpolation. If the middle node is a “hanging” node, the continuity of the displacement is achieved simply by setting the corresponding degrees of freedom to zero.

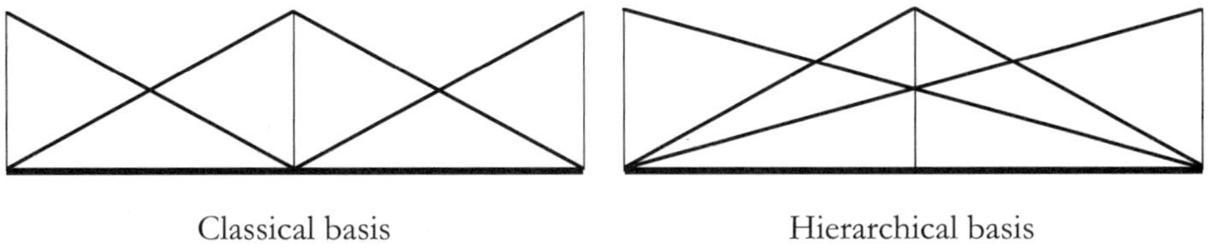

Figure 5.5. *Hierarchical elements in 1D.*

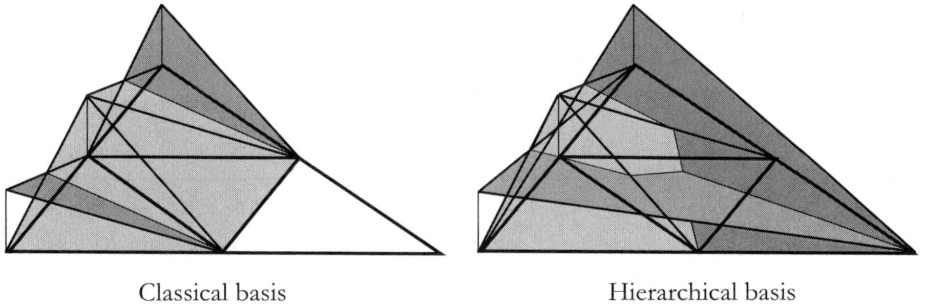

Figure 5.6. *Hierarchical elements in 2D.*

The only drawback of this technique is that it requires a special organization of the calculation which is not available in today's classical industrial codes. Nevertheless, research software has been constructed on this principle, e.g. the code developed by AUBRY [AUBRY - TIE, 1992], which also includes a method of error estimation.

REMARK. It is also possible, in order to avoid the creation of hanging nodes, to modify the mesh locally to reestablish appropriate connections compatible with the finite element displacement technique. For example, in 2D, simple and effective techniques have been proposed by [RIVARA, 1984, 1989].

h-Remeshing

In this method, the mesh is entirely redefined. The initial mesh serves only as the support for defining the sizes of the elements of the new mesh (Figure 5.7). Global *h*-remeshing came into use much later because it requires the use of automatic mesh generators capable of respecting the required mesh sizes correctly.

To our knowledge, the first uses were reported in [CARNET - LADEVEZE - LEGUILLON, 1981], [LADEVEZE - LEGUILLON, 1983] for 2D thermal problems, in [LADEVEZE - PELLE, 1984], [PELLE, 1985] for vibration problems using the automatic mesh generator developed by CARNET [CARNET, 1978], then in [COFFIGNAL - LADEVEZE, 1983] for 2D elasticity and plasticity problems using the automatic mesh-generator ARAIGNÉE [COFFIGNAL, 1987]. Today, this is the most widely used mesh adaptivity method for 2D problems because numerous robust 2D automatic mesh generators have been developed [GEORGE, 1991].

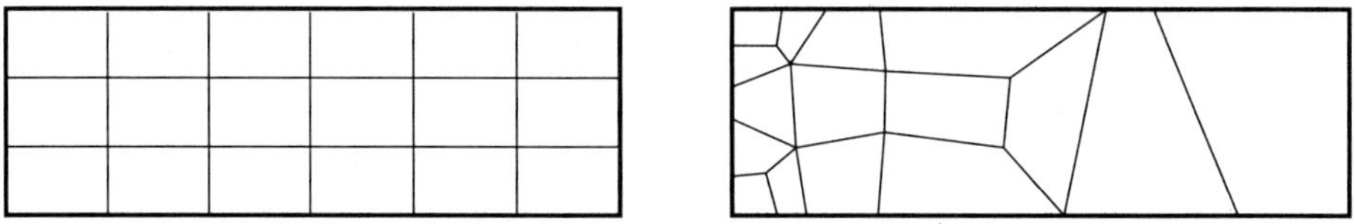

Figure 5.7. *h-remeshing.*

In 3D, such generators exist as yet only in the realm of research; this method continues to pose some implementation difficulties, even though progress has been made recently [COOREVITS - DUMEAU - PELLE, 1995, 1999]. The current investigations on generators capable of matching a size map for 3D problems and skewed surfaces should soon provide tools as robust as there are now in 2D.

5.2.3 The *p*-version

In the strict sense, the *p*-version is more a method of adaptivity of the type of interpolation than a method of adaptivity of the mesh. Indeed, this method consists in keeping the initial mesh and increasing in certain zones the degree of interpolation being used [BABUSKA - SZABO - KATZ, 1981], [BABUSKA - SURI, 1990], [SZABO, 1986, 1990], [PAPADRAKAKIS - BABILIS - BRAOUZI, 1997]. This method has some very attractive aspects, particularly concerning the cost of constructing the mesh. Indeed, a single mesh suffices; moreover, in the applications usually presented, this mesh is composed of very few elements and, therefore, is very easy to construct (Figure 5.8). On the other hand, high-degree interpolation functions must be used at least in some zones. However, this "coarse" mesh must not be constructed indiscriminately if one wants to be able to use high values of p. In particular, one must avoid elements with high aspect ratios; for somewhat complex 3D components, one ends up with meshes which are much more refined than one would have hoped a priori. Moreover, when the degree of interpolation changes locally, this method also introduces "hanging" nodes. Once again, the use of hierarchical bases [ZIENKIEWICZ - CRAIG, 1986] provides a remedy for this difficulty.

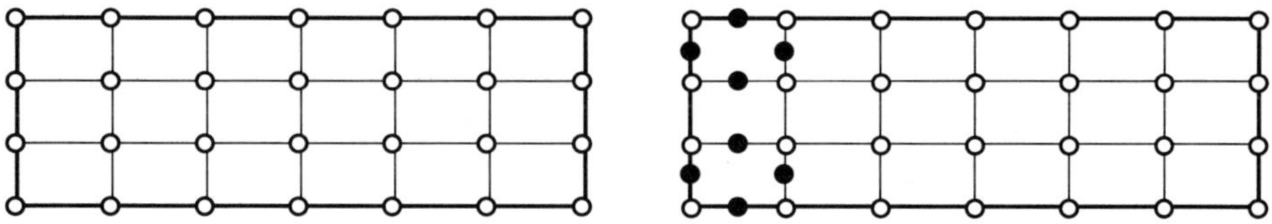

Figure 5.8. *The p-version.*

Today, this method is not yet widely used in industry because industrial codes (except some made especially for the p-version, such as [PROBE, 1985], [FIESTA, 1986]) rarely possess elements of degree greater than 2. Another difficulty which has not yet been resolved properly resides in the prediction at minimum computational cost.

REMARKS

1. Let us mention that studies on the simultaneous change of element sizes and interpolation functions have also been made [BABUSKA, 1986], [ZIENKIEWICZ - ZHU - GONG, 1989], [ODEN - DEMKOWICZ, 1991], [ODEN - PRUDHOMME, 1998]. This technique, called the h-p-version, can lead to very high convergence rates. However, as for the p-version, it is difficult to predict the satisfactory parameters to achieve a given quality.

2. This type of hierarchical approach has been extended to the calculation of plates and shells [ACTIS - SZABO - SCHWAB, 1999].

3. An adaptive strategy for equilibrium elements and hybrid elements has been developed [PEREIRA - ALMEIDA - MAUNDER, 1999].

5.3 Mesh adaptation

The objective of an adaptive procedure is to ensure a given precision level ε_0 while minimizing computational costs. We will address this subject in the framework of the h-version, which is the most widely used and the most effective technique today. We assume that we have at our disposal an estimate of the global relative error $\boldsymbol{\varepsilon}$ and an estimate of the corresponding element contribution ε_E, which verify

$$\varepsilon^2 = \sum_E \varepsilon_E^2$$

The adaptive techniques are different depending on whether one uses h-refinement or global h-remeshing.

5.3.1 Adaptation by local refinement

This method is iterative and very simple. Once the initial finite element calculation and error calculations have been carried out, one compares the level of contribution of each element to the average of all contributions

$$\varepsilon_{moy} = \frac{1}{N} \sum_E \varepsilon_E$$

Let α be a coefficient greater than 1 defined by the user. Each element E for which $\varepsilon_E \geq \alpha \varepsilon_{moy}$ gets subdivided. Then, one performs a new finite element calculation, a new estimate of the element contributions, and reiterates the procedure. This technique tends to make the element contributions uniform, but it is not obvious that it minimizes computational costs. We will return to this question later.

REMARKS

1. This procedure can be easily implemented in 2D with triangular or quadrilateral elements and in 3D with bricks, particularly if one uses hierarchical techniques.

2. This technique is not widely used in 3D with tetrahedral elements because there is no obvious method to subdivide a tetrahedron into subtetrahedra iteratively without running the risk of increasing the aspect ratios.

5.3.2 Adaptation by global remeshing

Here, the idea consists of using the results of an initial finite element analysis and the associated error estimates to determine an optimum mesh, i.e. a mesh which provides the desired precision while minimizing the calculation costs. The principle can be outlined as follows:

- one carries out an initial calculation on a relatively coarse mesh $\mathcal{T}$;
- on this mesh $\mathcal{T}$, one calculates the global relative error ε and the local contributions ε_E;
- one uses this information to determine the characteristics of the optimum mesh $\mathcal{T}^*$.

Then, one constructs the mesh $\mathcal{T}^*$ using an automatic mesh generator, and carries out a second finite element analysis.

The optimum mesh

The notion of optimum mesh was introduced for 1D problems in [LADEVEZE, 1977] and was subsequently developed in 2D in [LADEVEZE - LEGUILLON, 1981]. One can also mention, among the first works on mesh adaptation, [GAGO - KELLY - ZIENKIEWICZ - BABUSKA, 1983]. The criterion used consists in defining as the optimum a mesh $\mathcal{T}^*$ such that

$$\begin{aligned} &\varepsilon^* = \varepsilon_0 \quad \text{(the desired accuracy)} \\ &\varepsilon_E^* \ \text{ uniform on } \mathcal{T}^* \end{aligned} \tag{5.1}$$

Definition (5.1) is equivalent to saying that the best mesh corresponds to the discretization errors being uniformly distributed. However, as we will show below, this criterion does not always lead to the lowest computational costs.

More recently, a criterion which does not have this shortcoming was introduced in [PELLE, 1985] and [LADEVEZE - COFFIGNAL - PELLE, 1986]. It consists in adopting the following new definition:

Definition

For a given level of error ε_0, *a mesh* $\mathcal{T}^*$ *is optimum with respect to an error measure* ε *if:*

$$\begin{aligned} &\varepsilon^* = \varepsilon_0 \quad \textit{(the desired accuracy)} \\ &N^* \quad \textit{(number of elements of } \mathcal{T}^*\textit{)} \ \textit{is minimum} \end{aligned} \tag{5.2}$$

This criterion obviously leads to the lowest computational costs.

REMARK. For problems whose solutions are sufficiently regular, the two definitions are equivalent. This is not the case if the exact solution of the problem presents singularities (Section 5.4).

Determination of the optimum mesh

The principle of the determination of the characteristics of the optimum mesh $\mathcal{T}^*$ consists in calculating for each element E of mesh $\mathcal{T}$ the size modification coefficient

$$r_E = h_E^*/h_E \tag{5.3}$$

where h_E is the current size of element E and h_E^* is the (unknown) size which must be prescribed for the elements of $\mathcal{T}^*$ in the vicinity of element E in order to achieve optimality (Figure 5.9).

Thus, the determination of the optimum mesh is equivalent to the determination on the initial mesh $\mathcal{T}$ of the map of the size modification coefficients.

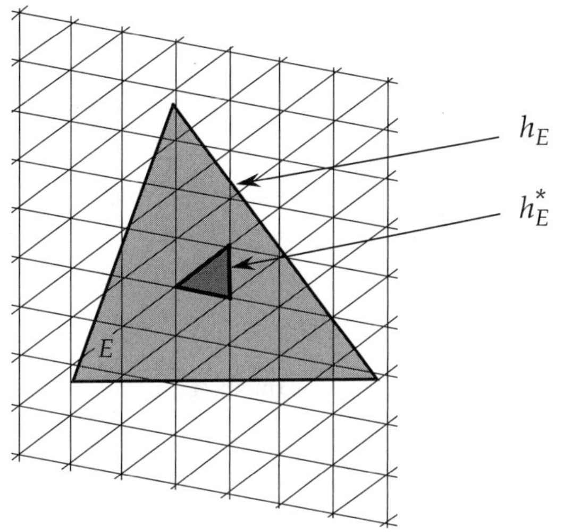

Figure 5.9. *Definition of the sizes.*

Calculation of the coefficients

The calculation of the coefficients r_E is based on the convergence rate of the error

$$\varepsilon = O(h^q) \tag{5.4}$$

where q depends not only on the element type being used, but also on the smoothness of the exact solution of the problem studied. For an error estimator which verifies Inequalities (1.34), the convergence rate is equal to the order of convergence of the finite element solution.

Initially, we assume that the exact solution is sufficiently smooth for the value of q to depend solely on the type of finite elements being used.

For example, one has in this case [CIARLET, 1978]:

- $q = 1$ for 3-node triangles and 4-node tetrahedral;
- $q = 2$ for 6-node triangles and 10-node tetrahedra.

In this case, in order to predict the optimum sizes, one writes that the ratio of the sizes and the ratio of the error contributions are related by

$$\frac{\varepsilon_E^*}{\varepsilon_E} = \left[\frac{h_E^*}{h_E}\right]^q = r_E^q \tag{5.5}$$

where ε_E^* is the contribution of the elements of $\mathcal{T}^*$ located in the vicinity of E,

$$\varepsilon_E^* = \left[\sum_{E^* \subset E} \varepsilon_{E^*}^2\right]^{1/2} \tag{5.6}$$

Thus, the square of the error for mesh $\mathcal{T}^*$ can be estimated by

$$\left(\varepsilon^*\right)^2 = \sum_E \left(\varepsilon_E^*\right)^2 = \sum_E \left[\sum_{E^* \subset E} \varepsilon_{E^*}^2\right] = \sum_E r_E^{2q} \varepsilon_E^2 \tag{5.7}$$

and the number of elements of mesh $\mathcal{T}^*$ by

$$N^* = \sum_E \frac{1}{r_E^d} \tag{5.8}$$

where d is the dimension of the space (in practice, $d = 1, 2$ or 3).

Therefore, one must solve the following problem:

$$\text{Minimize} \quad N^* = \sum_E \frac{1}{r_E^d} \quad \text{with} \quad \sum_E r_E^{2q} \varepsilon_E^2 = \varepsilon_0^2 \tag{5.9}$$

Property

The solution of the minimization problem (5.9) *determines the map of the sizes* $\{r_E\}_{E \in \mathcal{T}}$ *with*

$$r_E = \frac{\varepsilon_0^{1/q}}{\varepsilon_E^{2/(2q+d)} \left[\sum_E \varepsilon_E^{2d/(2q+d)}\right]^{1/2q}} \tag{5.10}$$

Proof

Introducing a LAGRANGE multiplier denoted A, Problem (5.9) is equivalent to making the Lagrangian stationary

$$L(\{r_E\}_{E\in\mathcal{T}}\ ;A)=\sum_E \frac{1}{r_E^d}+A\left(\sum_E r_E^{2q}\varepsilon_E^2-\varepsilon_0^2\right) \tag{5.11}$$

The extremum conditions yield

$$-\frac{d}{r_E^{d+1}}+A(2q\varepsilon_E^2)r_E^{2q-1}=0,\quad \forall E\in\mathcal{T} \tag{5.12}$$

from which one deduces

$$r_E=\left[\frac{d}{2Aq\varepsilon_E^2}\right]^{1/(2q+d)} \tag{5.13}$$

which, inserted into the second equation of (5.9), yields

$$\left[\frac{d}{2Aq}\right]^{2q/(2q+d)}=\frac{\varepsilon_0^2}{\sum_E \varepsilon_E^{2d/(2q+d)}} \tag{5.14}$$

Finally, one obtains the announced property by substituting (5.14) into (5.13).

REMARKS

1. As shown by Relations (5.7) and (5.8), the previous calculation is based on the following hypothesis:

On a given element of the initial mesh, the subelements of the optimized mesh are uniform (i.e. they have the same size).

We will see in Section 5.4 that in more severe situations, such as high-gradient zones, it is often preferable to use a more refined hypothesis.

2. The prediction of the sizes of the optimized mesh assumes that the reference element is always regular:

- equilateral triangle or square in 2D;
- regular tetrahedron or cube in 3D.

In other terms, the mesh we are seeking to construct is assumed a priori to be geometrically isotropic. For certain applications, particularly in dynamics (passing of a shock wave), or in the presence of shear strips, it may be preferable to use elements whose reference is not isotropic: relatively flat right-angle triangles, bricks with one dimension small compared to the others.... One refers to such a case as an *anisotropic mesh.*

Relation with the uniformization of the local contributions

Property

The size map defined by (5.10) *corresponds to uniform local contributions and is identical to that obtained by the solution of* (5.1).

Proof

The error contribution of an element E^* of $\boldsymbol{\tau}^*$ is given by

$$\varepsilon_{E^*}^2 = r_E^d(\varepsilon_E^*)^2 = r_E^{2q+d}\varepsilon_E^2 = \frac{\varepsilon_0^{(2q+d)/q}}{\left[\sum_E \varepsilon_E^{2d/(2q+d)}\right]^{(2q+d)/2q}} \tag{5.15}$$

which proves the first part of the property.

Therefore, in order to prove the second part, let us assume that on the mesh $\boldsymbol{\tau}^*$ being sought the contributions to the error are uniform. Thus, for any element E^* of $\boldsymbol{\tau}^*$, one has

$$\varepsilon_{E^*}^2 = \varepsilon_0^2 / N^* \tag{5.16}$$

Taking into account $\varepsilon_{E^*}^2 = r_E^{2q+d}\varepsilon_E^2$, one gets

$$r_E^{2q+d}\varepsilon_E^2 = \frac{\varepsilon_0^2}{N^*} \tag{5.17}$$

then

$$r_E = \left[\frac{\varepsilon_0}{\varepsilon_E}\right]^{2/2q+d)} \frac{1}{(N^*)^{1/(2q+d)}} \tag{5.18}$$

which, substituted into (5.8), yields an equation which enables one to calculate N^*,

$$N^* = \frac{\left[\sum_E \varepsilon_E^{2d/(2q+d)}\right]^{(2q+d)/2q}}{\varepsilon_0^{d/q}} \tag{5.19}$$

Finally, one finds (5.10) again

$$r_E = \frac{\varepsilon_0^{1/q}}{\varepsilon_E^{2/(2q+d)}\left[\sum_E \varepsilon_E^{2d/(2q+d)}\right]^{1/2q}} \tag{5.20}$$

In conclusion, provided the exact solution is sufficiently smooth, the two criteria (5.1) and (5.2) are equivalent.

5.3.3 Construction of an optimum mesh

To construct mesh $\mathcal{T}^*$, one must have access to a mesh generator capable of respecting the size specifications (5.10). This problem, while relatively simple in the 1D case, is much more complex in 2D and 3D.

2D meshes

It seems, at least in the field of structural mechanics, that the first 2D automatic mesh generator matching a size map was presented by CARNET [CARNET, 1978]. Today, for plane geometries, one can consider the problem solved. Indeed, among research teams as well as commercial software vendors, many automatic mesh generators which respect size specifications correctly have been developed using various techniques (the front propagation method, the VORONOÏ - DELAUNAY method, recursive decomposition method, quadtrees...).

The examples presented at the end of this chapter were treated with the mesh generator ARAIGNÉE developed at LMT-Cachan under the impulse of COFFIGNAL.

All these automatic 2D mesh generators require a continuous size map defined on the domain to be meshed. Starting from the map $\{r_E\}_{E \in \mathcal{T}}$ we already determined, a continuous size map can be easily constructed. Indeed, given r_E, one knows the optimum size on each element of the initial mesh: $h_E^* = r_E\, h_E$. Then, it suffices to define at each vertex i of the mesh an average size, for example, by letting

$$h_i^* = \frac{1}{n_i} \sum_{E \subset \Omega_i} h_E^* \qquad (5.21)$$

where Ω_i is the union of the n_i elements E connected to node i.

The continuous representation is obtained by using the shape functions of the 3-node triangles or 4-node quadrilaterals.

REMARKS

1. This technique can be easily extended to the 3D case.

2. In Section 5.4, we will give a method of calculation of the optimum sizes which enables the direct determination of the nodal sizes.

3. Most 2D automatic mesh generators construct a triangulation of the domain. There are very simple algorithms to transform triangles into quadrilaterals, either by combining adjacent triangles [RANK - SCHWEINGRUBER - SOMMER, 1993], or by subdividing each triangle into three quadrilaterals [COOREVITS - PELLE - ROUGEOT, 1994].

3D meshes

3D automatic mesh generators

In 3D, there are much more serious difficulties. Today, the most powerful automatic mesh generators are based either on the VORONOÏ - DELAUNAY method (e.g. the mesh generator GHS3D developed by GEORGE's team at INRIA) or on the front propagation method (e.g. the mesh generator SYSMESH developed by RASSINNEUX [RASSINNEUX - KROMER - GUEURY, 1997]). These two types of mesh generators construct solely tetrahedra and require as input a triangulation of the surface of the domain to be meshed.

Mesh generators using octrees have also been presented [SHEPARD - GEORGE, 1991]. A mesh generator based on the same principle, but using a subdivision into regular tetrahedra and octahedra, was developed in [PERRONNET, 1993, 1995]. Let us also mention the mesh generator known as the *topological mesh generator* developed by COUPEZ from a totally different principle, which presents the advantage of meshing the surface and the volume simultaneously [COUPEZ, 1991].

Matching a 3D size map

Even if these mesh generators yield topologically correct meshes and avoid generating elements with excessively high aspect ratios, none of them is yet truly able to respect the size specifications given by a 3D size map. Steering procedures for existing automatic mesh generators have recently been proposed in order to generate matching 3D meshes [COOREVITS - DUMEAU - PELLE, 1995, 1996, 1997]. These were the techniques used in the examples presented at the end of this chapter.

Due to the significant increase in the number of 3D calculations carried out in industrial environments, one may conjecture that specialists will soon come up with automatic mesh generators matching 3D size maps which will be as effective as today's 2D mesh generators.

Mesh generation on skewed surfaces

Most 3D mesh generators presuppose the existence of a mesh of the outer surface of the volume to be meshed. If this surface is an assembly of plane surfaces, it is relatively easy to construct the mesh with the help of a 2D automatic mesh generator and, if necessary, to match a size map.

Conversely, for a complex skewed surface generally defined as an assembly of patches by CAD software, the construction of the mesh is a difficult problem. To obtain such a mesh, one goes through the following steps (Figure 5.10):

- definition of nodes at the boundaries of the patches in order to enforce conformity;

• for each patch, transfer of the discretization from the boundary to the parameter space;

• creation of a mesh in the parameter space;

• transfer of the mesh onto the actual surface.

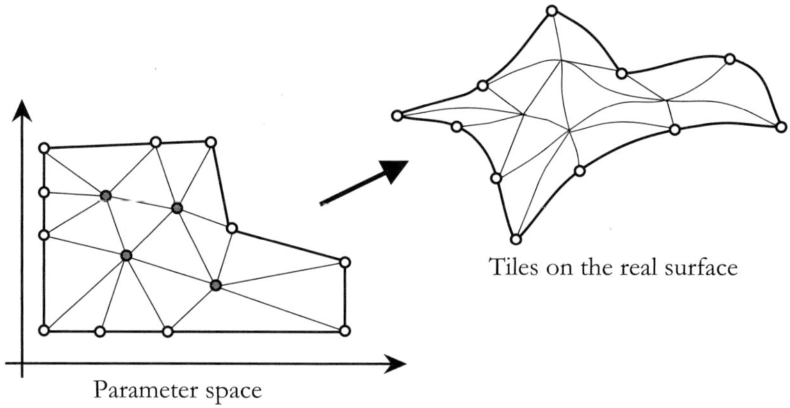

Figure 5.10. *Meshing of a patch by the transfer procedure.*

This method does yield a conforming mesh of a complex surface provided that the definition of the geometry is correct (no patch overlapping, no holes...). However, the adaptation of such a mesh is a difficult problem which requires the constraints of the boundaries between the patches to be removed. A semi-interactive adaptive procedure for meshing skewed surfaces, whose performance seems very promising, can be found in [NOËL, 1994]; an application to the adaptation of 3D meshes can be found in [COOREVITS - DUMEAU - NOËL - PELLE, 1997].

5.3.4 Verification of optimality

Once the optimum sizes have been determined and the optimized mesh $\mathcal{T}^*$ has been constructed, a new finite element calculation is carried out. In order to verify the effectiveness of the different procedures used, one must control the optimality of mesh $\mathcal{T}^*$.

One verification consists simply in calculating the discretization error ε^* on mesh $\mathcal{T}^*$. If this error is not close to the desired accuracy ε_0, the new mesh certainly is not optimum.

However, even if ε^* is close to ε_0, this is not sufficient proof of the optimality of $\mathcal{T}^*$.

To check optimality, a simple technique consists in determining a new map of optimum sizes for a desired accuracy identical to the accuracy ε^* which has been obtained.

If the mesh is truly optimum for accuracy ε^*, the procedure should yield size modification coefficients such that

$$\forall E^* \in \mathcal{T}^*, \quad r_{E^*} = 1 \tag{5.22}$$

In actual analyses, such mesh quality is never achieved for several reasons:

• As we have already pointed out, the prediction of the size calculation assumes that the reference element is always perfectly regular (equilateral triangle, regular tetrahedron…).

• The mesh generator does not always conform to the prescribed size map, particularly in zones of pronounced size variation. Indeed, it is necessary, in order to connect small elements to large elements, to insert elements of intermediate size!

Nevertheless, examination of the maps of the coefficients r_{E^*} enables one to control the quality of the mesh obtained. In practice, as a consequence of the previous remarks, one considers that the element sizes are satisfactory in all the zones where the following condition is satisfied

$$\frac{2}{3} \leq r_{E^*} \leq \frac{3}{2} \tag{5.23}$$

In order to facilitate this analysis, particularly in 3D, one can also consider a size match histogram constructed as follows:

One defines a number N_c of classes among the set of the values of r_{E^*} of the form

$[\, R_i, R_{i+1} [\quad$ *for* $i = 1, 2, \ldots, N_c$

with

$$0 < R_1 < 1, \quad R_{N_c+1} = \frac{1}{R_1}, \quad R_{i+1} = R_i\,(R_1)^{-2/N_c}$$

On a logarithmic scale for r_{E^*}, such a representation enables one:

• to give the same importance to coefficients greater than 1 (sizes too small) and to coefficients smaller than 1 (sizes too large);

• to represent classes which correspond to comparable relative size ratios by classes of the same width (Figure 5.11).

In particular, with these conditions, the class considered to be valid is represented by an interval centered on 1.

On the *y*-coordinate, one plots the relative number of elements which corresponds to each class.

The example of Figure 5.11 corresponds to the values $N_c = 19$ and $R_1 = 1/3$.

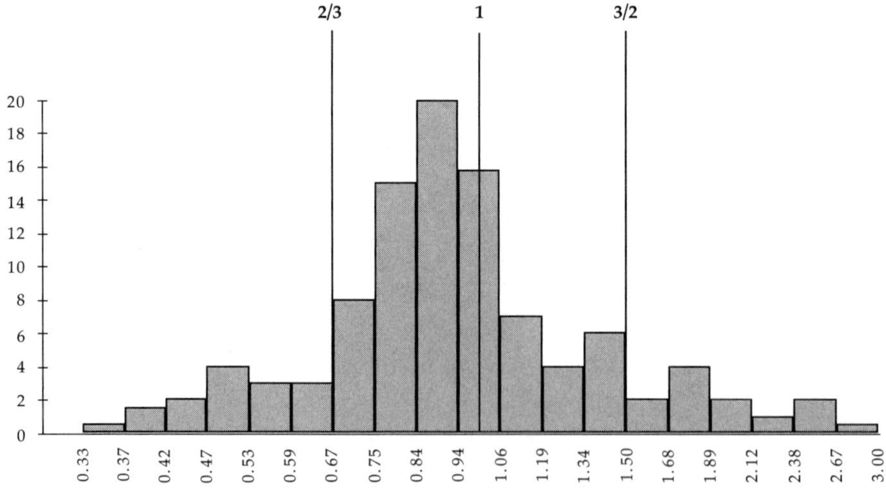

Figure 5.11. *Optimality histogram.*

5.4 Treatment of high-gradient zones

5.4.1 High-gradient zones

When the exact solution of the problem being studied has singularities, it is well-known that the convergence rate of the finite element solution is modified. Consequently, if the error estimator verifies (1.34), its convergence rate is also modified. Let us consider, for example, a plane elasticity problem discretized with triangular elements of degree p.

If the exact solution $\underline{U}_{ex}$ is smooth, one knows [STRANG - FIX, 1976], [CIARLET, 1978] that

$$\| \underline{e}_h \|_{u,\Omega} \leq C\, h^p \tag{5.24}$$

Conversely, if the exact solution has a singularity, such as if in the vicinity of a point M_0 (with r and θ the polar coordinates in the vicinity of point M_0) the displacement field is of the form

$$\underline{U}_{ex} = r^{\alpha}\, \underline{V}(\theta) + \underline{W} \quad \text{with } 0 < \alpha < 1 \tag{5.25}$$

then, one can show [STRANG - FIX, 1976] that

$$\| \underline{e}_h \|_{u,\Omega} \leq C\, h^{\alpha}$$

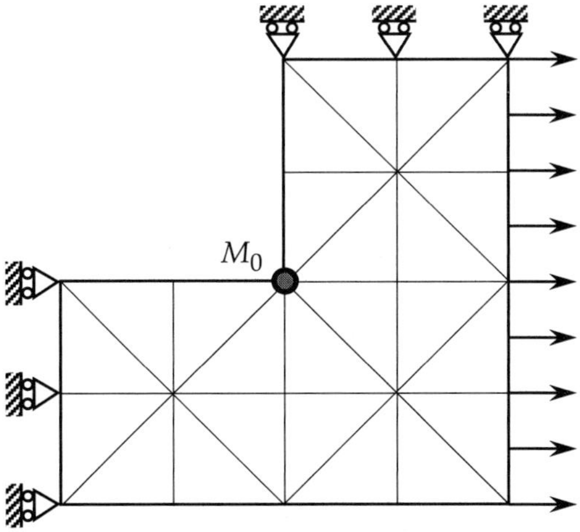

Figure 5.12. *A 2D example with a singularity.*

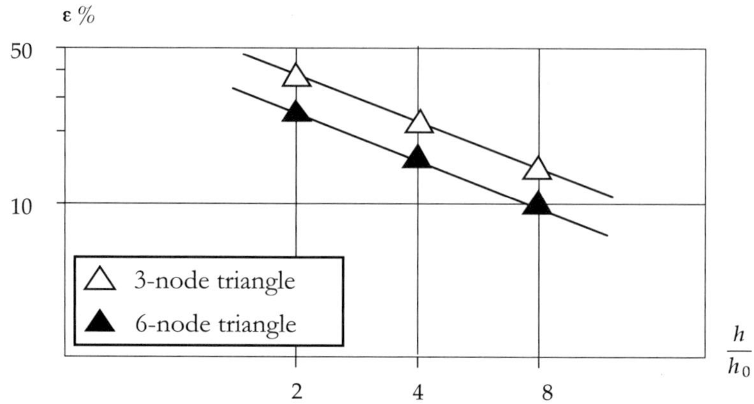

Figure 5.13. *Convergence rate of the error in constitutive relation.*

It follows that the convergence rate of the global error in energy is independent of the order of the finite elements being used, and so is the convergence rate of the error measure. These theoretical results are in perfect agreement with numerical tests.

Let us consider the problem described in Figure 5.12. For uniform meshes obtained by subdivision of the initial mesh of size h_0, the convergence rate of the error in constitutive relation ε as a function of h is given in Figure 5.13. One can see that these convergence rates are practically the same for 3-node and 6-node triangles.

Similar results obtained with different error estimators can be found in [ZIENKIEWICZ - CRAIG, 1986].

One should note that this type of result also occurs in problems with only zones of very high stress concentration, which, in a mathematical sense, are not singularities. But first, we will deal with the case of singularities.

5.4.2 Study of a test example

Let us consider the typical example of a cracked plate under mode-I loading (Figure 5.14) which has an $r^{1/2}$ singularity at the crack tip. We consider an initial mesh consisting of 124 6-node triangles (Figure 5.15).

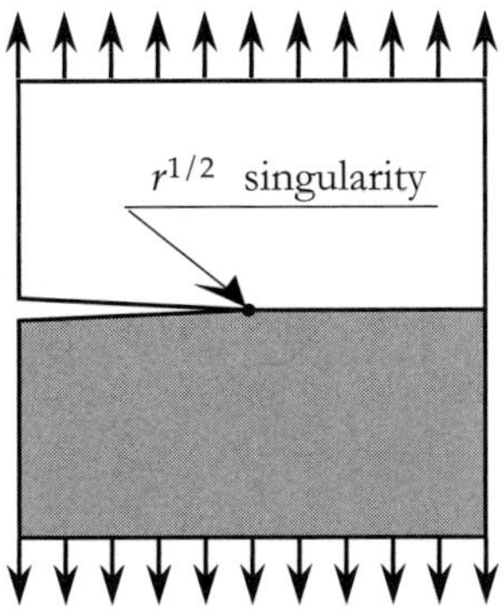

Figure 5.14. *Cracked plate in mode I.*

The error in constitutive relation calculated for this mesh is 11.4 %.

If for this problem one uses the procedure described in Section 5.3.2 with $q = 2$ for a desired precision $\varepsilon_0 = 2\%$, one obtains a mesh with 448 elements (Figure 5.16). However, if one calculates the error in constitutive relation for this new mesh, one finds $\varepsilon = 4.2\%$. Therefore, the mesh is not optimum!

This example confirms that in the presence of singularities the convergence rate of the global error is not directly related to the degree of interpolation being used.

Because of the theoretical results reviewed above, one might think that in order to obtain a properly optimized mesh it suffices to use the procedure described in Section 5.3.2 with $q = \alpha = 0.5$. However, one then obtains a mesh with 53,907 elements! There is no need to carry out the calculation to perceive clearly that the corresponding mesh is not optimum.

Let us consider the plate shown in Figure 5.17 with a crack of length 2*a*, subjected to a mode-I loading case with a parabolic profile. The calculations were carried out with 6-node triangles. Figure 5.18 shows the convergence rate of the error in constitutive relation for various crack lengths.

For $a = 0$, i.e. the plate with no crack, one finds a numerical convergence rate of 1.8, which is very close to the theoretical value $p = 2$.

For $a = b/12$, the numerical convergence rate varies between 1.33 and 0.79, while for $a = b/2$ it varies between 0.85 and 0.52, a value close to $\alpha = 0.5$.

This example shows that the theoretical convergence rate is achieved only for values of h that are sufficiently small, and that the numerical convergence rate depends on the influence of the singularity (through the stress intensity factors).

Therefore, it cannot be determined a priori on the basis of the theoretical results alone.

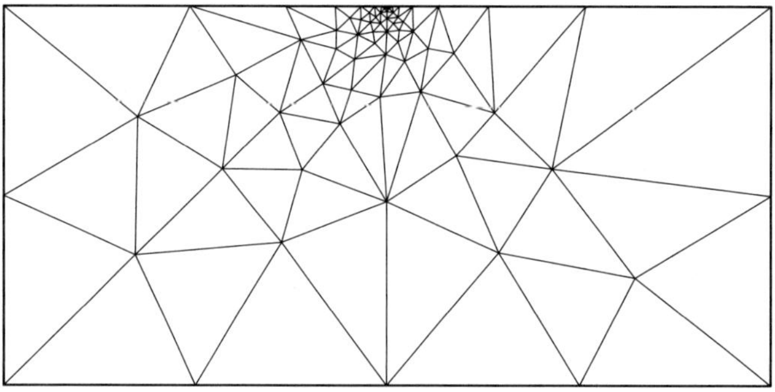

Figure 5.15. *Initial mesh - 124 elements - 273 nodes 6-node triangles - error 11.4%.*

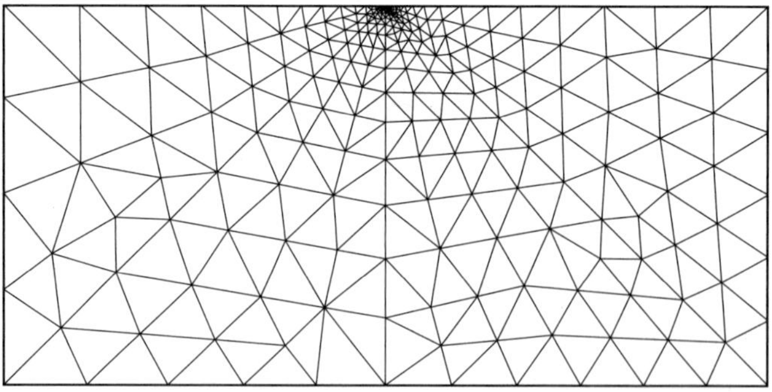

Figure 5.16. *Optimized mesh with* $q = 2$ *- 448 elements - 953 nodes 6-node triangles - error 4.2%.*

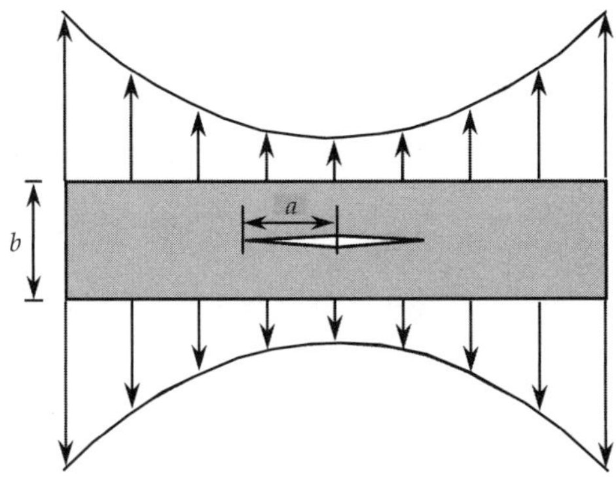

Figure 5.17. *Plate with a central crack under mode-I loading.*

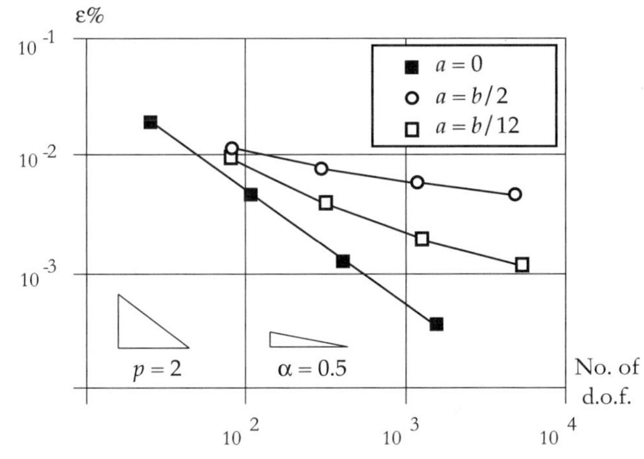

Figure 5.18. *Numerical convergence rates.*

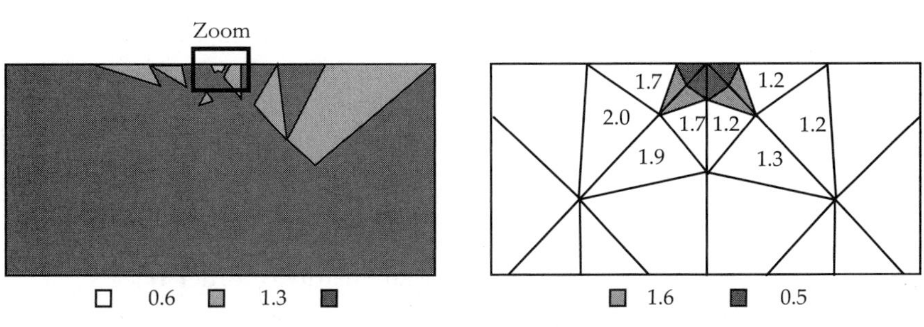

Figure 5.19. *Convergence rates of the local contributions.*

Besides, an examination of the convergence rates of the local contributions ε_E (Figure 5.19) shows that for the elements connected to the crack tip the rate is close to α, while for the other elements it is close to p.

5.4.3 Taking into account high-gradient zones

Thus, a good prediction of the optimized meshes requires nonuniform convergence rates. Initially, this approach was applied to test cases with singularities by using the position and the theoretical order of these singularities, which resulted in significant improvement of the quality of the resulting optimized meshes [PELLE - ROUGEOT, 1988, 1989], [ROUGEOT, 1989], [LADEVEZE - PELLE - ROUGEOT, 1991].

In practice, besides singularities, there can be zones where the solution presents very high gradients without being singular.

A fully automated method for the optimization of meshes in such situations was developed in [COOREVITS - LADEVEZE - PELLE - ROUGEOT, 1992], [COOREVITS - LADEVEZE - PELLE, 1994]; this method requires no a priori knowledge of the high-gradient zones or even of the singularities (position and order). However, it uses information obtained from the finite element solution already calculated.

This procedure was extended to 3D problems in [COOREVITS - DUMEAU - PELLE, 1996].

Use of local convergence coefficients

In order to account for singularities properly, one uses a convergence rate per element q_E such that

$$\varepsilon_E = O(h^{q_E}) \tag{5.26}$$

A simple way of defining these local coefficients consists in taking:

- $q_E = \alpha$ if element E is connected to a singularity of order α;
- $q_E = p$ for all other elements.

Modification of the calculation of the size map

Now, in order to calculate the map of the $\{r_E\}_{E \in \mathcal{T}}$, one must have

$$\frac{\varepsilon_E^*}{\varepsilon_E} = \left[\frac{h_E^*}{h_E}\right]^{q_E} = r_E^{q_E}$$

As in Section 5.3.2, one can show that the r_E are the solutions of

$$\text{Minimize} \quad N^* = \sum_E \frac{1}{r_E^d} \quad \text{with} \quad \sum_E r_E^{2q_E} \varepsilon_E^2 = \varepsilon_0^2 \tag{5.27}$$

Property

The solution of Problem (5.27) *is given by*

$$r_E = \left[\frac{d}{2Aq_E\varepsilon_E^2}\right]^{1/(2q_E+d)} \tag{5.28}$$

where A is the solution of the nonlinear equation

$$\sum_E \left[\left(\frac{d}{2Aq_E\varepsilon_E^2}\right)^{2q_E/(2q_E+d)} \varepsilon_E^{2d/(2q_E+d)}\right] = \varepsilon_0^2 \tag{5.29}$$

The proof is not given here because it is similar to that of Section 5.3.2. To obtain the size map, it suffices to solve (5.29) numerically by the NEWTON method.

REMARK. Using (5.28), one gets

$$\varepsilon_{E^*}^2 = r_E^d\,(\varepsilon_E^*)^2 = r_E^{2q_E+d}\varepsilon_E^2 = d\,/\,(2\,A\,q_E) \tag{5.30}$$

which proves that on the optimized mesh the local contributions are not uniform. It is the products $q_E^{1/2}\varepsilon_{E^*}$ which are uniform on $\mathcal{T}^*$. In the presence of singularities, the criteria (5.1) and (5.2) are no longer equivalent.

Direct calculation of the nodal sizes

The above method for calculating sizes is based on the modification of sizes which are assumed to be locally uniform. For example, in a zone where the mesh should be refined, an element E of the initial mesh is subdivided into r_E^d elements E^* with the same measure. If one of the vertices of the element is a singular point for the solution, this method does not enable the size to vary rapidly as one moves away from the singularity.

To avoid this shortcoming, it is preferable to define the map of the optimum sizes directly through sizes h_i^* defined at the vertices i of the initial mesh. Let us consider the case of triangular $(d = 2)$ or tetrahedral $(d = 3)$ meshes. Assuming that the measures of the elements of mesh $\mathcal{T}^*$ vary linearly on an element E of the initial mesh, element E is subdivided into N_E^* elements and N_E^* can be evaluated by

$$N_E^* = \frac{h_E^d}{\text{mes}(E)} \int_E \left[\sum_{k=1}^{d+1} \lambda_k (h_k^*)^d \right]^{-1} dE \tag{5.31}$$

where λ_k are the barycentric coordinates in E.

Similarly, ε_E^* can be evaluated by

$$(\varepsilon_E^*)^2 = \frac{\varepsilon_E^2}{h_E^{2q_E}\,\text{mes}(E)} \int_E \left[\sum_{k=1}^{d+1} \lambda_k (h_k^*)^d \right]^{2q_E/d} dE \tag{5.32}$$

which leads to the resolution of the optimization problem

$$\text{Minimize} \quad N^* = \sum_E N_E^* \quad \text{with} \quad \sum_E (\varepsilon_E^*)^2 = \varepsilon_0^2 \tag{5.33}$$

a problem which, this time, must be solved numerically.

The application of the techniques described above results in adapted meshes of excellent quality.

For example, in the crack problem, starting from the initial mesh illustrated in Figure 5.15, one obtains for 2% prescribed accuracy an error of 1.6%.

The corresponding mesh is shown in Figure 5.20 and the optimality histogram in Figure 5.21.

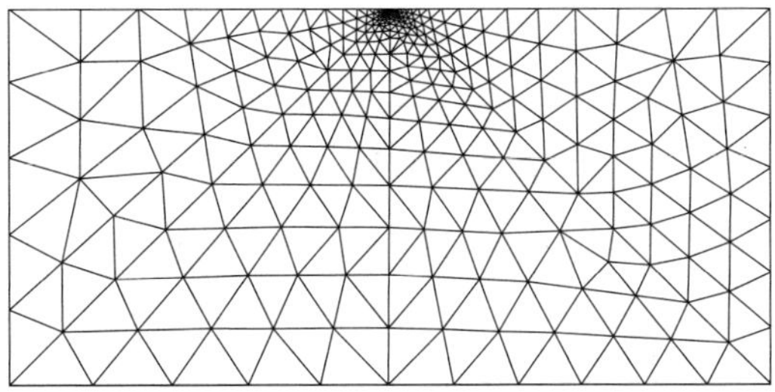

Figure 5.20. *Mode-I crack - Optimized mesh - 800 elements - 1,699 nodes 6-node triangles - error 1.6%.*

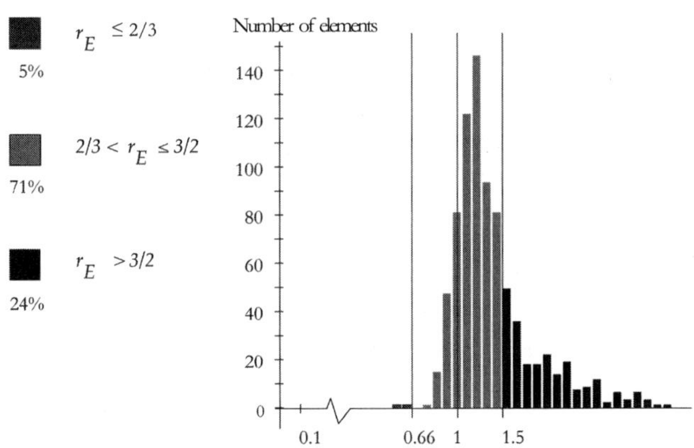

Figure 5.21. *Optimality histogram.*

5.4.4 Automatic detection of high-gradient zones

The objective is to determine both the "singular" nodes and the corresponding q_E coefficients automatically.

Detection of singular nodes

The idea consists in using the local errors

$$\bar{\varepsilon}_E = \left[\frac{\text{mes}(\Omega)}{\text{mes}(E)}\right]^{1/2} \varepsilon_E \tag{5.34}$$

Indeed, numerical tests show that these local errors have a peak value in the vicinity of a singularity.

Thus, for every node i of the mesh, one compares the average $\bar{m}_i$ of the $\bar{\varepsilon}_E$ for the elements E connected to node i to the average $\bar{M}$ of the $\bar{\varepsilon}_E$ for the entire structure. Node i is considered to be singular if

$$\bar{m}_i \geq \beta \bar{M} \tag{5.35}$$

where β is a coefficient greater than 1.

REMARK. Numerical tests showed that singular nodes are correctly detected by setting $\beta = 2$ in 2D and $\beta = 3$ in 3D.

Estimation of the order of the singularity

For every singular node i detected, the order of the singularity, i.e. the value of q_E which is to be used for the element connected to node i, is determined by identifying the value of the energy density of the finite element solution with its theoretical value in the vicinity of i.

The 2D case

In this case, one calculates the average finite element energy in disks A centered on i and with radius r,

$$\bar{e}_h(r) = \frac{1}{\text{mes}(A)} \int_A \text{Tr}[\sigma_h K^{-1} \sigma_h] \, dA$$

By identifying this average energy, using a least-squares method, with the theoretical value in the vicinity of a singularity of order α,

$$e(r) = ar^{2(\alpha-1)} + b \tag{5.36}$$

one obtains numerically a value $\bar{\alpha}$ close to α (Figure 5.22).

In practice, numerical tests showed that it is sufficient to carry out the identification in a zone corresponding to three element layers around the singular point and to evaluate $\bar{e}_h(r)$ for five to eight values of r uniformly distributed in this zone.

For example, in the crack problem, this technique applied to the mesh of Figure 5.15 yields $\bar{\alpha} = 0.52$, which is quite satisfactory.

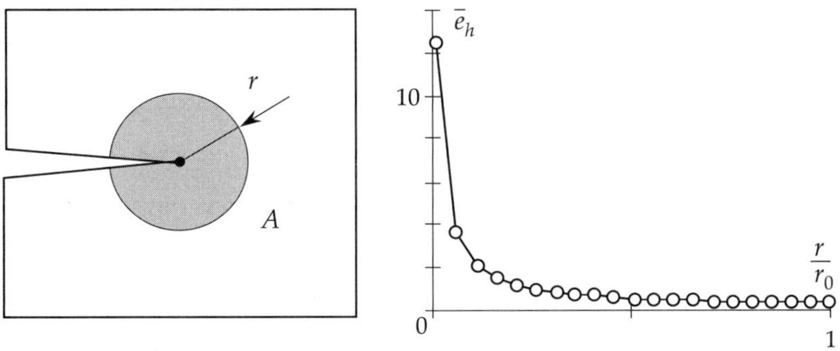

Figure 5.22. *Numerical evaluation of* α.

The 3D case

In 3D, the situation is more complex. The singular points are rarely isolated, and one must often deal with singular edges.

For example, for a built-in cube in tension, all the points of the edges of the built-in side are singular (Figure 5.23).

In this situation, the evaluation of the average energy in spheres A of increasing radius centered at a singular node does not enable the identification of $\bar{\alpha}$. Indeed, as the radius increases, the extent of the singular zone within sphere A increases and one does not get a rapid decrease of $\bar{e}_h(r)$ (Figure 5.24).

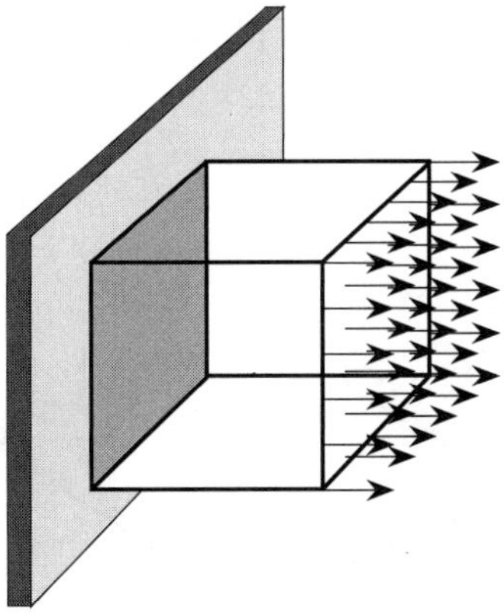

Figure 5.23. *Built-in cube in tension.*

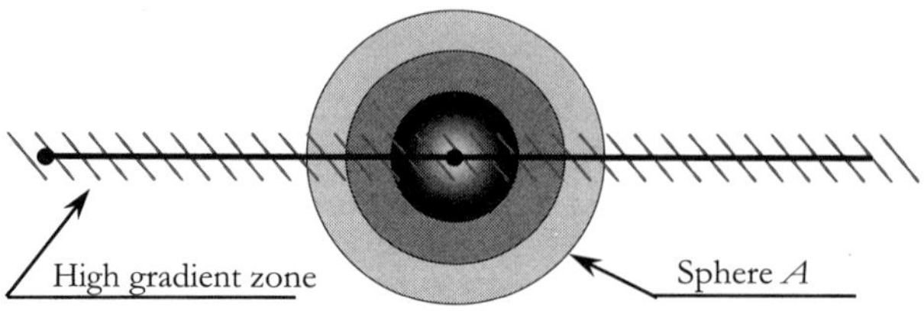

Figure 5.24. *Energy in concentric spheres.*

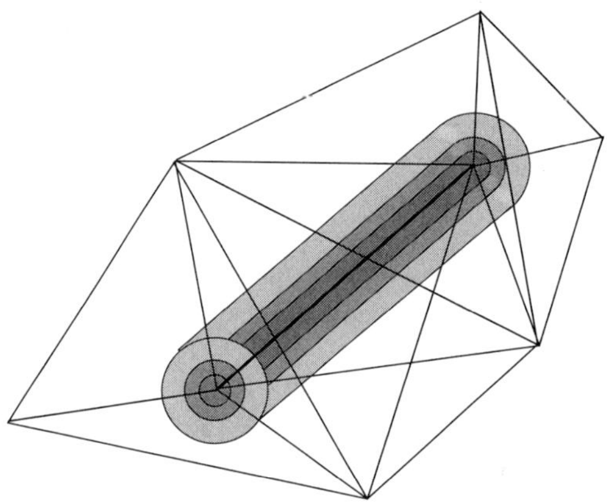

Figure 5.25. *Energy in coaxial cylinders.*

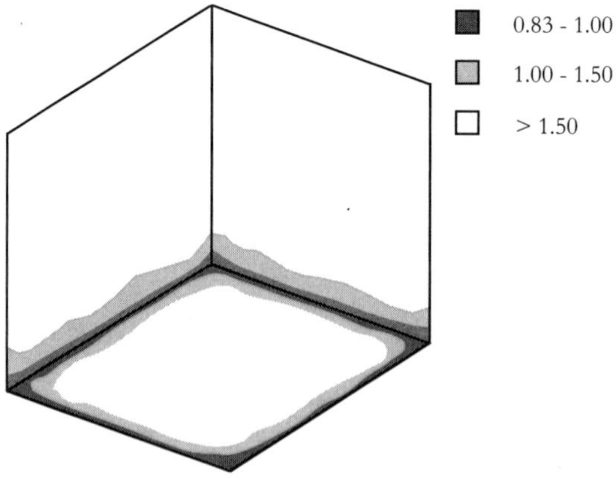

Figure 5.26. *Evaluation of* $\bar{\alpha}$.

When the singular points are not isolated, the coefficient must be identified by calculating the energy density in coaxial cylinders defined around the edges whose end points are considered to be singular (Figure 5.25).

The results obtained for the cube of Figure 5.23 are given in Figure 5.26.

5.4.5 Extension to stress concentration zones

The method used to take into account high-gradient zones automatically is identical to that used in situations presenting mathematical singularities. The coefficient a of (5.36) characterizes the intensity of the gradient.

Numerical examples show that this technique also works quite well in zones of high stress gradients, even though mathematically these zones do not correspond to singularities.

REMARKS

1. In statics, it is well-known that if the interior loads f_d are regular (particularly if they are zero, which is often the case in industrial applications), the high-gradient zones are necessarily the "boundary" zones. In this case, in order to cut costs, one can limit the use of the automatic procedure to the nodes located on the boundary.

2. In 2D, a very similar technique is currently being used by BECKERS et al. The only difference is that the identification of the singularity coefficient is carried out from the average values of the energy of the finite element solution in disks centered on the singularity [DUFEU, 1997].

5.4.6 A 3D example of adaptation

Let us give a simple example of mesh adaptation in 3D.

It concerns a piece, called a pressure pot, used in certain apparatus for very high-pressure experiments (Figure 5.27).

Figure 5.28 describes the mechanical problem and, after symmetries have been taken into account, the zone actually being calculated.

Figure 5.29 shows two views of the initial mesh which was used (4-node tetrahedra), and

Figure 5.30 shows the optimum mesh obtained for a desired 5% accuracy. The histogram of Figure 5.31 shows that the mesh was properly optimized.

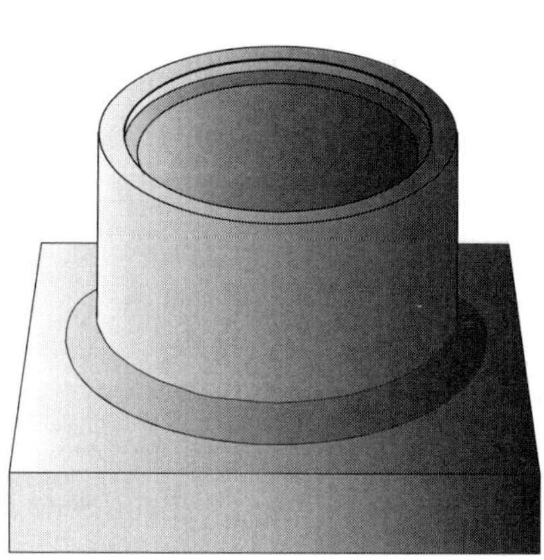

Figure 5.27. *Geometry of the pressure pot.*

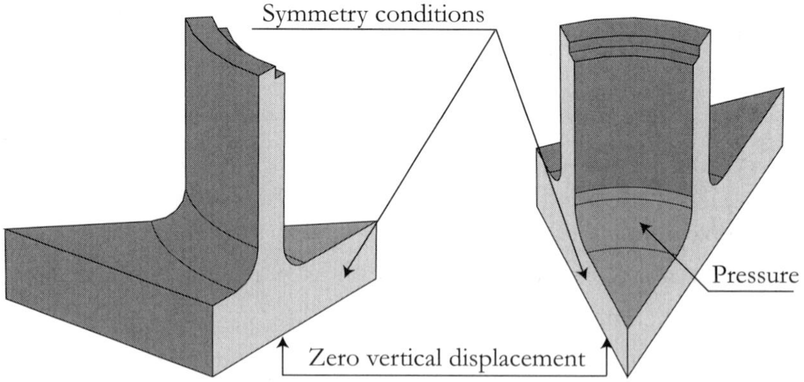

Figure 5.28. *Pressure pot under internal pressure - Calculation zone.*

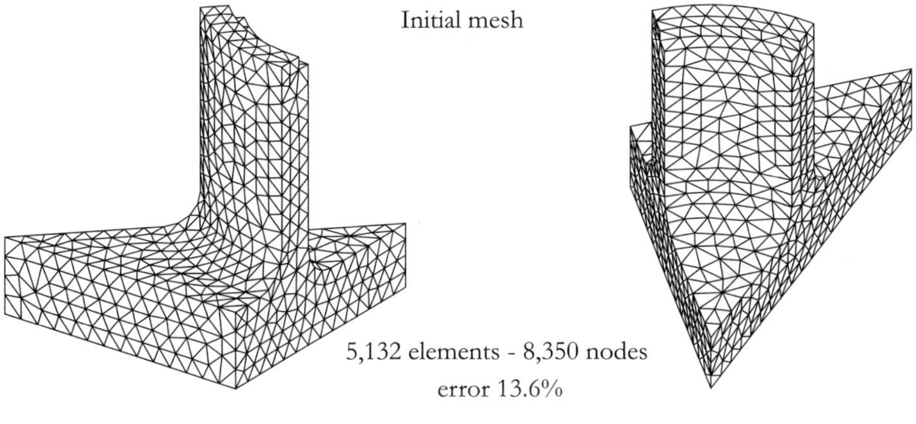

Figure 5.29. *Initial mesh: 4-node tetrahedra - 5% desired error.*

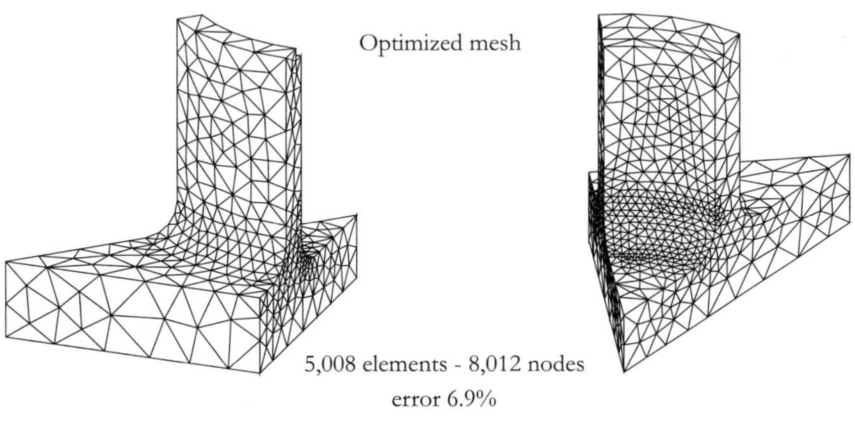

Figure 5.30. *Optimized mesh.*

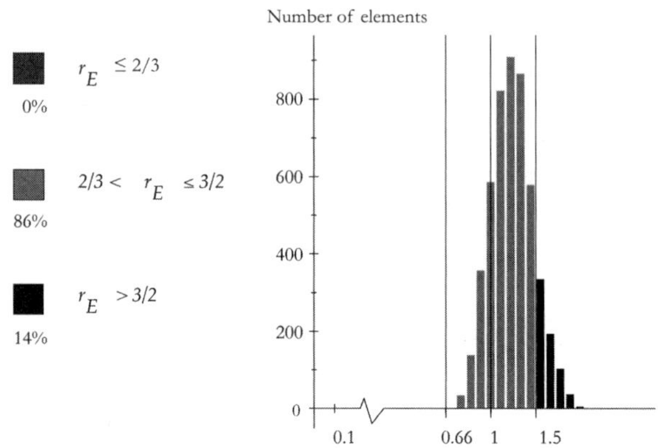

Figure 5.31. *Optimality histogram.*

5.5 Toward the automation of finite element analyses

5.5.1 Objectives

The procedures described above – reliable error estimator, mesh adaptation technique, automatic mesh generator – open the possibility of automating finite element analyses and controlling the quality of the solutions obtained while minimizing costs.

To cut costs, one must naturally reduce the cost of the computation itself, but one must also reduce the cost of human intervention. More precisely, the objective is to limit the operations by the user merely to the description of the problem being treated (geometry of the structure and mechanical data) and to the choice of the target accuracy level for the finite element analysis. All other steps of the calculation, including the choice of the initial mesh and the construction of adapted meshes, should be completely automated.

5.5.2 An automation algorithm

First, the user describes the problem being considered: description of the geometry of the structure by CAD software; description of the mechanical problem: prescribed displacements, applied loads, characteristics of the materials. Then, he chooses the accuracy ε_0 with which he wishes his analysis to be performed.

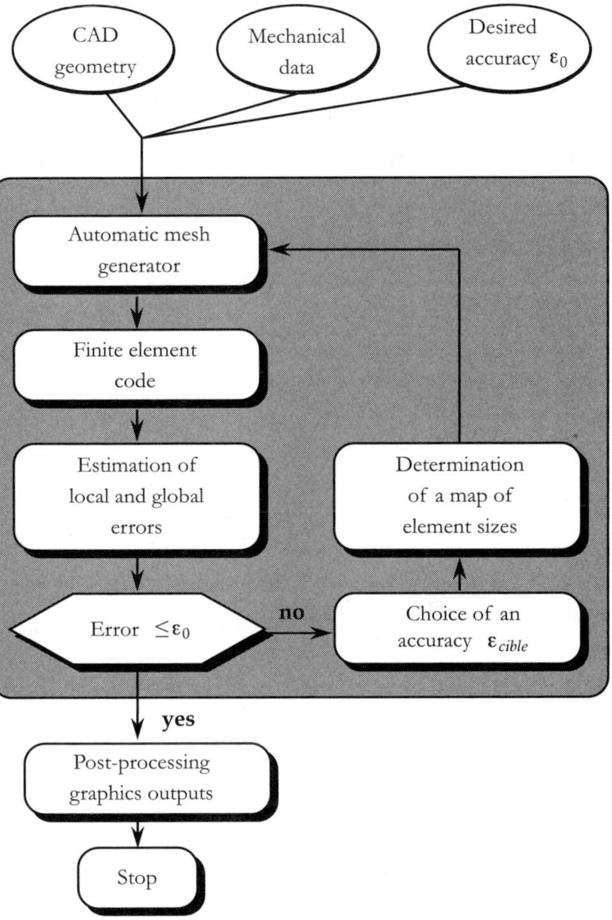

Figure 5.32. *Algorithm of automation of finite element analyses.*

Starting from an initial mesh $\mathcal{T}_0$, the algorithm described in Figure 5.32 yields a suitable adapted mesh to achieve the desired accuracy ε_0 in a few iterations.

For the choice of the initial mesh $\mathcal{T}_0$, two procedures can be used:

• Choice (1): The user describes the geometry through supertriangles and superquadrilaterals, which then constitute the initial mesh $\mathcal{T}_0$.

• Choice (2): Alternatively, the user may specify the approximate number of elements he wants for the initial mesh (which is equivalent to setting the cost of the initial finite element analysis). In this case, based on the measure of the structure, the average size to be used is calculated and the mesh generator constructs, based on this specification, a quasi-uniform initial mesh $\mathcal{T}_0$.

Once the initial coarse mesh is constructed, the initial finite element analysis is carried out and the corresponding accuracy ε is calculated. This accuracy is compared with the desired accuracy ε_0.

Usually, since mesh $\mathcal{T}_0$ is coarse, one has $\varepsilon \geq \varepsilon_0$. Therefore, at least one optimization iteration is required to reach the accuracy ε_0.

If ε is close to ε_0, i.e. in practice if

$$\varepsilon \leq \theta\, \varepsilon_0 \tag{5.37}$$

where θ is a coefficient ($\theta \geq 1$), one sets the target value as

$$\varepsilon_{target} = \varepsilon_0 \tag{5.38}$$

Otherwise, one sets the target value as

$$\varepsilon_{target} = \varepsilon / d \tag{5.39}$$

where d is a coefficient ($d \geq 1$).

Then, a size map is calculated taking ε_{target} as the required accuracy. The new mesh is constructed using an automatic mesh generator and one starts a new iteration by performing a new finite element analysis.

In a few iterations (between 1 and 5 depending on the initial mesh and the desired level of accuracy), the accuracy ε_0 is achieved.

The coefficients θ and d can be determined from numerical tests. For 2D or axisymmetric elasticity calculations, the values $\theta = 4$ and $d = 3$ give good results, whether one uses 3-node or 6-node triangles, or 4-node or 8-node quadrilaterals [COOREVITS - LADEVEZE - PELLE, 1994], [COOREVITS -PELLE - ROUGEOT, 1994].

For 3D elasticity calculations, it is preferable to take $\theta = 3$ and $d = 2$.

5.5.3 Comments

For 2D calculations, this automation procedure is operational at present and works very well if the mesh generator used is sufficiently robust for meshing complex domains while respecting the size specifications.

In 3D, the first examples of automation of calculations have been treated on geometries which are not too complex [COOREVITS - DUMEAU - PELLE, 1996]. The most significant difficulties are presently connected with the creation of the mesh:

- difficulties in creating a mesh of a complex skewed surface while respecting the size specifications;
- difficulties in controlling the sizes in the interior of the volume correctly.

5.6 Examples

5.6.1 First example

The first example is a 2D elastic test problem.

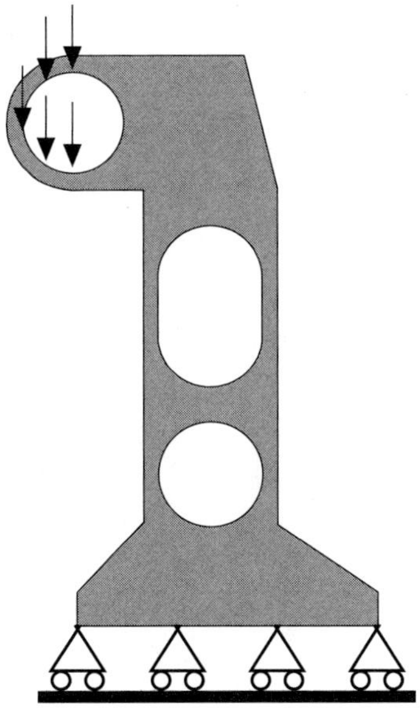

Figure 5.33. *Test problem 1 – Discretization: 3-node triangles.*

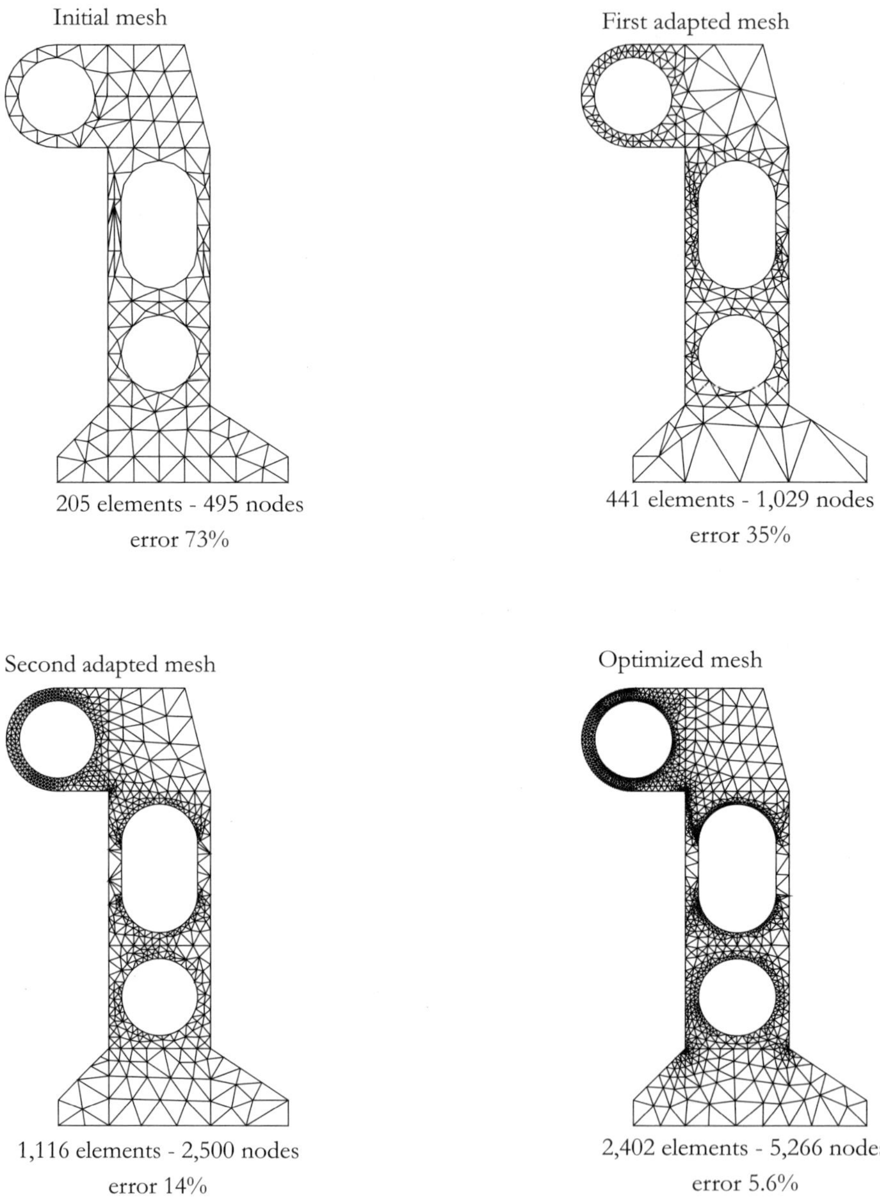

Figure 5.34. *Mesh automation starting from a coarse mesh. Prescribed accuracy 5%.*

This is quite a severe case, since the structure being considered presents stress concentration zones as well as relatively slender parts in which it is usually difficult to achieve good size predictions (particularly if one uses an indicator based on a stress smoothing technique). The discretization was performed using 6-node triangles.

The mechanical problem is described in Figure 5.33. The adaptation was carried out starting from a relatively coarse mesh which led to a global error of 73%.

The first adapted mesh took some stress concentration zones into account; then, because of the poor quality of the initial mesh in the slender zones, one of these zones came out too finely meshed.

The next two steps corrected this situation gradually, ending up with a very good optimized mesh (Figure 5.34).

5.6.2 Second example

The second example is a less complex test case (Figure 5.35) consisting of a brace loaded in bending. In all the calculations, the finite element discretization used 6-node triangles.

In Figure 5.36 the automation of the calculations was carried out for 2% prescribed accuracy, starting from a very coarse mesh which consisted of only 10 elements. In Figure 5.37, the adaptation was performed based on a moderately coarse mesh consisting of 69 elements.

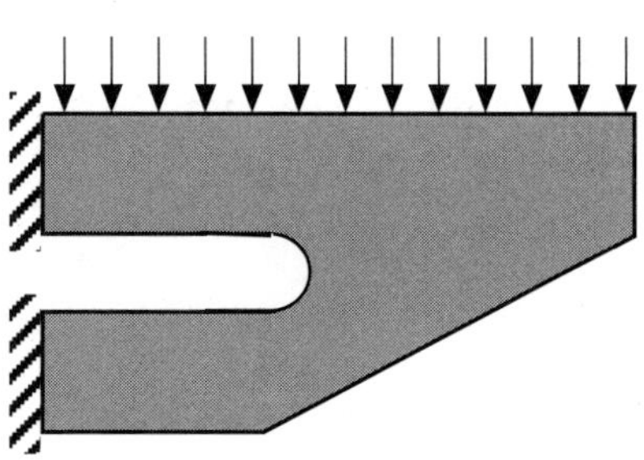

Figure 5.35. *Test problem 2 - Discretization : 6-node triangles.*

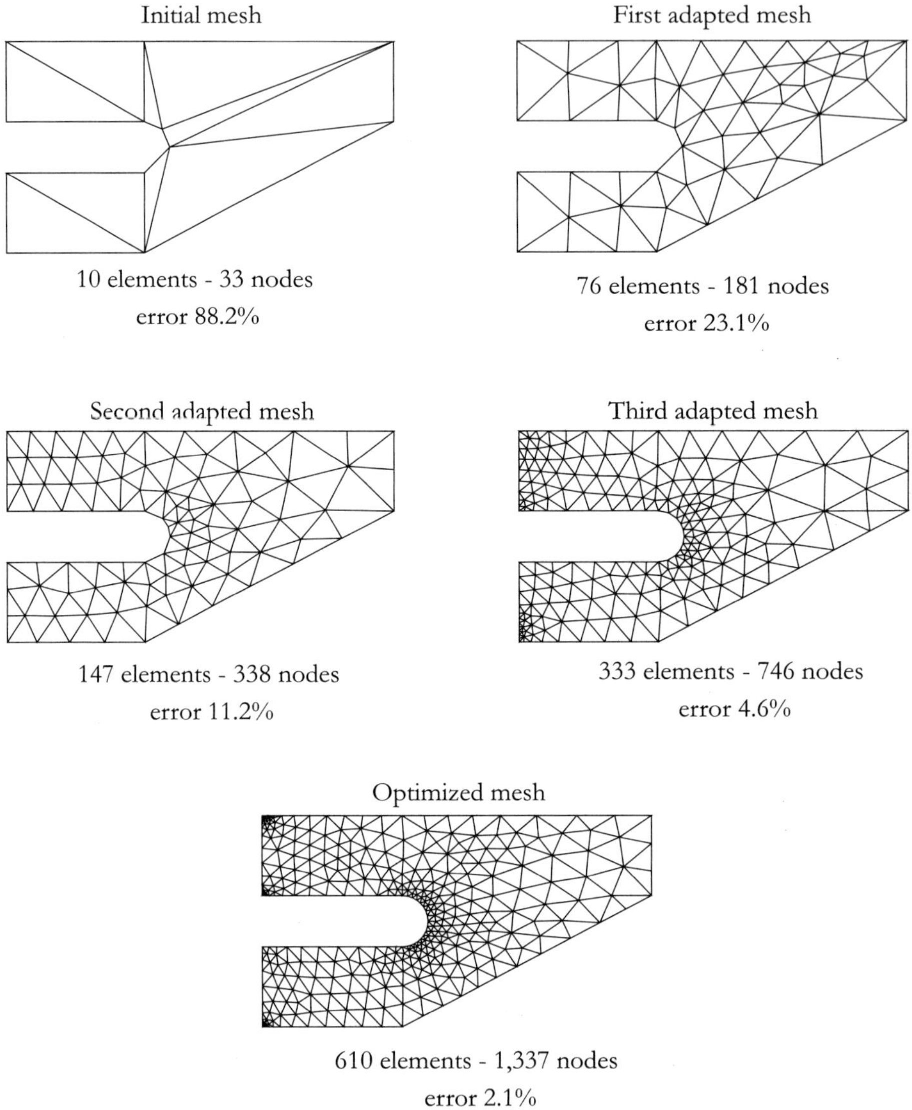

Figure 5.36. *Mesh automation starting from a very coarse mesh.*

Prescribed accuracy 2%.

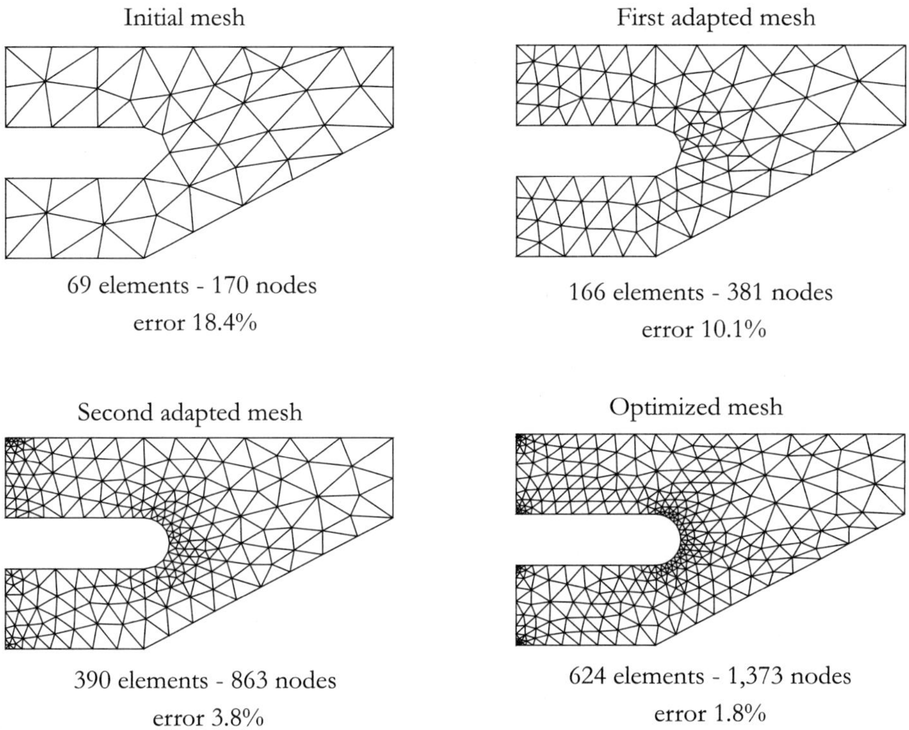

Figure 5.37. *Mesh automation starting from a coarse mesh.*

Prescribed accuracy 2%.

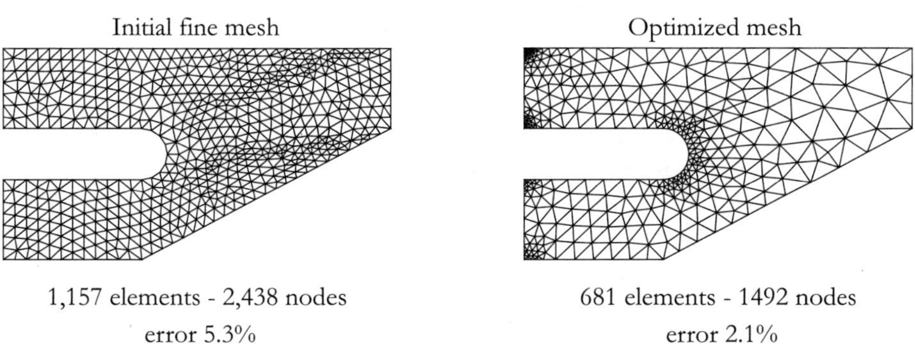

Figure 5.38. *Mesh automation starting from a very fine mesh. Prescribed accuracy 2%.*

The adapted meshes obtained in the two cases are very similar. This demonstrates the stability of our method. This stability was confirmed by the third adaptation presented in Figure 5.38 which, starting from a finer mesh, led again to an optimized mesh close to the two obtained previously.

5.6.3 Third example

This third example (Figure 5.39) concerns a dome of a pressure vessel with manifold piping.

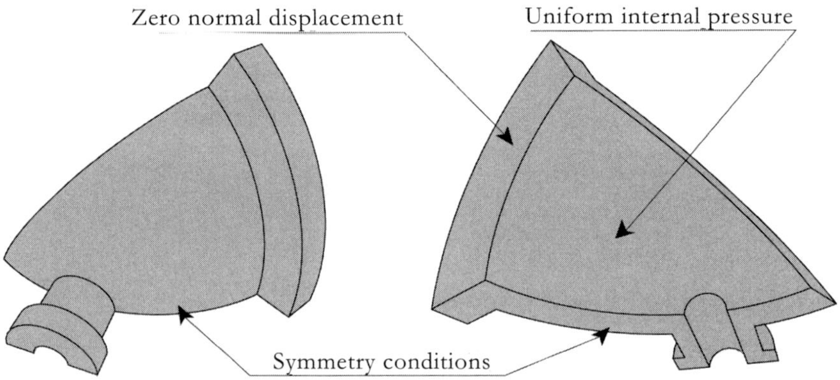

Figure 5.39. *Reservoir under internal pressure.*

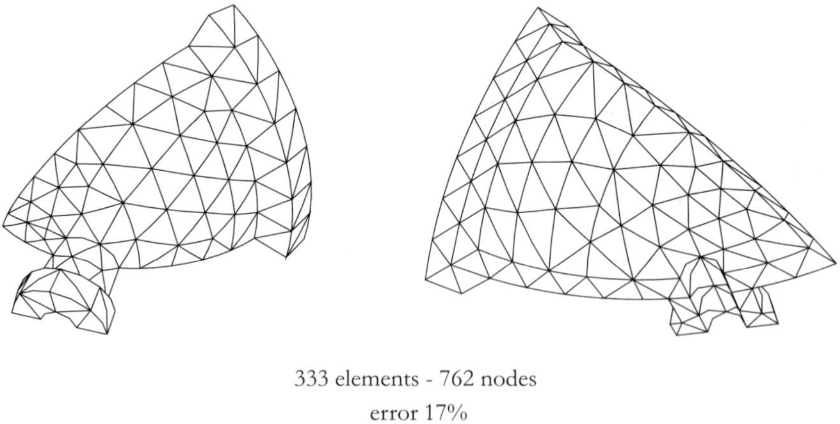

Figure 5.40. *Initial mesh.*

Due to the symmetries, only one-eighth of the piece was meshed. The adaptation for 5% prescribed error was performed starting from a relatively coarse mesh consisting of only 333 10-node tetrahedra and 762 nodes (Figure 5.40). Three adaptation steps were necessary to achieve the desired accuracy (Figure 5.41, Figure 5.42 and Figure 5.43). The map of the error contributions for the final optimized mesh is given in Figure 5.44.

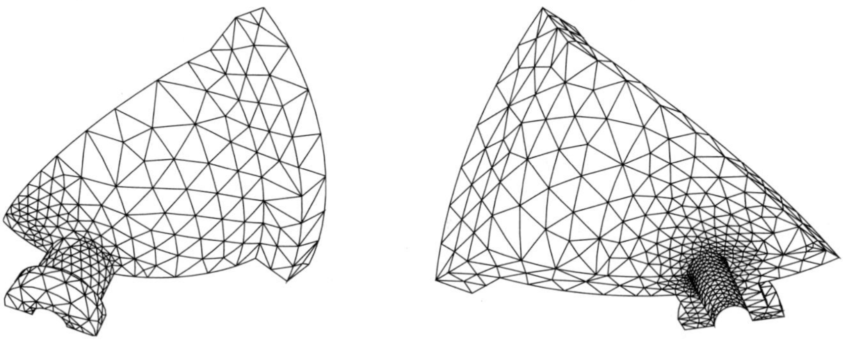

3,740 elements - 6,264 nodes
errror 8.2%

Figure 5.41. *First adapted mesh.*

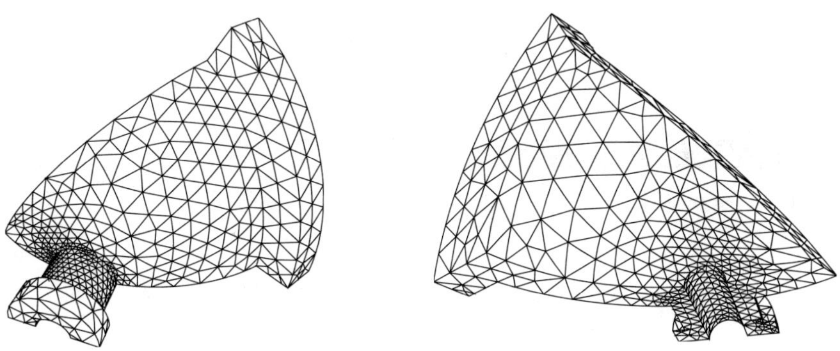

4,653 elements - 7,889 nodes
error 6.1%

Figure 5.42. *Second adapted mesh.*

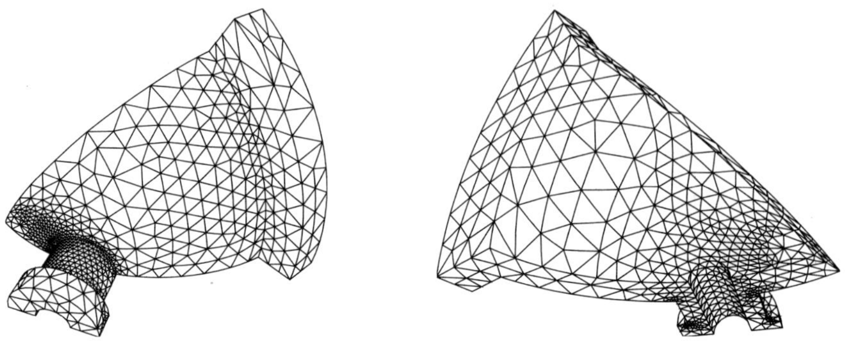

6,679 elements - 11,110 nodes
error 5.5%

Figure 5.43. *Optimized mesh.*

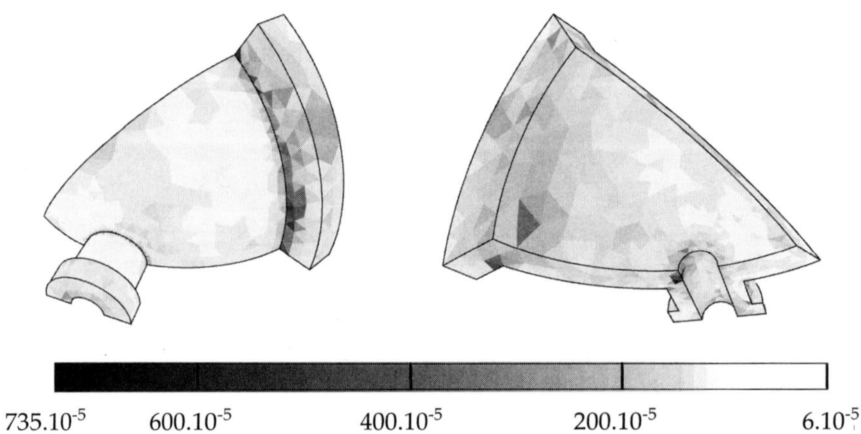

Figure 5.44. *Local contributions for the optimized mesh.*

Chapter 6

The constitutive relation error method for nonlinear evolution problems

6.1 Introduction

As we saw in the previous chapters, there have been many works proposing error estimators for linear problems. However, there have been clearly far fewer for nonlinear problems. For nonlinear elasticity or HENCKY-type plasticity problems, in which time plays no part, estimators based on the equilibrium residuals were proposed, for example, in [BABUSKA - RHEINBOLDT, 1982], [JOHNSON - HANSBO, 1991, 1992]. For problems in which history intervenes, a common approach consists in using, at each nodal instant of the incremental algorithm, the techniques developed for the linear case [BASS - ODEN, 1987], [ZIENKIEWICZ - LIU - HUANG, 1988], [JIN - BERNSPANG - LARSSON - WIBERG, 1990], [AUBRY - TIE, 1991, 1992], [FOURMENT - CHENOT, 1995], [BOROOMAND - ZIENKIEWICZ, 1998], [RANNACHER - SUTTMEIER, 1998],

[WUNDERLICH - CRAMER - STEINL, 1998], [HUERTA - DIEZ, 2000]. Other indicators introducing specific quantities related to plasticity have also been proposed [PERIC - YU - OWEN, 1994], [BARTHOLD - SCHMIDT - STEIN, 1997], [ANTUNEZ, 1998], [STEIN - BARTHOLD - OHNIMUS - SCHMIDT, 1998].

These types of methods cannot account for errors due to the time discretization and, therefore, do not yield a complete and reliable estimate of the discretization errors.

The objective of this chapter is to show that for nonlinear evolution problems the concept of error in constitutive relation enables the easy construction of error measures which have a strong mechanical meaning and which take into account all the approximation errors. In order to simplify the presentation, this approach will be explained in detail for quasi-static problems with small perturbations. We will make comments concerning its extension to problems with large transformations.

An error measure called *error in DRUCKER's sense* was proposed in [LADEVEZE, 1985]. It was developed in [COFFIGNAL - LADEVEZE, 1983], [LADEVEZE - COFFIGNAL - PELLE, 1986], [GALLIMARD - LADEVEZE - PELLE, 1996, 1997a, 1997b] for plasticity and viscoplasticity problems with small strains solved by the incremental method.

Another error measure called *dissipation error*, which fully exploits the formulation of the constitutive relations through internal variables, was proposed in [LADEVEZE, 1996]. For this error measure, a property similar to the PRAGER - SYNGE theorem in elasticity was proved. Its implementation was carried out in [LADEVEZE - MOES, 1998a, 1998b, 1999] for classical incremental methods, and in [PELLE - RYCKELYNCK, 1998] for the nonincremental "LATIN" method.

For each of these two error measures, specific indicators enable one to separate the various sources of error: space discretization errors, time discretization errors, errors due to the use of iterative algorithms.... Thus, it is possible to develop methods of control and adaptation of the parameters for these simulations: adaptation of the mesh, of the time representations and of the iterative resolution scheme over each time increment.

Almost all the error estimators presented here are designed for standard, normal material models [LADEVEZE, 1989]. In the last section, we will show how one can take into account other constitutive models, in particular, by utilizing the concept of bipotential recently introduced by DE SAXCE [DE SAXCE, 1992]. As an example, the application of this last method will be detailed for problems of contact among elastic solids, with or without friction.

6.2 Plasticity and viscoplasticity in small perturbations

6.2.1 Reconsideration of the reference problem

Let us consider the nonlinear problem [(1.9) - (1.12)] which describes the evolution of a structure whose constitutive relation is nonlinear and time-dependent. We assume small strains and neglect the acceleration terms:

Find $\underline{U}(M, t)$ *and* $\sigma(M, t)$ *defined on* $\Omega \times [0, T]$ *such that:*

- *Kinematic constraint equations and initial conditions*

$$\begin{aligned} &\underline{U} \in \boldsymbol{\mathcal{U}}^{[0,T]}, \\ &\forall t \in [0, T], \quad \underline{U}_{|\partial_1\Omega} = \underline{U}_d \\ &\forall M \in \Omega, \quad \underline{U}_{|t=0} = 0 \end{aligned} \tag{6.1}$$

- *Equilibrium equations*

$$\begin{aligned} &\sigma \in \boldsymbol{\mathcal{S}}^{[0,T]}, \\ &\forall t \in [0, T], \quad \forall \underline{U}^* \in \boldsymbol{\mathcal{U}}_0 \\ &\int_\Omega \mathrm{Tr}[\sigma \varepsilon(\underline{U}^*)]\, d\Omega = \int_\Omega \underline{f}_d \cdot \underline{U}^*\, d\Omega + \int_{\partial_2\Omega} \underline{F}_d \cdot \underline{U}^*\, dS \end{aligned} \tag{6.2}$$

- *Constitutive relation*

$$\forall t \in [0, T], \quad \forall M \in \Omega, \quad \sigma_{|t} = \mathbf{A}\big(\varepsilon(\dot{\underline{U}})_{|\tau};\, \tau \in [0,t]\big) \tag{6.3}$$

REMARKS

1. We will always assume that the quantities $\underline{U}_d$, $\underline{f}_d$ and $\underline{F}_d$ representing the environment are zero at $t = 0$. Displacements, stresses and, more generally, all state variables are zero at that instant.

2. Generally, in small perturbations, Problem [(6.1) - (6.3)] has a unique solution. The uniqueness or nonuniqueness depends on the type of constitutive relation being used. For plasticity or viscoplasticity problems, the classical conditions leading to uniqueness are the monotony conditions. A very general way of bringing the results of the literature together is to introduce DRUCKER's stability condition (Section 6.3.2). If the constitutive relation is stable in DRUCKER's sense, then uniqueness is guaranteed. Proof of this property can be found in [LADEVEZE, 1996]. However, in situations where degradation of the material is modeled by a damage law, one may lose this uniqueness property. For example, critical points – in particular instability points – can appear.

3. For a proof of the existence of solutions in plasticity and viscoplasticity problems, the reader is referred to [JOHNSON, 1976], [NECAS - HLAVACEK, 1981], [SUQUET, 1981], [LABORDE - NGUYEN, 1990], [LE TALLEC, 1990].

6.2.2 The constitutive relation

In (6.3), the constitutive relation was expressed in functional form. Today, it is most often formulated using internal variables. Models with internal variables were introduced a fairly long time ago; the basic idea consists in summing up the history of the material being studied until time t using some variables called *internal variables*. These models are required to verify the principles of thermodynamics, particularly the second principle. An elegant way to verify the second principle is to introduce the concept of dissipation pseudopotential, a notion which was clearly formulated in the 1970s [RICE, 1971]. Since this pseudopotential is generally a convex function, the works of MOREAU [MOREAU, 1966] and ROCKAFELLAR [ROCKAFELLAR, 1970] on the theory of convex functions have had a very significant impact on the theory of constitutive relations. The formulation based on a pair of dual functions and the determination of the plasticity pseudopotential are major contributions from MOREAU [MOREAU, 1966, 1974]. Various applications were given by NAYROLES [NAYROLES, 1973]. The most concise formulation, known as *formulation with two potentials* is due to NGUYEN [NGUYEN, 1984, 2000]; it came after [HALPHEN - NGUYEN, 1975]. Let us also mention the influence of GERMAIN in the popularization of this approach [GERMAIN, 1973], a recent illustration of which is the book by [LEMAÎTRE - CHABOCHE, 1985] on the mechanics of materials. Another presentation can be found in [WATANABE - ALTURI, 1986]. A perhaps more physical version which brings into play the various scales of observation is given in [FRANÇOIS - PINEAU - ZAOUI, 1991]. Regarding the difficult problem of identification, which falls into the general category of inverse problems, the reader can refer, for example, to [BUI, 1992].

Formulation with internal variables

At a point M, the state of the material is defined by the total strain $\mathbb{\varepsilon}$, the plastic strain $\mathbb{\varepsilon}^p$ and internal variables, globally denoted $\mathbb{X}$, belonging to $\mathbb{R}^m$. Introducing the conjugate variables $\mathbb{\sigma}$ and $\mathbb{Y}$, the mechanical dissipation can be written ($\circ$ denoting the inner product in $\mathbb{R}^m$) as

$$\mathrm{Tr}(\mathbb{\sigma}\dot{\mathbb{\varepsilon}}^p) - \mathbb{Y} \circ \dot{\mathbb{X}}$$

More precisely, the space $\boldsymbol{e}$ of the pairs $(\dot{\mathbb{\varepsilon}}^p, -\dot{\mathbb{X}})$ and the space $\boldsymbol{f}$ of the pairs $(\mathbb{\sigma}, \mathbb{Y})$ are put in duality by the bilinear form

$$\boldsymbol{e} \times \boldsymbol{f} \;\rightarrow\; \mathbb{R}, \qquad \left(\begin{bmatrix} \dot{\mathbb{\varepsilon}}^p \\ -\dot{\mathbb{X}} \end{bmatrix}, \begin{bmatrix} \mathbb{\sigma} \\ \mathbb{Y} \end{bmatrix} \right) \mapsto \mathrm{Tr}(\mathbb{\sigma}\dot{\mathbb{\varepsilon}}^p) - \mathbb{Y} \circ \dot{\mathbb{X}}$$

The elastic strain is defined by $\mathbb{\varepsilon}^e = \mathbb{\varepsilon} - \mathbb{\varepsilon}^p$.

The constitutive behavior is entirely described by the free energy $\rho\psi(\mathbb{\varepsilon}^e, \mathbb{X})$ and an operator $\mathbf{B}$ such that:

- *State law*

$$\mathbb{\sigma} = \rho \frac{\partial \psi}{\partial \mathbb{\varepsilon}^e} \quad \textit{and} \quad \mathbb{Y} = \rho \frac{\partial \psi}{\partial \mathbb{X}} \tag{6.4}$$

- *Evolution law*

$$\begin{bmatrix} \dot{\mathbb{\varepsilon}}^p \\ -\dot{\mathbb{X}} \end{bmatrix} = \mathbf{B} \left(\begin{bmatrix} \mathbb{\sigma} \\ \mathbb{Y} \end{bmatrix} \right) \tag{6.5}$$

REMARKS

1. In (6.4) the notations $\partial\psi/\partial\mathbb{X}$ and $\partial\psi/\partial\varepsilon^e$ must be interpreted as gradients. It would be more correct to use

$$\sigma = \rho\,\partial_{\varepsilon^e}\psi \quad \text{and} \quad \mathbb{Y} = \rho\,\partial_{\mathbb{X}}\psi \quad \text{or}$$

$$\sigma = \rho\,\text{grad}_{\varepsilon^e}\psi \quad \text{and} \quad \mathbb{Y} = \rho\,\text{grad}_{\mathbb{X}}\psi$$

In other terms, this is the same as considering $\rho\,\delta\,\psi$ to be given by

$$\rho\delta\psi = \text{Tr}(\sigma\delta\varepsilon^e) + \mathbb{Y} \circ \delta\mathbb{X}$$

2. In many applications, elasticity is linear and one has the decoupling property

$$\rho\psi(\varepsilon^e, \mathbb{X}) = \frac{1}{2}\text{Tr}(\varepsilon^e\mathbf{K}\varepsilon^e) + \rho\tilde{\psi}(\mathbb{X}) \tag{6.6}$$

Then

$$\sigma = \mathbf{K}\varepsilon^e \quad \text{and} \quad \mathbb{Y} = \mathbf{G}(\mathbb{X}) \tag{6.7}$$

where $\mathbf{K}$ and $\mathbf{G}$ are material-dependent operators.

3. For the second principle of thermodynamics to be verified

$$\text{Tr}(\sigma\dot{\varepsilon}^p) - \mathbb{Y} \circ \dot{\mathbb{X}} \geq 0$$

it suffices that operator **B** be chosen positive.

4. For certain types of constitutive behavior, in particular for plasticity, operator **B** may be multivalued.

5. A state-of-the-art review of works related to the estimation of anelastic strains is given in [MAIER - COMI - CORRIGLIANO - PEREGO, 1993]. Up to now, these works do not appear to have been used to develop error estimators.

Dissipation pseudopotential

A standard approach for defining operator **B** is to specify a convex dissipation pseudopotential $\varphi^*(\sigma, \mathbb{Y})$. In such a case, the material model is said to be *standard* [HALPHEN - NGUYEN, 1975].

One has

$$\dot{\varepsilon}^p \in \partial_\sigma \varphi^* \quad \text{and} \quad -\dot{X} \in \partial_Y \varphi^* \tag{6.8}$$

where $(\partial_\sigma \varphi^*, \partial_Y \varphi^*)$ is the subdifferential of φ^* in (σ, Y) [MOREAU, 1966].

If the convex function $\varphi^*(\sigma, Y)$ is differentiable, the subdifferential coincides with the common gradient (as is the case, for example, in viscoplasticity).

Then, one gets

$$\dot{\varepsilon}^p = \partial_\sigma \varphi^* = \operatorname{grad}_\sigma \varphi^* \quad \text{and} \quad -\dot{X} = \partial_Y \varphi^* = \operatorname{grad}_Y \varphi^* \tag{6.9}$$

One has the following classical property:

Property

For the second principle to be verified, it suffices to choose $\varphi^*(\sigma, Y)$ such that

$$\varphi^* \text{ is convex}, \quad \forall (\sigma, Y) \in \boldsymbol{f}, \quad \varphi^*(\sigma, Y) \geq 0, \quad \varphi^*(0, 0) = 0 \tag{6.10}$$

Proof of this property can be found in [LADEVEZE, 1996].

Let us introduce the convex function φ, dual of φ^*, defined by

$$\varphi(\dot{\varepsilon}^p, -\dot{X}) = \sup_{(\sigma, Y) \in \boldsymbol{f}} \left[\operatorname{Tr}(\sigma \dot{\varepsilon}^p) - Y \circ \dot{X} - \varphi^*(\sigma, Y) \right]$$

Under very broad regularity conditions, one can prove that the dual function of φ is the function φ^* itself [MOREAU, 1966]

$$\varphi^*(\sigma, Y) = \sup_{(\dot{\varepsilon}^p, -\dot{X}) \in \boldsymbol{e}} \left[\operatorname{Tr}(\sigma \dot{\varepsilon}^p) - Y \circ \dot{X} - \varphi(\dot{\varepsilon}^p, -\dot{X}) \right]$$

Then, one has the following theorems whose proof is classical:

Theorems

1. LEGENDRE - FENCHEL *inequality*

$$\forall (\dot{\varepsilon}^p, -\dot{X}) \in \boldsymbol{e}, \quad \forall (\sigma, Y) \in \boldsymbol{f},$$

$$\varphi(\dot{\varepsilon}^p, -\dot{X}) + \varphi^*(\sigma, Y) - \left[\operatorname{Tr}(\sigma \dot{\varepsilon}^p) - Y \circ \dot{X} \right] \geq 0 \tag{6.11}$$

2. *The following three properties are equivalent:*

- *Property* 1

$$\varphi(\dot{\varepsilon}^p, -\dot{X}) + \varphi^*(\sigma, Y) - \left[\mathrm{Tr}(\sigma\dot{\varepsilon}^p) - Y \circ \dot{X}\right] = 0 \tag{6.12}$$

- *Property* 2

$$\dot{\varepsilon}^p \in \partial_\sigma \varphi^* \quad and \quad -\dot{X} \in \partial_Y \varphi^* \tag{6.13}$$

- *Property* 3

$$\sigma \in \partial_{\dot{\varepsilon}^p}\varphi \quad and \quad Y \in \partial_{(-\dot{X})}\varphi \tag{6.14}$$

In Relations [(6.12) - (6.14)], φ and φ^* denote a pair of convex dual functions.

The constitutive relations most commonly used in plasticity or in viscoplasticity can be defined in terms of a pair of convex dual functions; thus, they correspond to standard formulations.

Examples of plastic and viscoplastic constitutive relations

Below we give some classical examples of plasticity and viscoplasticity taken from [LADEVEZE, 1996].

Plasticity model P1 (PRANDTL - REUSS)

For this model, the internal variable X is the scalar p, which can be interpreted as the accumulated plastic strain. The conjugate variable Y is the scalar R which describes the size of the elastic convex (isotropic hardening).

One has

$$\rho\psi(\varepsilon^e, p) = \frac{1}{2}\mathrm{Tr}(\varepsilon^e \mathbf{K} \varepsilon^e) + g(p) \tag{6.15}$$

where $g(p)$ is strictly convex. This leads to the state laws

$$\sigma = \mathbf{K}\varepsilon^e \quad \text{and} \quad R = \frac{dg}{dp}(p) = g'(p) \tag{6.16}$$

The pseudopotential φ^* is defined by

$$\varphi^*(\sigma, R) = \begin{cases} 0 & \text{if } f = \|\sigma_D\| - R - R_0 \leq 0 \\ +\infty & \text{otherwise} \end{cases} \tag{6.17}$$

where

$$\sigma_D = \sigma - \frac{1}{3}\,\mathrm{Tr}(\sigma)\mathbf{I} \quad \text{and} \quad \|\sigma_D\| = \sqrt{\mathrm{Tr}(\sigma_D \sigma_D)}$$

R_0 is a given positive constant (initial elastic threshold).

A simple calculation shows that for this model

$$\varphi(\dot{\varepsilon}^p, -\dot{p}) = R_0\,\dot{p} + \chi_C(\dot{\varepsilon}^p, -\dot{p}) \tag{6.18}$$

where χ_C is the indicator function of the convex C defined by

$$C = \left\{(\dot{\varepsilon}^p, -\dot{p}) \in \boldsymbol{e} \setminus \|\dot{\varepsilon}^p\| - \dot{p} \leq 0 \text{ and } \mathrm{Tr}(\dot{\varepsilon}^p) = 0\right\} \tag{6.19}$$

From (6.8) and (6.17), one deduces

$$\dot{\varepsilon}^p = \dot{\lambda}\,\frac{\sigma_D}{R + R_0} \quad \text{and} \quad -\dot{p} = -\dot{\lambda}$$

$$\dot{\lambda} = \frac{\langle \mathrm{Tr}(\sigma_D \dot{\sigma}_D)\rangle_+}{(R + R_0)\, g''(p)} \quad \text{if } f = 0 \text{ and } \dot{\lambda} = 0 \text{ otherwise}$$

where $\langle \cdot \rangle_+$ denotes the positive part.

Viscoplasticity model VP1

This model is obtained very simply by starting from model P1 and modifying the pseudopotential φ^*,

$$\varphi^*(\sigma, R) = \frac{k}{n+1}\left[\langle f \rangle_+\right]^{n+1} \quad \text{with } f = \|\sigma_D\| - R - R_0 \tag{6.20}$$

where k and n are material constants.

The dual potential is given by

$$\varphi(\dot{\varepsilon}^p, -\dot{p}) = R_0\,\dot{p} + \frac{kn}{n+1}\left[\frac{\dot{p}}{k}\right]^{1+\frac{1}{n}} + \chi_C(\dot{\varepsilon}^p, -\dot{p}) \tag{6.21}$$

and the evolution law by

$$\dot{\varepsilon}^p = k\left[\langle f\rangle_+\right]^n \frac{\sigma_D}{\|\sigma_D\|} \quad \text{and} \quad -\dot{p} = -k\left[\langle f\rangle_+\right]^n \tag{6.22}$$

Plasticity model P2

This is a standard variation of the MARQUIS - CHABOCHE model. The internal variables are the accumulated plastic strain p and a linear operator α with zero trace. The conjugate variables are a scalar R and an operator β describing the position of the elasticity convex (kinematic hardening). One has

$$\rho\psi(\varepsilon^e, p, \alpha) = \frac{1}{2}\mathrm{Tr}[\varepsilon^e \mathbf{K} \varepsilon^e] + \frac{1}{2}\, c\, \mathrm{Tr}[\alpha\alpha] + g(p) \tag{6.23}$$

where $g(p)$ is strictly convex and c is a strictly positive constant.

One deduces the state laws

$$\sigma = \mathbf{K}\varepsilon^e, \quad R = g'(p) \quad \text{and} \quad \beta = c\alpha \tag{6.24}$$

The pseudopotential φ^* is defined (a being a given positive constant) by

$$\varphi^*(\sigma, R) = \begin{cases} 0 & \textit{if } f = \|\sigma_D - \beta\| + \dfrac{a}{2c}\mathrm{Tr}[\beta\beta] - R - R_0 \le 0 \\ +\infty & \textit{otherwise} \end{cases} \tag{6.25}$$

In this case, the dual potential is

$$\begin{aligned} &\text{if } \dot{p} \neq 0: \quad \varphi(\dot{\varepsilon}^p; -\dot{p}, -\dot{\alpha}) = R_0\,\dot{p} + \frac{c}{2a}\frac{\|\dot{\varepsilon}^p - \dot{\alpha}\|^2}{\dot{p}} + \chi_C(\dot{\varepsilon}^p, -\dot{p}) \\ &\text{if } \dot{p} = 0: \quad \varphi(\dot{\varepsilon}^p; -\dot{p}, -\dot{\alpha}) = 0 \end{aligned} \tag{6.26}$$

where C is still defined by (6.19).

The evolution laws are given by

$$\dot{\varepsilon}^p = \lambda \frac{\sigma_D - \beta}{\|\sigma_D - \beta\|}$$

$$-\dot{p} = -\dot{\lambda}$$
$$-\dot{\alpha} = \dot{\lambda}\left[-\frac{\sigma_D - \beta}{\|\sigma_D - \beta\|} + \frac{a}{c}\beta\right] \tag{6.27}$$

with

$$\dot{\lambda} \geq 0, \quad f \leq 0 \quad \text{and} \quad \dot{\lambda} f = 0$$

Viscoplasticity model VP2

This model is obtained very simply by starting from model P2 and modifying the pseudopotential φ^*,

$$\varphi^*(\sigma, R, \beta) = \frac{k}{n+1}[\langle f \rangle_+]^{n+1} \tag{6.28}$$

where k and n are material constants and

$$f = \|\sigma_D - \beta\| + \frac{a}{2c}\mathrm{Tr}[\beta\beta] - R - R_0 \tag{6.29}$$

The dual potential is given by

$$\text{if } \dot{p} \neq 0: \quad \varphi(\dot{\varepsilon}^p; -\dot{p}, -\dot{\alpha}) = R_0 \dot{p} + \frac{c}{2a}\frac{\|\dot{\varepsilon}^p - \dot{\alpha}\|^2}{\dot{p}} + \frac{kn}{n+1}\left[\frac{\dot{p}}{k}\right]^{1+\frac{1}{n}} + \chi_C(\dot{\varepsilon}^p, -\dot{p}) \tag{6.30}$$

$$\text{if } \dot{p} = 0: \quad \varphi(\dot{\varepsilon}^p; -\dot{p}, -\dot{\alpha}) = 0$$

and the evolution laws by

$$\dot{\varepsilon}^p = k[\langle f \rangle_+]^n \frac{\sigma_D - \beta}{\|\sigma_D - \beta\|}$$
$$-\dot{p} = -k[\langle f \rangle_+]^n \tag{6.31}$$
$$-\dot{\alpha} = k[\langle f \rangle_+]^n\left[-\frac{\sigma_D - \beta}{\|\sigma_D - \beta\|} + \frac{a}{c}\beta\right]$$

REMARK. Other constitutive models can be found in [LEMAÎTRE - CHABOCHE, 1985] for metallic materials and in [HERAKOVITCH, 1996] for composites.

Normal formulation

The way the great majority of constitutive relations are written is biased in favor of the variables $\mathbb{X}$ and against the variables $\mathbb{Y}$. In many cases, it is possible through a change of variables of the type [LADEVEZE, 1989, 1996]

$$\underset{\sim}{\mathbb{X}} = \mathbf{R}(\mathbb{X}) \quad \text{and} \quad \underset{\sim}{\mathbb{Y}} = \mathbf{S}(\mathbb{Y})$$

to write the state laws in the form $\underset{\sim}{\mathbb{Y}} = \Lambda \underset{\sim}{\mathbb{X}}$, where Λ is a constant operator which is linear, symmetric and positive. In such a formulation, called *normal*, the variables $\mathbb{X}$ and $\mathbb{Y}$ are treated equally.

For example, for models P1, VP1, P2 and VP2, one obtains the associated normal formulation very easily using the change of variable

$$\underset{\sim}{p} = \frac{1}{\sqrt{\lambda}} \int_0^p [g''(q)]^{1/2}\, dq = \mathbf{S}(p) \quad \text{and} \quad \underset{\sim}{R} = \lambda\, \mathbf{S}\left[g'^{-1}(R)\right]$$

where λ is a positive constant.

The nonlinear part of the state laws $R = g'(p)$ becomes $\underset{\sim}{R} = \lambda \underset{\sim}{p}$.

Letting $R = \ell(\underset{\sim}{R}) = g'[\mathbf{S}^{-1}(\underset{\sim}{R}/\lambda)]$, one obtains the normal formulations corresponding to models P1, VP1, P2 and VP2 by replacing:

- in the state laws $R = g'(p)$ by $\underset{\sim}{R} = \lambda \underset{\sim}{p}$;
- in the expression of the threshold f, R by $\ell(\underset{\sim}{R})$.

From here on, whenever we use a standard and normal formulation, we will always write the linear state laws in the form

$$\mathbb{\sigma} = \mathbf{K}\, \mathbb{\varepsilon}^e \quad \text{and} \quad \mathbb{Y} = \Lambda\, \mathbb{X} \tag{6.32}$$

dropping $\underset{\sim}{\cdot}$ in order to simplify the notations and, thus, replacing R by $\ell(R)$ in the expression of the threshold function f.

6.3 Error in DRUCKER's sense

The error measure in DRUCKER's sense uses the functional formulation of the constitutive relation. One assumes that the material is in a virgin state at the initial time $t = 0$.

This error measure was introduced in [LADEVEZE, 1985] in the framework of the "LATIN" method. Its first application to the control of finite element calculations was carried out in [COFFIGNAL - LADEVEZE, 1983] and in [LADEVEZE - COFFIGNAL - PELLE, 1986]. More recently, in [GALLIMARD - LADEVEZE - PELLE, 1996, 1997a, 1997b], specific indicators for the time discretization and for the space discretization were introduced, enabling these discretizations to be controlled simultaneously.

6.3.1 Admissible fields

The notion of admissible field is similar to that used for linear problems.

Definition

A pair $(\underline{U}_{KA}, \boldsymbol{\sigma}_{SA})$ *is called admissible if:*

- $\underline{U}_{KA}$ *verifies the kinematic constraints;*
- $\boldsymbol{\sigma}_{SA}$ *verifies the equilibrium equations;*
- $(\underline{U}_{KA}, \boldsymbol{\sigma}_{SA})_{|t=0} = 0$.

6.3.2 Definition of the error measure

The error measure in DRUCKER's sense is based on DRUCKER's stability inequality [DRUCKER, 1964], which is verified by many constitutive models and, in particular, by the models presented above.

DRUCKER's stability inequality

Let us consider two stress-strain pairs $(\boldsymbol{\varepsilon}, \boldsymbol{\sigma})$ and $(\tilde{\boldsymbol{\varepsilon}}, \tilde{\boldsymbol{\sigma}})$ which verify the constitutive relation (6.3)

$$\forall t \in [0, T], \quad \boldsymbol{\sigma}_{|t} = \mathbf{A}\left(\dot{\boldsymbol{\varepsilon}}_{|\tau}; \tau \in [0, t]\right), \quad \tilde{\boldsymbol{\sigma}}_{|t} = \mathbf{A}\left(\dot{\tilde{\boldsymbol{\varepsilon}}}_{|\tau}; \tau \in [0, t]\right) \tag{6.33}$$

and are zero at the initial time $t = 0$

$$(\varepsilon, \sigma)_{|t=0} = 0, \quad (\underset{\sim}{\varepsilon}, \underset{\sim}{\sigma})_{|t=0} = 0$$

Definitions

1. *A material is said to be stable in* DRUCKER*'s sense if for any* (ε, σ) *and for any* $(\underset{\sim}{\varepsilon}, \underset{\sim}{\sigma})$ *verifying* (6.33) *one has*

$$\forall t \in [0, T], \quad \int_0^t \mathrm{Tr}[(\underset{\sim}{\sigma} - \sigma)(\dot{\underset{\sim}{\varepsilon}} - \dot{\varepsilon})]\, d\tau \geq 0 \tag{6.34}$$

2. *It is said to be strictly stable if under the same conditions* (6.34) *is verified and*

$$\left[\forall t \in [0, T], \quad \int_0^t \mathrm{Tr}[(\underset{\sim}{\sigma} - \sigma)(\dot{\underset{\sim}{\varepsilon}} - \dot{\varepsilon})]\, d\tau = 0\right] \Rightarrow \left[\underset{\sim}{\varepsilon} = \varepsilon \textit{ and } \underset{\sim}{\sigma} = \sigma\right] \tag{6.35}$$

Let us consider a standard constitutive relation defined by the free energy $\rho\,\psi(\varepsilon^e, \mathbb{X})$ and the dissipation pseudopotential $\varphi^*(\sigma, \mathbb{Y})$.

Theorem

The material is stable in DRUCKER*'s sense if:*

- *the free energy is of the following form, with* $\tilde{\psi}$ *strictly convex*

$$\rho\psi(\varepsilon^e, \mathbb{X}) = \frac{1}{2}\mathrm{Tr}(\varepsilon^e \mathbf{K} \varepsilon^e) + \rho\tilde{\psi}(\mathbb{X})\,;$$

- $\varphi^*(\sigma, \mathbb{Y})$ *is convex and verifies* (6.10);
- *the function* $\Phi^*(\sigma, \mathbb{X}) = \varphi^*(\sigma, \mathbf{G}(\mathbb{X}))$ *with* $\mathbb{Y} = \mathbf{G}(\mathbb{X})$ *is convex*;
- $\mathbf{G}(0) = 0$.

Proof

Let us denote $\Phi(\dot{\varepsilon}^p, -\dot{\mathbb{Y}})$ the conjugate function of $\Phi^*(\sigma, \mathbb{X})$,

$$\Phi(\dot{\varepsilon}^p, -\dot{\mathbb{Y}}) = \underset{(\sigma, \mathbb{X})}{\mathrm{Sup}}\left[\mathrm{Tr}(\sigma\dot{\varepsilon}^p) - \mathbb{X} \circ \dot{\mathbb{Y}} - \Phi^*(\sigma, \mathbb{X})\right] \tag{6.36}$$

Let us consider two stress-strain pairs (ε, σ) and $(\underset{\sim}{\varepsilon}, \underset{\sim}{\sigma})$ which verify the constitutive relation (6.3), and introduce the associated pairs of conjugate internal variables $(\mathbb{X}, \mathbb{Y})$ and $(\underset{\sim}{\mathbb{X}}, \underset{\sim}{\mathbb{Y}})$. Starting from

$$\mathbb{Y} = \mathbf{G}(\mathbb{X}), \quad \dot{\varepsilon}^p = \partial_\sigma \varphi^* \quad \text{and} \quad -\dot{\mathbb{X}} = \partial_{\mathbb{Y}} \varphi^*$$

a simple calculation of composite derivatives shows that

$$\dot{\varepsilon}^p = \partial_\sigma \Phi^* \quad \text{and} \quad -\dot{\mathbb{Y}} = \partial_{\mathbb{X}} \Phi^* \tag{6.37}$$

which proves that the constitutive relation can be defined indifferently starting from the conjugate potentials φ and φ^* for the conjugate variables $(-\dot{\mathbb{X}}, \mathbb{Y})$ or starting from the conjugate potentials Φ and Φ^* for the conjugate variables $(-\dot{\mathbb{Y}}, \mathbb{X})$.

Introducing plastic strains, one has

$$\begin{aligned}
\mathrm{Tr}[(\underset{\sim}{\sigma} - \sigma)(\dot{\underset{\sim}{\varepsilon}} - \dot{\varepsilon})] &= \mathrm{Tr}[(\underset{\sim}{\sigma} - \sigma)(\dot{\underset{\sim}{\varepsilon}}^e - \dot{\varepsilon}^e)] \\
&\quad + \mathrm{Tr}[(\underset{\sim}{\sigma} - \sigma)(\dot{\underset{\sim}{\varepsilon}}^p - \dot{\varepsilon}^p)] \\
&= \frac{1}{2}\frac{d}{dt}\Big[\mathrm{Tr}[(\underset{\sim}{\sigma} - \sigma)\mathbf{K}^{-1}(\underset{\sim}{\sigma} - \sigma)]\Big] \\
&\quad + \mathrm{Tr}[(\underset{\sim}{\sigma} - \sigma)(\dot{\underset{\sim}{\varepsilon}}^p - \dot{\varepsilon}^p)]
\end{aligned} \tag{6.38}$$

Using the properties of the conjugate functions φ and φ^*, one obtains

$$\mathrm{Tr}[\sigma\dot{\varepsilon}^p] = \varphi(\dot{\varepsilon}^p, -\dot{\mathbb{X}}) + \varphi^*(\sigma, \mathbb{Y}) + \mathbb{Y} \circ \dot{\mathbb{X}}$$

$$\mathrm{Tr}[\underset{\sim}{\sigma}\dot{\underset{\sim}{\varepsilon}}^p] = \varphi(\dot{\underset{\sim}{\varepsilon}}^p, -\dot{\underset{\sim}{\mathbb{X}}}) + \varphi^*(\underset{\sim}{\sigma}, \underset{\sim}{\mathbb{Y}}) + \underset{\sim}{\mathbb{Y}} \circ \dot{\underset{\sim}{\mathbb{X}}}$$

and therefore

$$\begin{aligned}
\mathrm{Tr}[(\underset{\sim}{\sigma} - \sigma)(\dot{\underset{\sim}{\varepsilon}}^p - \dot{\varepsilon}^p)] &= \varphi(\dot{\varepsilon}^p, -\dot{\mathbb{X}}) + \varphi^*(\sigma, \mathbb{Y}) + \mathbb{Y} \circ \dot{\mathbb{X}} \\
&\quad + \varphi(\dot{\underset{\sim}{\varepsilon}}^p, -\dot{\underset{\sim}{\mathbb{X}}}) + \varphi^*(\underset{\sim}{\sigma}, \underset{\sim}{\mathbb{Y}}) + \underset{\sim}{\mathbb{Y}} \circ \dot{\underset{\sim}{\mathbb{X}}} \\
&\quad - \mathrm{Tr}[\sigma\dot{\underset{\sim}{\varepsilon}}^p] - \mathrm{Tr}[\underset{\sim}{\sigma}\dot{\varepsilon}^p]
\end{aligned}$$

One deduces

$$\begin{aligned}
\mathrm{Tr}[(\underset{\sim}{\sigma} - \sigma)(\dot{\underset{\sim}{\varepsilon}}^p - \dot{\varepsilon}^p)] &= \Big[\varphi(\dot{\varepsilon}^p, -\dot{\mathbb{X}}) + \varphi^*(\underset{\sim}{\sigma}, \underset{\sim}{\mathbb{Y}}) - \mathrm{Tr}[\underset{\sim}{\sigma}\dot{\varepsilon}^p] + \underset{\sim}{\mathbb{Y}} \circ \dot{\mathbb{X}}\Big] \\
&\quad + \Big[\varphi(\dot{\underset{\sim}{\varepsilon}}^p, -\dot{\underset{\sim}{\mathbb{X}}}) + \varphi^*(\sigma, \mathbb{Y}) - \mathrm{Tr}[\sigma\dot{\underset{\sim}{\varepsilon}}^p] + \mathbb{Y} \circ \dot{\underset{\sim}{\mathbb{X}}}\Big] \\
&\quad + (\mathbb{Y} - \underset{\sim}{\mathbb{Y}}) \circ (\dot{\mathbb{X}} - \dot{\underset{\sim}{\mathbb{X}}})
\end{aligned}$$

Taking into account (6.37), the pairs (ε, σ) and $(\underset{\sim}{\varepsilon}, \underset{\sim}{\sigma})$ also verify

$$\mathrm{Tr}[\sigma\dot{\varepsilon}^{p}] = \Phi(\dot{\varepsilon}^{p}, -\dot{\mathbb{Y}}) + \Phi^{*}(\sigma, \mathbb{X}) + \dot{\mathbb{Y}} \circ \mathbb{X}$$

$$\mathrm{Tr}[\underset{\sim}{\sigma}\dot{\underset{\sim}{\varepsilon}}^{p}] = \Phi(\dot{\underset{\sim}{\varepsilon}}^{p}, -\dot{\underset{\sim}{\mathbb{Y}}}) + \Phi^{*}(\underset{\sim}{\sigma}, \underset{\sim}{\mathbb{X}}) + \dot{\underset{\sim}{\mathbb{Y}}} \circ \underset{\sim}{\mathbb{X}}$$

and therefore

$$\begin{aligned}\mathrm{Tr}[(\underset{\sim}{\sigma} - \sigma)(\dot{\underset{\sim}{\varepsilon}}^{p} - \dot{\varepsilon}^{p})] = &\left[\Phi(\dot{\varepsilon}^{p}, -\dot{\mathbb{Y}}) + \Phi^{*}(\underset{\sim}{\sigma}, \underset{\sim}{\mathbb{X}}) - \mathrm{Tr}[\underset{\sim}{\sigma}\dot{\varepsilon}^{p}] + \underset{\sim}{\mathbb{X}} \circ \dot{\mathbb{Y}}\right] \\ &+ \left[\Phi(\dot{\underset{\sim}{\varepsilon}}^{p}, -\dot{\underset{\sim}{\mathbb{Y}}}) + \Phi^{*}(\sigma, \mathbb{X}) - \mathrm{Tr}[\sigma\dot{\underset{\sim}{\varepsilon}}^{p}] + \mathbb{X} \circ \dot{\underset{\sim}{\mathbb{Y}}}\right] \\ &+ (\mathbb{Y} - \underset{\sim}{\mathbb{Y}}) \circ (\dot{\mathbb{X}} - \dot{\underset{\sim}{\mathbb{X}}})\end{aligned}$$

Using these results in Relation (6.38), one finally has

$$\int_0^t \mathrm{Tr}[(\underset{\sim}{\sigma} - \sigma)(\dot{\underset{\sim}{\varepsilon}} - \dot{\varepsilon})]\, d\tau =$$

$$\frac{1}{2}\int_0^t \Gamma\, d\tau + \frac{1}{2}\left[\mathrm{Tr}[(\underset{\sim}{\sigma} - \sigma)\mathbf{K}^{-1}(\underset{\sim}{\sigma} - \sigma)] + (\mathbb{Y} - \underset{\sim}{\mathbb{Y}}) \circ (\mathbb{X} - \underset{\sim}{\mathbb{X}})\right]_{|t} \qquad (6.39)$$

where

$$\begin{aligned}\Gamma = &\left[\varphi(\dot{\varepsilon}^{p}, -\dot{\mathbb{X}}) + \varphi^{*}(\underset{\sim}{\sigma}, \underset{\sim}{\mathbb{Y}}) - \mathrm{Tr}[\underset{\sim}{\sigma}\dot{\varepsilon}^{p}] + \underset{\sim}{\mathbb{Y}} \circ \dot{\mathbb{X}}\right] \\ &+ \left[\varphi(\dot{\underset{\sim}{\varepsilon}}^{p}, -\dot{\underset{\sim}{\mathbb{X}}}) + \varphi^{*}(\sigma, \mathbb{Y}) - \mathrm{Tr}[\sigma\dot{\underset{\sim}{\varepsilon}}^{p}] + \mathbb{Y} \circ \dot{\underset{\sim}{\mathbb{X}}}\right] \\ &+ \left[\Phi(\dot{\varepsilon}^{p}, -\dot{\mathbb{Y}}) + \Phi^{*}(\underset{\sim}{\sigma}, \underset{\sim}{\mathbb{X}}) - \mathrm{Tr}[\underset{\sim}{\sigma}\dot{\varepsilon}^{p}] + \underset{\sim}{\mathbb{X}} \circ \dot{\mathbb{Y}}\right] \\ &+ \left[\Phi(\dot{\underset{\sim}{\varepsilon}}^{p}, -\dot{\underset{\sim}{\mathbb{Y}}}) + \Phi^{*}(\sigma, \mathbb{X}) - \mathrm{Tr}[\sigma\dot{\underset{\sim}{\varepsilon}}^{p}] + \mathbb{X} \circ \dot{\underset{\sim}{\mathbb{Y}}}\right]\end{aligned} \qquad (6.40)$$

The four bracketed terms of Γ are positive due to the LEGENDRE - FENCHEL inequality.

Similarly, the second term of (6.39) is positive because, on the one hand, of **K** being positive and, on the other hand, of $\tilde{\psi}(\mathbb{X})$ being strictly convex, which implies that the derivation operator **G** is monotonic

$$(\mathbb{Y} - \underset{\sim}{\mathbb{Y}}) \circ (\mathbb{X} - \underset{\sim}{\mathbb{X}}) = \left[\mathbf{G}(\mathbb{X}) - \mathbf{G}(\underset{\sim}{\mathbb{X}})\right] \circ (\mathbb{X} - \underset{\sim}{\mathbb{X}}) \geq 0$$

The verification of strict stability is trivial.

REMARK. Materials verifying the assumptions of the above theorem, already mentioned in [HALPHEN - NGUYEN, 1975], are called *bistandard materials* or *materials with two potentials.*

Corollary

If function $g'(p)$ is strictly concave, the constitutive models $P1$, $VP1$, $P2$ and $VP2$ are strictly stable in DRUCKER's sense.

Proof

All the assumptions of the previous theorem are obviously verified, except perhaps the convexity of Φ^* which follows from the concavity of $g'(p)$.

For example, for P1, one has $\mathbb{X} = p$, $\mathbb{Y} = R$, $R = g'(p)$ and

$$\Phi^*(\sigma, p) = \varphi^*(\sigma, g'(p)) = \begin{cases} 0 & \text{if } \tilde{f} \leq 0 \\ +\infty & \text{otherwise} \end{cases}$$

The concavity of $g'(p)$ yields the convexity of $\tilde{f} = \|\sigma_D\| - g'(p) - R_0$ and, thus, that of Φ^*. The proof for the other models is similar.

Error measure

Let us consider an admissible pair $(\underline{U}_{KA}, \sigma_{SA})$ which is zero at the initial time $t = 0$. This pair does not verify a priori the constitutive relation (6.3). It is therefore an approximate solution of the reference problem [(6.1) - (6.3)]. Through the constitutive relation, one associates the stress σ_{KA} with the field $\varepsilon_{KA} = \varepsilon(\underline{U}_{KA})$. Similarly, through the inverse constitutive relation, one associates the strain ε_{SA} with the stress field σ_{SA}. For $M \in \Omega$ and $t \in [0, T]$, let

$$\eta(M, t) = \int_0^t \mathrm{Tr}[(\sigma_{SA} - \sigma_{KA})(\dot{\varepsilon}_{SA} - \dot{\varepsilon}_{KA})]\, d\tau \tag{6.41}$$

Property

For a strictly stable constitutive model in DRUCKER's sense, one has

$\forall t \in [0, T], \quad \eta(M, t) \geq 0$

and

$$[\forall t \in [0, T],\ \eta(M, t) = 0] \Rightarrow [\varepsilon_{CA} = \varepsilon_{SA} \text{ and } \sigma_{CA} = \sigma_{SA}] \tag{6.42}$$

The proof of this property follows immediately from (6.34) and (6.35). One deduces the theorem:

Theorem

$(\underline{U}_{KA}, \sigma_{SA})$ is the exact solution of the reference problem if and only if

$$\forall M \in \Omega, \quad \forall t \in [0, T], \quad \eta(M, t) = 0 \tag{6.43}$$

These properties lead us to give the following definition

Definition

One calls error in constitutive relation in DRUCKER's sense associated with the pair $(\underline{U}_{KA}, \sigma_{SA})$ the real quantity

$$\mathbf{e}_{RdC}(\underline{U}_{KA}, \sigma_{SA}) = \left[\underset{t \in [0, T]}{\text{Sup}} \int_\Omega \eta(M, t)\, d\Omega \right]^{1/2} \tag{6.44}$$

One associates with this absolute error the relative error

$$\varepsilon = \mathbf{e}_{RdC}/D \tag{6.45}$$

where the denominator D is defined by

$$D^2 = 2 \underset{t \in [0, T]}{\text{Sup}} \int_\Omega \int_0^t \text{Tr}[\sigma_{SA} \dot{\varepsilon}_{SA} + \sigma_{KA} \dot{\varepsilon}_{KA}]\, dt\, d\Omega \tag{6.46}$$

REMARKS

1. D is a quantity which is always positive or zero, and which is nonzero when one of the fields σ_{SA} or σ_{KA} is nonzero. Indeed, it is sufficient to observe that

$$\int_0^t \text{Tr}[\sigma_{SA} \dot{\varepsilon}_{SA} + \sigma_{KA} \dot{\varepsilon}_{KA}]\, dt = \int_0^t \text{Tr}[(\sigma_{SA} - 0)(\dot{\varepsilon}_{SA} - 0)]\, dt$$
$$+ \int_0^t \text{Tr}[(\sigma_{KA} - 0)(\dot{\varepsilon}_{KA} - 0)]\, dt$$

From (6.34) each of these terms is positive or zero and from (6.35) these two terms are zero if and only if σ_{SA} and σ_{KA} are zero.

2. Using (6.39), one shows that the error at time T can be divided into two terms:

- a term which depends only on the state at time T,

$$\frac{1}{2}\int_{\Omega} \mathrm{Tr}\Big[(\sigma_{SA}-\sigma_{KA})\mathbf{K}^{-1}(\sigma_{SA}-\sigma_{KA})\Big]_{|t=T}\, d\Omega$$

$$+\frac{1}{2}\int_{\Omega}\Big[(\mathbb{Y}_{SA}-\mathbb{Y}_{KA})\circ(\mathbb{X}_{SA}-\mathbb{X}_{KA})\Big]_{|t=T}\, d\Omega \tag{6.47}$$

- a term involving the loading history over $[0, T]$

$$\frac{1}{2}\int_{\Omega}\int_0^T \Gamma \, dt\, d\Omega \tag{6.48}$$

3. Another possibility consists in defining the error in constitutive relation over $[0, T]$ by

$$\tilde{\mathbf{e}}_{RdC}(\underline{U}_{KA}, \sigma_{SA}) = \left[\frac{1}{T}\int_0^T \int_{\Omega} \eta(M, t)\, d\Omega\, dt\right]^{1/2} \tag{6.49}$$

Relationship with the error in elasticity

Let us assume that the loading is such that the entire structure remains elastic and that the stresses σ_{SA} and σ_{KA} never exceed the elastic limit at any point.

Then, with these hypotheses, one obtains

$$\sigma_{SA} = \mathbf{K}\varepsilon_{SA} \quad \text{and} \quad \sigma_{KA} = \mathbf{K}\varepsilon_{KA} \tag{6.50}$$

and by integrating (6.41) with respect to time

$$\eta(t, M) = \frac{1}{2}\mathrm{Tr}[(\sigma_{SA}-\sigma_{KA})\mathbf{K}^{-1}(\sigma_{SA}-\sigma_{KA})] \tag{6.51}$$

One has

$$\mathbf{e}^2_{RdC} = \sup_{t\in[0,T]} \int_{\Omega} \eta(t, M)\, d\Omega = \frac{1}{2}\sup_{t\in[0,T]} \Big\|\big[\sigma_{SA}-\mathbf{K}\varepsilon_{KA}\big]\Big\|^2_{\sigma,\Omega} \tag{6.52}$$

Similarly, one can show that

$$D^2 = \underset{t \in [0,T]}{\text{Sup}} \left[\left\| \boldsymbol{\sigma}_{SA} \right\|^2_{\sigma,\Omega} + \left\| \mathbf{K}\boldsymbol{\varepsilon}_{KA} \right\|^2_{\sigma,\Omega} \right] \tag{6.53}$$

Therefore, one has

$$\varepsilon^2 = \frac{\underset{t \in [0,T]}{\text{Sup}} \left\| \boldsymbol{\sigma}_{SA} - \mathbf{K}\boldsymbol{\varepsilon}_{KA} \right\|^2_{\sigma,\Omega}}{2 \underset{t \in [0,T]}{\text{Sup}} \left[\left\| \boldsymbol{\sigma}_{SA} \right\|^2_{\sigma,\Omega} + \left\| \mathbf{K}\boldsymbol{\varepsilon}_{KA} \right\|^2_{\sigma,\Omega} \right]} \tag{6.54}$$

If one compares this error with the errors defined in (2.8) for elasticity, one can see that the expressions are very similar. The denominator contains the sum of the norms

$$2 \left[\left\| \boldsymbol{\sigma}_{SA} \right\|^2_{\sigma,\Omega} + \left\| \mathbf{K}\boldsymbol{\varepsilon}_{KA} \right\|^2_{\sigma,\Omega} \right] \tag{6.55}$$

instead of

$$\left\| \boldsymbol{\sigma}_{SA} + \mathbf{K}\boldsymbol{\varepsilon}_{KA} \right\|^2_{\sigma,\Omega} \tag{6.56}$$

REMARKS

1. When $\boldsymbol{\sigma}_{SA}$ and $\mathbf{K}\boldsymbol{\varepsilon}_{KA}$ are nearly equal, i.e. when the error is small, the difference between (6.55) and (6.56) is truly negligible.

2. In many common situations where the loading is monotonic, one observes that in (6.54) the "Sup"s are achieved at $t = T$.

6.3.3 Construction of the admissible fields

Calculations in plasticity and viscoplasticity are most often carried out using the incremental method, whose principle was reviewed in Section 1.3.2. In order to outline the construction of the admissible fields, let us first elaborate on the resolution of Problem (1.28):

Find $\mathbf{q}_{n+1} = \mathbf{q}(t_{n+1})$ *and* $\boldsymbol{\sigma}_{h,n+1} = \boldsymbol{\sigma}_h(t_{n+1})$ *such that*

$$\int_\Omega \mathbf{B}^T \boldsymbol{\sigma}_{h,n+1} \, d\Omega = \mathbf{F}_{n+1} \quad \text{and} \quad \boldsymbol{\sigma}_{h,n+1} = \mathbf{A}\left[\mathbf{B}\dot{\mathbf{q}}(\tau)_{\,|\tau \le t_{n+1}} \right] \tag{6.57}$$

with

$$\mathbf{F}_{n+1} = \int_{\Omega} \mathbf{N}^T \underline{f}_d(t_{n+1})\, d\Omega + \int_{\partial_2\Omega} \mathbf{N}^T \underline{F}_d(t_{n+1})\, dS$$

$$\underline{U}_h(M, t) = \mathbf{N}(M)\mathbf{q}(t) \quad \textit{and} \quad \varepsilon(\underline{U}_h) = \mathbf{B}(M)\mathbf{q}(t)$$

$$\mathbf{q}(t) = \left[\mathbf{q}(t_{n+1}) - \mathbf{q}(t_n)\right] \frac{t - t_n}{t_{n+1} - t_n} + \mathbf{q}(t_n)$$

Determination of $\mathbf{q}_{n+1}$

To determine $\mathbf{q}_{n+1}$ one uses an iterative method which we outline here in a form applicable to the different solution techniques. Usually, an initial approximation $\mathbf{q}_{n+1}^{(0)}$ of $\mathbf{q}_{n+1}$ is obtained by assuming elastic behavior over the increment $[t_n, t_{n+1}]$ in order to treat cases of elastic unloading correctly (Figure 6.1)

$$\mathbf{q}_{n+1}^{(0)} = \mathbf{q}_n + \Delta\mathbf{q}^{(0)}$$

with

$$\int_{\Omega} \mathbf{B}^T \tilde{\sigma}_{h,\,n+1}^{(0)}\, d\Omega = \mathbf{F}_{n+1} \quad \text{and} \quad \tilde{\sigma}_{h,\,n+1}^{(0)} = \sigma_{h,\,n} + \mathbf{KB}\Delta\mathbf{q}^{(0)}$$

i.e.

$$\mathbb{K}\Delta\mathbf{q}^{(0)} = \mathbf{F}^{(0)}$$

with

$$\mathbf{F}^{(0)} = \mathbf{F}_{n+1} - \int_{\Omega} \mathbf{B}^T \sigma_{h,\,n}\, d\Omega$$

Let us describe the *k*th iteration. At iteration $k-1$, one has previously constructed $\mathbf{q}_{n+1}^{(k-1)}$.

Starting from the strain history defined over $[0, t_n]$ by the calculations on previous increments and over $[t_n, t_{n+1}]$ by

$$\mathbf{Bq}_n + \frac{\mathbf{Bq}_{n+1}^{(k-1)} - \mathbf{Bq}_n}{t_{n+1} - t_n}(t - t_n)$$

one obtains through the integration of the constitutive relation the stress field $\sigma_{h,\,n+1}^{(k-1)}$.

Then, one calculates the residual

$$\mathbf{R}^{(k-1)} = \mathbf{F}_{n+1} - \int_{\Omega} \mathbf{B}^T \sigma_{h,\,n+1}^{(k-1)}\, d\Omega \tag{6.58}$$

If this residual is considered to be sufficiently small, one ends the iterations; otherwise, one seeks a new correction $\mathbf{\Delta q}^{(k)}$ which is the solution of

$$\left[\int_{\Omega} \mathbf{B}^T \left[\frac{\partial \sigma}{\partial \mathbf{q}} \right]_{|\mathbf{q} = \mathbf{q}_{n+1}^{(k-1)}} d\Omega \right] \mathbf{\Delta q}^{(k)} = \mathbf{R}^{(k-1)}$$

In the vicinity of the state associated with $(\mathbf{Bq}_{n+1}^{(k-1)}, \sigma_{h,\,n+1}^{(k-1)})$ one can, to make things clear, consider that

$$\left[\frac{\partial \sigma}{\partial \mathbf{q}} \right]_{|\mathbf{q} = \mathbf{q}_{n+1}^{(k-1)}} = \mathbf{K}_{tg}\mathbf{B}$$

where $\mathbf{K}_{tg}$ is the tangent constitutive operator.

Therefore, the problem to be solved is

$$\mathbb{K}_{tg}\, \mathbf{\Delta q}^{(k)} = \mathbf{R}^{(k-1)} \tag{6.59}$$

with

$$\mathbb{K}_{tg} = \int_{\Omega} \mathbf{B}^T \mathbf{K}_{tg} \mathbf{B}\, d\Omega$$

Thus, one obtains a new approximation of the displacement field

$$\mathbf{N}(M)\mathbf{q}_{n+1}^{(k)}$$

and an equilibrated stress field in the finite element sense

$$\tilde{\sigma}_{h,\,n+1}^{(k)} = \sigma_{h,\,n+1}^{(k-1)} + \mathbf{K}_{tg}\mathbf{B}\mathbf{\Delta q}^{(k)} \tag{6.60}$$

The calculation of the residual $\mathbf{R}^{(k)}$ yields a stress field $\sigma_{h,\,n+1}^{(k)}$ related to the displacement field by the constitutive relation.

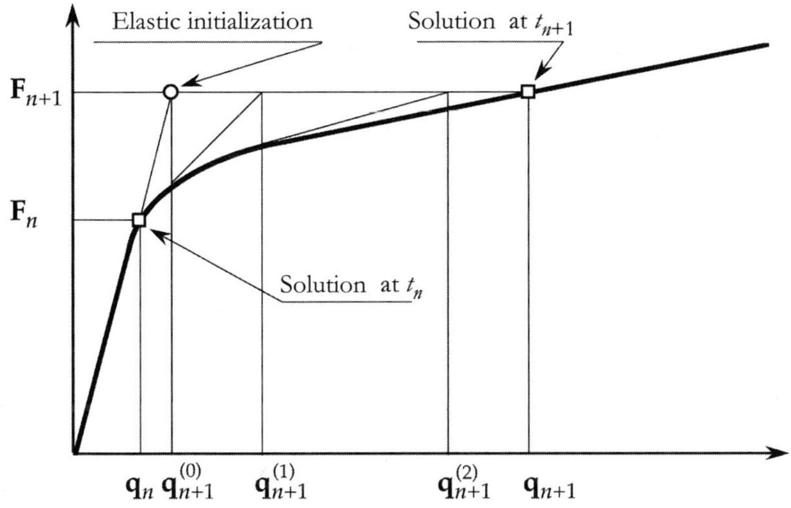

Figure 6.1. *Determination of* $\mathbf{q}_{n+1}$.

REMARKS

1. Usually, the results supplied by the finite element code are:

• nodal values $\mathbf{q}(t_n)$ at each nodal instant t_n of the time discretization;

• values of the stress field $\sigma_h(G_j, t_n) = \sigma_{h,n}(G_j)$ at a number of GAUSS points G_j and at each nodal instant t_n.

2. There are many variants of the above algorithm. These differ by:

• the choice of the numerical scheme for integrating the constitutive relation over the known strain history;

• the use of operators other than $\mathbf{K}_{tg}$ to make the calculation less costly or to accelerate the convergence (quasi-NEWTON methods, BFGS method…).

For these aspects, one can, for example, refer to [DENNIS - MORE, 1977], [OWEN - HINTON, 1980], [GERADIN - IDELSOHN - HOGGE, 1981], [BATHE, 1982], [MAIER - NAPPI, 1983], [SIMO - TAYLOR, 1985], [ORTIZ - SIMO, 1986], [HUGHES, 1987], [ZIENKIEWICZ - TAYLOR, 1988], [CORIGLIANO - PEREGO, 1990], [GEAR, 1971], [CRISFIELD, 1991], [NOOR - PETER, 1991].

Construction of an admissible displacement

At each instant t_n, one constructs the field $\underline{U}_{KA}$ by letting

$$\underline{U}_{KA}(M, t_n) = \mathbf{N}(M)\mathbf{q}(t_n)$$

then $\underline{U}_{KA}$ is determined at any time by linear interpolation. Thus, with the assumption that the data $\underline{U}_d$ is also linear in time over each time interval $[t_n, t_{n+1}]$, which is not very restrictive, one obtains a kinematically admissible displacement field at every instant. The field $\boldsymbol{\sigma}_{KA}$ is obtained by integrating the constitutive relation.

Construction of an admissible stress

The techniques developed in elasticity can be easily adapted, as long as one starts from a field which verifies the equilibrium in the finite element sense, to the construction at each nodal instant t_n of a field $\hat{\boldsymbol{\sigma}}_{h,n}(M)$ exactly admissible at time t_n. If one has access inside the finite element code, it suffices to take the field $\tilde{\boldsymbol{\sigma}}_h$ introduced for calculating Residual (6.58). Otherwise, one simply needs, starting from $\boldsymbol{\sigma}_h$, to solve a linear problem similar to (6.59). If one assumes that the time variation of the loads is linear over $[t_n, t_{n+1}]$, one obtains a field $\boldsymbol{\sigma}_{SA}$ admissible at each instant by letting

$$\boldsymbol{\sigma}_{SA}(M, t) = \frac{\hat{\boldsymbol{\sigma}}_{h,n+1}(M) - \hat{\boldsymbol{\sigma}}_{h,n}(M)}{t_{n+1} - t_n}(t - t_n) + \hat{\boldsymbol{\sigma}}_{h,n}(M)$$

The strain field $\boldsymbol{\varepsilon}_{SA}$ is obtained by integrating the inverse constitutive relation.

REMARK. The construction of $\boldsymbol{\sigma}_{SA}$ poses no special difficulty if the strain hardening coefficient is always positive or, in other terms, if any stress verifies the plasticity criterion for certain values of the internal variables which can be calculated.

Conversely, in perfect plasticity, the stress thus constructed may be outside the domain of elasticity, in which case it is no longer possible to calculate the corresponding anelastic strain. Adjustments in the construction of equilibrated $\hat{\boldsymbol{\sigma}}_{h,n}(M)$ can be made so that $\boldsymbol{\sigma}_{SA}$ remains legal for the highest possible solicitations.

6.3.4 Error indicators in time and in space

The measure of the error in constitutive relation in DRUCKER's sense takes into account all the approximations made between the reference problem and the numerical model being calculated: discretization error in time, discretization errors in space, errors due to the use of iterative algorithms for solving the nonlinear problem over each increment....

In order to adapt the different calculation parameters, particularly the mesh and the size of the time steps, one must be capable of identifying these different contributions in the global error.

We will show that the concept of error in constitutive relation applied to different "intermediate" models enables one to achieve this objective.

Error indicator in time

Let us consider Problem [(1.24) - (1.26)], which is discretized in space, but continuous in time:

Find $\underline{U}_h$ of the finite element type and σ_h such that:

$$\forall t \in [0, T], \quad \underline{U}_h \text{ is KA and satisfies the initial conditions} \tag{6.61}$$

$$\forall t \in [0, T] \quad \text{and} \quad \forall \underline{U}_h^* \in \boldsymbol{\mathcal{U}}_{h0},$$

$$\int_\Omega \operatorname{Tr}\left[\sigma_h \varepsilon(\underline{U}_h^*)\right] d\Omega = \int_\Omega \underline{f}_d \cdot \underline{U}_h^* \, d\Omega + \int_{\partial_2\Omega} \underline{F}_d \cdot \underline{U}_h^* \, dS \tag{6.62}$$

$$\forall t \in [0, T], \ \forall M \in \Omega, \ \sigma_{h|t} = \mathbf{A}\left(\varepsilon(\dot{\underline{U}}_h)_{|\tau}; \tau \in [0, t]\right) \tag{6.63}$$

One can use the same approach with this problem as with the reference problem [(6.1) - (6.3)].

Definition

A pair $(\underline{U}_{h,KA}, \sigma_{h,SA})$ is said to be admissible with respect to Problem [(6.61) - (6.63)] *if:*

- $\underline{U}_{h,KA}$ *verifies the kinematic constraints* (6.61);
- $\sigma_{h,SA}$ *verifies the equilibrium in the weak finite element sense* (6.62);
- $(\underline{U}_{h,KA}, \sigma_{h,SA})_{|t=0} = 0$.

Using the constitutive relation, one associates a stress field $0\sigma_{h,KA}$ with the field $\varepsilon_{h,KA} = \varepsilon(\underline{U}_{h,KA})$. Similarly, using the inverse constitutive relation, one associates a strain field $\varepsilon_{h,SA}$ with each admissible stress field $\sigma_{h,SA}$. The quality of the pair $(\underline{U}_{h,KA}, \sigma_{h,SA})$ as an approximate solution of Problem [(6.61) - (6.63)] may be evaluated by the error in constitutive relation built on DRUCKER's inequality

$$\mathbf{e}_{h,RdC}(\underline{U}_{h,KA}\,,\,\sigma_{h,SA}) =$$

$$\left[\operatorname*{Sup}_{t\in[0,\,T]} \int_\Omega \int_0^t \mathrm{Tr}[(\sigma_{h,SA}-\sigma_{h,KA})\,(\dot{\varepsilon}_{h,SA}-\dot{\varepsilon}_{h,KA})]\,dt\,d\Omega\right]^{1/2} \qquad (6.64)$$

For a given space discretization, this error measure enables one to evaluate the errors due to the use of the incremental method and to the use of a NEWTON-type iterative method over each increment.

Its implementation is very simple. Indeed, a pair $(\underline{U}_{h,KA}, \sigma_{h,SA})$ can be easily constructed from the results provided by the incremental method.

For the displacement field, one simply takes

$$\underline{U}_{h,KA} = \underline{U}_h$$

and for the stress field $\sigma_{h,SA}$, one takes the field defined over $[t_n, t_{n+1}]$ by

$$\tilde{\sigma}_h(M, t) = \frac{\tilde{\sigma}_{h,\,n+1}(M) - \tilde{\sigma}_{h,\,n}(M)}{t_{n+1} - t_n}\,(t - t_n) + \tilde{\sigma}_{h,\,n}(M)$$

where the fields $\tilde{\sigma}_{h,\,n}$ are defined by (6.60).

Indeed, this field is admissible in the weak finite element sense.

Then, one lets

$$\mathbf{i}_{time} = \mathbf{e}_{h,RdC}(\underline{U}_h\,,\,\tilde{\sigma}_h)$$

A zero value of $\mathbf{i}_{time}$ means that $(\underline{U}_h\,,\,\tilde{\sigma}_h)$ is the exact solution of Problem [(6.61) - (6.63)]. This measure of the error in constitutive relation between the numerical model and the problem discretized in space, but continuous in time, is used as an indicator of the error in time between the numerical model and the reference model [(6.1) - (6.3)].

Error indicator in space

The approach is similar to that used for the error indicator in time. Now one considers a problem continuous in space, but discretized in time, whose unknowns are a priori fields defined on Ω at each of the nodal instants t_n $(n = 0, 1, 2, \dots, p)$ of the time discretization chosen. Many variations are possible. Here, we follow the approach of [LADEVEZE - MOES, 1998b].

Let us introduce the notations

$$\Delta = \{t_0 = 0, t_1, t_2, \dots, t_n, \dots, t_p = T\}$$

$$\underline{U}^\Delta = \{\underline{U}^0, \underline{U}^1, \dots, \underline{U}^n, \dots, \underline{U}^p\}$$

$$\varepsilon^\Delta = \{\varepsilon(\underline{U}^0), \varepsilon(\underline{U}^1), \dots, \varepsilon(\underline{U}^n), \dots, \varepsilon(\underline{U}^p)\} = \varepsilon^\Delta(\underline{U}^\Delta)$$

$$\sigma^\Delta = \{\sigma^0, \sigma^1, \dots, \sigma^n, \dots, \sigma^p\}$$

where $\underline{U}^n$ denotes a displacement field and σ^n a stress field defined on Ω at time t_n.

The problem can be formulated as follows:

Find $\underline{U}^\Delta$ *and* σ^Δ *such that:*

$$\forall t_n \in \Delta, \quad \underline{U}^n \text{ is KA} \tag{6.65}$$

$$\forall t_n \in \Delta \quad \text{and} \quad \forall \underline{U}^* \in \boldsymbol{\mathcal{U}}_0,$$

$$\int_\Omega \mathrm{Tr}\left[\sigma^n \varepsilon(\underline{U}^*)\right] d\Omega = \int_\Omega \underline{f}_d(t_n) \cdot \underline{U}^* \, d\Omega + \int_{\partial_2\Omega} \underline{F}_d(t_n) \cdot \underline{U}^* \, dS \tag{6.66}$$

$$\sigma^\Delta = \mathbf{A}^\Delta(\varepsilon^\Delta(\underline{U}^\Delta)) \tag{6.67}$$

where $\mathbf{A}^\Delta$ denotes the discretized constitutive operator.

Numerous variations are possible for defining the discretized operator $\mathbf{A}^\Delta$ which is not necessarily the operator used in the calculation.

Here, to define this operator $\mathbf{A}^\Delta$, we will use the internal variable formulation $\forall t_n \in \Delta$

$$\sigma^n = \mathbf{K}\left[\varepsilon(\underline{U}^n) - \varepsilon^{p,n}\right] \quad \text{and} \quad \mathbb{Y}^n = \mathbf{\Lambda}\, \mathbb{X}^n$$

$$\dot{\varepsilon}^{p,n} \in \partial_\sigma \varphi^*(\sigma^{\theta n}, \mathbb{Y}^{\theta n}) \quad \text{and} \quad -\dot{\mathbb{X}}^{p,n} \in \partial_{\mathbb{Y}} \varphi^*(\sigma^{\theta n}, \mathbb{Y}^{\theta n}) \tag{6.68}$$

where the superscript n indicates that the value is taken at t_n and where, for any quantity z,

$$\dot{z}^n = \frac{z^n - z^{n-1}}{t_n - t_{n-1}} \quad \text{and for } 0 \le \theta \le 1,\ z^{\theta n} = (1-\theta)z^{n-1} + \theta z^n$$

Let us consider two pairs $(\varepsilon^\Delta, \sigma^\Delta)$ and $(\tilde{\varepsilon}^\Delta, \tilde{\sigma}^\Delta)$ verifying (6.68),

$$\sigma^\Delta = \mathbf{A}^\Delta(\varepsilon^\Delta) \quad \text{and} \quad \tilde{\sigma}^\Delta = \mathbf{A}^\Delta(\tilde{\varepsilon}^\Delta)$$

and such that

$$(\varepsilon^0, \sigma^0) = (\tilde{\varepsilon}^0, \tilde{\sigma}^0) = 0$$

Then, one has the following property, which is a stability property of the discrete constitutive relation:

Property

For $\frac{1}{2} \le \theta \le 1$ *and for* $m \in \{1, 2, \ldots, p\}$ *one has*

$$\sum_{n=1}^{m} (t_n - t_{n-1}) \operatorname{Tr}[(\tilde{\sigma}^{\theta n} - \sigma^{\theta n})(\dot{\tilde{\varepsilon}}^n - \dot{\varepsilon}^n)] \ge 0$$

and

$$\left[\forall m \in \{1, 2, \ldots, p\},\ \sum_{n=1}^{m} (t_n - t_{n-1}) \operatorname{Tr}[(\tilde{\sigma}^{\theta n} - \sigma^{\theta n})(\dot{\tilde{\varepsilon}}^n - \dot{\varepsilon}^n)] = 0\right]$$
$$\Rightarrow \left[\forall n \in \{1, 2, \ldots, p\},\ \varepsilon^n = \tilde{\varepsilon}^n \text{ and } \sigma^n = \tilde{\sigma}^n\right]$$

Proof

As in Section 6.3.2., one first shows that

$$\begin{aligned}
&\operatorname{Tr}[(\tilde{\sigma}^{\theta n} - \sigma^{\theta n})(\dot{\tilde{\varepsilon}}^n - \dot{\varepsilon}^n)] \\
&\quad = \left[\varphi(\dot{\varepsilon}^{p,n}, -\dot{\mathbb{X}}^n) + \varphi^*(\tilde{\sigma}^{\theta n}, \tilde{\mathbb{Y}}^{\theta n}) - \operatorname{Tr}[\tilde{\sigma}^{\theta n}\dot{\varepsilon}^{p,n}] + \tilde{\mathbb{Y}}^{\theta n} \circ \dot{\mathbb{X}}^n\right] \\
&\quad + \left[\varphi(\dot{\tilde{\varepsilon}}^{p,n}, -\dot{\tilde{\mathbb{X}}}^n) + \varphi^*(\sigma^{\theta n}, \mathbb{Y}^{\theta n}) - \operatorname{Tr}[\sigma^{\theta n}\dot{\tilde{\varepsilon}}^{p,n}] + \mathbb{Y}^{\theta n} \circ \dot{\tilde{\mathbb{X}}}^n\right] \\
&\quad + \operatorname{Tr}\left[(\tilde{\sigma}^{\theta n} - \sigma^{\theta n})\mathbf{K}^{-1}(\dot{\tilde{\sigma}}^n - \dot{\sigma}^n)\right] + (\mathbb{Y}^{\theta n} - \tilde{\mathbb{Y}}^{\theta n}) \circ \Lambda^{-1}(\dot{\mathbb{Y}}^n - \dot{\tilde{\mathbb{Y}}}^n)
\end{aligned}$$

Observing that

$$z^{\theta n} = \frac{1}{2}(z^n + z^{n-1}) + \frac{2\theta - 1}{2}(z^n - z^{n-1})$$

one gets

$$\mathrm{Tr}\Big[(\underset{\sim}{\sigma}^{\theta n} - \sigma^{\theta n})\mathbf{K}^{-1}(\dot{\underset{\sim}{\sigma}}^n - \dot{\sigma}^n)\Big]$$

$$= \frac{\| \underset{\sim}{\sigma}^n - \sigma^n \|_{\mathbf{K}}^2 - \| \underset{\sim}{\sigma}^{n-1} - \sigma^{n-1} \|_{\mathbf{K}}^2}{2(t_n - t_{n-1})} + \frac{(2\theta - 1)(t_n - t_{n-1}) \| \dot{\underset{\sim}{\sigma}}^n - \dot{\sigma}^n \|_{\mathbf{K}}^2}{2}$$

$$(\mathbb{Y}^{\theta n} - \underset{\sim}{\mathbb{Y}}^{\theta n}) \circ \Lambda^{-1}(\dot{\mathbb{Y}}^n - \dot{\underset{\sim}{\mathbb{Y}}}^n)$$

$$= \frac{\| \underset{\sim}{\mathbb{Y}}^n - \mathbb{Y}^n \|_{\Lambda}^2 - \| \underset{\sim}{\mathbb{Y}}^{n-1} - \mathbb{Y}^{n-1} \|_{\Lambda}^2}{2(t_n - t_{n-1})} + \frac{(2\theta - 1)(t_n - t_{n-1}) \| \dot{\underset{\sim}{\mathbb{Y}}}^n - \dot{\mathbb{Y}}^n \|_{\Lambda}^2}{2}$$

where

$$\| \sigma \|_{\mathbf{K}}^2 = \mathrm{Tr}[\sigma \mathbf{K}^{-1} \sigma] \text{ and } \| \mathbb{Y} \|_{\Lambda}^2 = \mathbb{Y} \circ \Lambda^{-1} \mathbb{Y} .$$

Summing over all the time increments, one deduces

$$\sum_{n=1}^{m} (t_n - t_{n-1})\, Tr[(\underset{\sim}{\sigma}^{\theta n} - \sigma^{\theta n})(\dot{\underset{\sim}{\varepsilon}}^n - \dot{\varepsilon}^n)]$$

$$= \sum_{n=1}^{m} (t_n - t_{n-1}) \Big[\varphi(\dot{\varepsilon}^{p,n}, -\dot{\mathbb{X}}^n) + \varphi^*(\underset{\sim}{\sigma}^{\theta n}, \underset{\sim}{\mathbb{Y}}^{\theta n}) - \mathrm{Tr}[\underset{\sim}{\sigma}^{\theta n} \dot{\varepsilon}^{p,n}] + \underset{\sim}{\mathbb{Y}}^{\theta n} \circ \dot{\mathbb{X}}^n \Big]$$

$$+ \sum_{n=1}^{m} (t_n - t_{n-1}) \Big[\varphi(\dot{\underset{\sim}{\varepsilon}}^{p,n}, -\dot{\underset{\sim}{\mathbb{X}}}^n) + \varphi^*(\sigma^{\theta n}, \mathbb{Y}^{\theta n}) - \mathrm{Tr}[\sigma^{\theta n} \dot{\underset{\sim}{\varepsilon}}^{p,n}] + \mathbb{Y}^{\theta n} \circ \dot{\underset{\sim}{\mathbb{X}}}^n \Big]$$

$$+ \sum_{n=1}^{m} \frac{(2\theta - 1)(t_n - t_{n-1})^2}{2} \Big[\| \dot{\underset{\sim}{\sigma}}^n - \dot{\sigma}^n \|_{\mathbf{K}}^2 + \| \dot{\underset{\sim}{\mathbb{Y}}}^n - \dot{\mathbb{Y}}^n \|_{\Lambda}^2 \Big]$$

$$+ \frac{\| \underset{\sim}{\sigma}^m - \sigma^m \|_{\mathbf{K}}^2 + \| \underset{\sim}{\mathbb{Y}}^m - \mathbb{Y}^m \|_{\Lambda}^2}{2}$$

From this point on, the proof of the property is immediate.

In order to define the error indicator in space, one uses a similar approach.

Definition

A pair $(\underline{U}_{KA}^{\Delta}, \sigma_{SA}^{\Delta})$ *is said to be admissible with respect to Problem* [(6.65) - (6.67)] *if:*

- $\underline{U}_{KA}^{\Delta}$ *verifies the kinematic constraints* (6.65);
- σ_{SA}^{Δ} *verifies the equilibrium* (6.66);
- $(\underline{U}_{KA}^{0}, \sigma_{SA}^{0}) = 0$.

Through the discretized constitutive relation, one associates the stresses σ_{KA}^{Δ} with the strains $\varepsilon_{KA}^{\Delta}$. Similarly, through the discretized inverse constitutive relation, one associates the strains $\varepsilon_{SA}^{\Delta}$ with the stresses σ_{SA}^{Δ}.

The quality of the pair $(\underline{U}_{KA}^{\Delta}, \sigma_{SA}^{\Delta})$ as an approximate solution of Problem [(6.65) - (6.67)] may be evaluated by an error in constitutive relation built on the stability property of the discretized constitutive relation.

Then, one defines the error by

$$\mathbf{e}_{RdC}^{\Delta}(\underline{U}_{KA}^{\Delta}, \sigma_{SA}^{\Delta}) = \left[\operatorname*{Sup}_{\substack{m \text{ such that} \\ 1 \le m \le p}} \int_{\Omega} \sum_{n=1}^{m} (t_n - t_{n-1})\, Tr[(\sigma_{SA}^{\theta n} - \sigma_{KA}^{\theta n})(\dot{\varepsilon}_{SA}^{n} - \dot{\varepsilon}_{KA}^{n})]\, d\Omega \right]^{1/2} \tag{6.69}$$

For a given time discretization, this error measure provides an evaluation of the errors due to the use of a space discretization of the finite element type. Its implementation is very simple. Indeed, a pair $(\underline{U}_{KA}^{\Delta}, \sigma_{SA}^{\Delta})$ can be easily constructed from the results given by the incremental method. For the displacement fields, one simply takes

$$\underline{U}_h^{\Delta} = \{\underline{U}_h(0), \underline{U}_h(t_1), \ldots, \underline{U}_h(t_n), \ldots, \underline{U}_h(T)\}$$

Regarding the stress fields, at each nodal instant t_n one knows a field $\tilde{\sigma}_{h,n}$ which is in equilibrium in the finite element sense, with which one associates, using the stress field construction technique, a field $\hat{\sigma}_{h,n}$ which is in equilibrium in the strong sense. One simply takes field σ_{SA}^{Δ} as

$$\hat{\sigma}^{\Delta} = \{\hat{\sigma}_{h,0}, \hat{\sigma}_{h,1}, \ldots, \hat{\sigma}_{h,n}, \ldots, \hat{\sigma}_{h,p}\}$$

Then, one lets

$$\mathbf{i}_{space} = \mathbf{e}^{\Delta}_{RdC}(\underline{U}^{\Delta}_{h}, \hat{\sigma}^{\Delta})$$

If $\mathbf{i}_{space} = 0$, then $(\underline{U}^{\Delta}_{h}, \hat{\sigma}^{\Delta})$ is the exact solution of Problem [(6.65) - (6.67)]. This measure of the error between the numerical model and the problem discretized in space and continuous in time is used as an indicator of the error in space between the numerical model and the reference model [(6.1) - (6.3)].

REMARK. Another approach for defining an indicator of the error in time consists, in principle, in taking as the indicator in space

$$\left[\mathbf{e}^{2}_{RdC}(\underline{U}_{KA}, \sigma_{SA}) - \mathbf{e}^{2}_{h,RdC}(\underline{U}_{h}, \tilde{\sigma}_{h})\right]^{1/2}$$

which is the same as considering that the square of the global error is the sum of the squares of the errors due to the time discretization and to the space discretization. Details of this variant can be found in [GALLIMARD - LADEVEZE - PELLE, 1995, 1996].

6.3.5 Global effectivity index

As in elasticity, the global effectivity index is an important criterion for evaluating the quality of an error estimator. Below, we give two simple examples of the study of the global effectivity index. These were taken from [GALLIMARD - LADEVEZE - PELLE, 2000]. These examples concern 2D plane stress calculations for the PRANDTL - REUSS plasticity model P1. The loading was a simple monotonically increasing solicitation.

The "true" error was evaluated numerically using a very fine mesh and a very fine time discretization.

First example

Let us consider a perforated square plate loaded in compression as in Figure 6.2.

Taking the symmetries into account, only one-fourth of the plate was modeled. The corresponding mesh is shown in Figure 6.3. The finite element discretization used 6-node triangles.

The characteristics of the PRANDTL - REUSS model were

$$E = 200{,}000 \text{ MPa}, \quad \nu = 0.3, \quad R = H p^{1/2}$$

where H is a constant which, in order to study the influence of strain hardening, was given different values in the interval $0.001E \leq H \leq 0.1E$

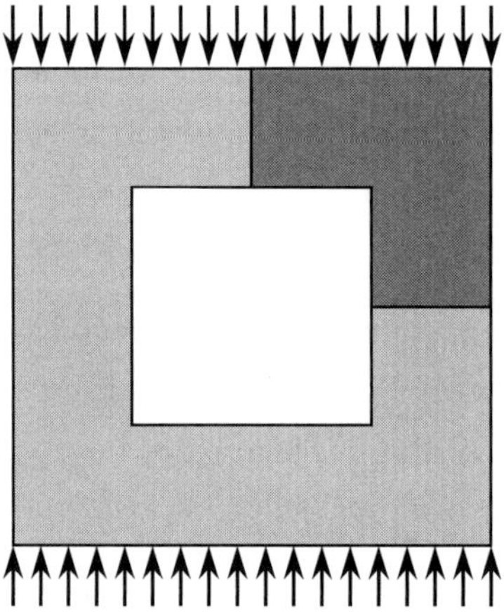

Figure 6.2. *Perforated plate and the calculation zone.*

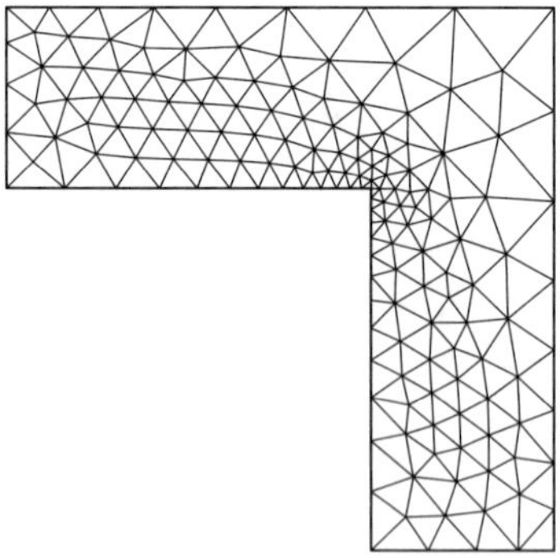

Figure 6.3. *Mesh.*

Figure 6.4 shows the global effectivity index obtained with two methods of construction of the admissible stress field σ_{SA}:

• the first method was that described in Section 6.3.3 which uses the classical construction of the densities $\hat{\underline{F}}$ (see Sections 2.4.3 and 8.3.1);

• the second uses an improved version which is an extension to plasticity of the method described in Section 8.6.

With the standard construction, the effectivity index varies between 3.5 and 4.5, provided that the ratio E/H is less than 100. Conversely, when this ratio is large, which corresponds to a constitutive behavior close to perfect plasticity, the effectivity index can reach values close to 10.

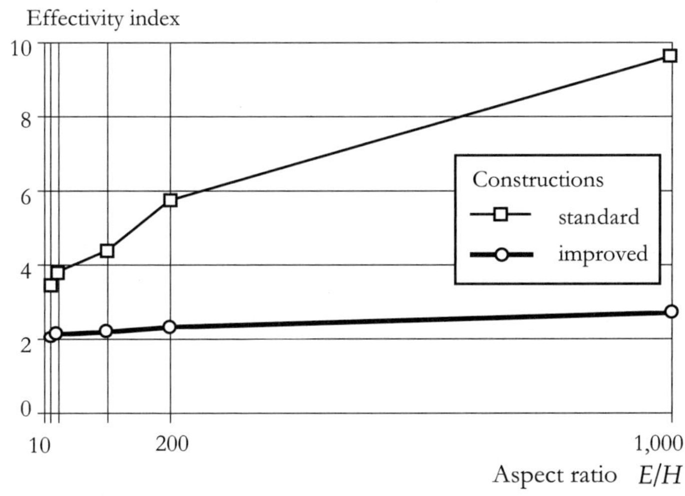

Figure 6.4. *Global effectivity index as a function of* E/H.

With the improved construction, the effectivity index is of much better quality:

• it varies between 2.1 and 2.8 for E/H going from 10 to 1,000;

• its evolution is much smoother.

Second example

The objective of this example is to study the influence of the aspect ratio of the elements on the quality of the estimator of the error in DRUCKER's sense. In order to do this, we consider a plane problem in a rectangular domain of dimensions a and b, subjected to a prestress of the form

$$f(t)\begin{bmatrix} -2xy & y^2-x^2 \\ y^2-x^2 & 2xy \end{bmatrix}$$

where $f(t)$ is a monotonically increasing function. The mesh used is shown in Figure 6.5.

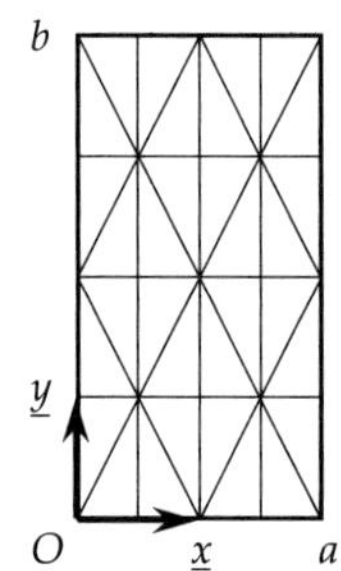

Figure 6.5. *Anisotropic regular mesh.*

The characteristics of the PRANDTL - REUSS model were

$$E = 200{,}000\ MPa, \qquad \nu = 0.3, \qquad R = 2{,}000\, p^{1/2}$$

We studied the effectivity index of DRUCKER's error as a function of the aspect ratio by making the value of a vary between $b/2$ and $b/16$.

The results for the standard and improved constructions are shown in Figure 6.6 in the case of the structure remaining elastic, then in Figure 6.7 in the case of extensive plasticity development.

We observe that at low aspect ratios the two estimators are very close to each other and that at high aspect ratios the improved estimator behaves extremely well.

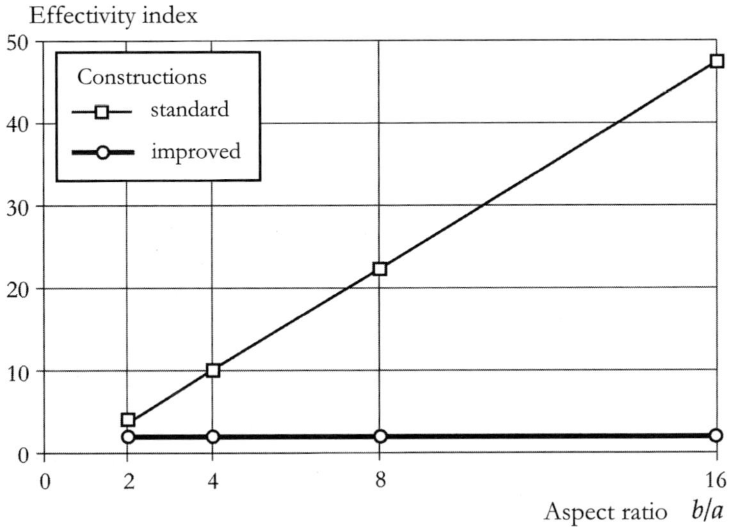

Figure 6.6. *Effectivity index as a function of b/a in the elastic case.*

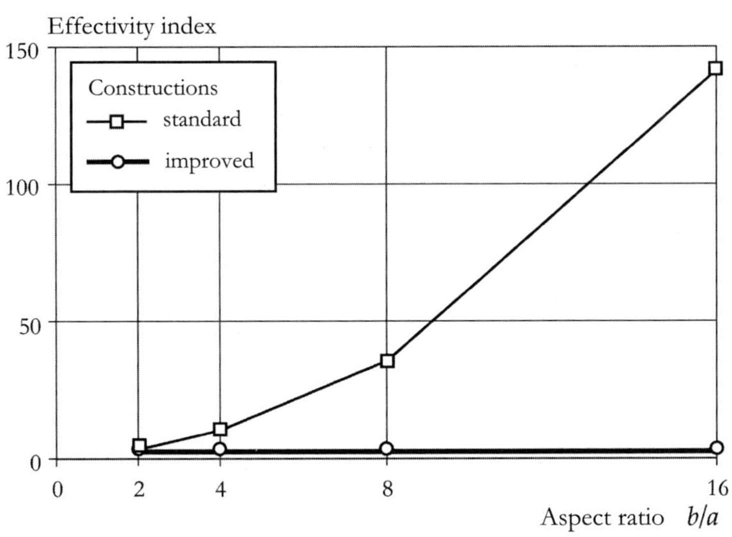

Figure 6.7. *Effectivity index as a function of b/a in the plastic case.*

6.4 Dissipation error

The measure of the dissipation error uses the standard formulation of constitutive models. The concept of dissipation error was introduced in [LADEVEZE, 1989]; its first applications to the control of finite element calculations were given in [LADEVEZE - MOES, 1998a, 1998b, 1999]. In order to simplify the presentation we will systematically assume that the formulation of the constitutive behavior is standard and normal, but the results can be generalized without difficulty [LADEVEZE, 1996].

Therefore, our framework is that described in Section 6.2.2 and we will use the notations (6.32) for the state laws. The definition of the new measure of the dissipation error also requires a redefinition of the notion of admissible solution and, consequently (because of the point of view chosen), of the notion of approximate solution of the reference problem.

6.4.1 Admissible fields

In the constitutive relation, the state laws are associated with the free energy, and the evolution laws (described by a pseudopotential) are associated with dissipative phenomena. As proposed in [LADEVEZE, 1989, 1996], the state laws are included in the admissibility conditions.

Definition

Fields $\underline{U}$, $\boldsymbol{\sigma}$, $\boldsymbol{\varepsilon}^p, \mathbb{X}, \mathbb{Y}$ *defined in* $\Omega \times [0, T]$ *are said to be admissible if:*

- $\underline{U}$ *verifies the kinematic constraints and the initial condition* (6.1);
- $\boldsymbol{\sigma}$ *verifies the equilibrium equations* (6.2) *and* $\boldsymbol{\sigma}$ *is zero at* $t = 0$;
- $\boldsymbol{\sigma}, \boldsymbol{\varepsilon}^p, \mathbb{X}, \mathbb{Y}$ *verify the state laws:*

$$\boldsymbol{\sigma} = \mathbf{K}(\boldsymbol{\varepsilon}(\underline{U}) - \boldsymbol{\varepsilon}^p) \quad \textit{and} \quad \mathbb{Y} = \Lambda\, \mathbb{X}.$$

With this definition, the admissibility conditions are related to the free energy. The *constitutive relation* is therefore reduced to the sole part of the constitutive model which is associated with dissipation, i.e. the relation which describes the evolution of the material's state.

More precisely, the admissible set $\underline{U}, \mathbb{\sigma}, \mathbb{\varepsilon}^p, \mathbb{X}, \mathbb{Y}$ is the exact solution of the reference problem if and only if:

$$\dot{\mathbb{\varepsilon}}^p = \partial_{\mathbb{\sigma}} \varphi^* \quad \textit{and} \quad -\dot{\mathbb{X}} = \partial_{\mathbb{Y}} \varphi^* \tag{6.70}$$

with $\mathbb{\varepsilon}^p|_{t=0} = 0$ and $\mathbb{X}|_{t=0} = 0$

Notations

If fields $\underline{U}, \mathbb{\sigma}, \mathbb{\varepsilon}^p, \mathbb{X}, \mathbb{Y}$ are *admissible*, we designate them globally by

$$s_{Ad} = (\underline{U}_{Ad}, \mathbb{\sigma}_{Ad}, \mathbb{\varepsilon}^p_{Ad}, \mathbb{X}_{Ad}, \mathbb{Y}_{Ad})$$

Because the evolution laws (and, consequently, the quantities $\dot{\mathbb{\varepsilon}}^p_{Ad}, \dot{\mathbb{X}}_{Ad}, \mathbb{\sigma}_{Ad}, \mathbb{Y}_{Ad}$) are going to play a special role in the following developments, very often we will also use the more concise notation

$$s_{Ad} = (\dot{\mathbb{\varepsilon}}^p_{Ad}, \dot{\mathbb{X}}_{Ad}, \mathbb{\sigma}_{Ad}, \mathbb{Y}_{Ad})$$

6.4.2 Measures of the dissipation error

An admissible solution s_{Ad} defined on $\Omega \times [0, T]$ is an approximate solution of Problem [(6.1) - (6.3)]. The quality of this approximation depends on the way that it verifies the constitutive relation (6.70). Figure 6.8 illustrates this situation at a point M and at a time t.

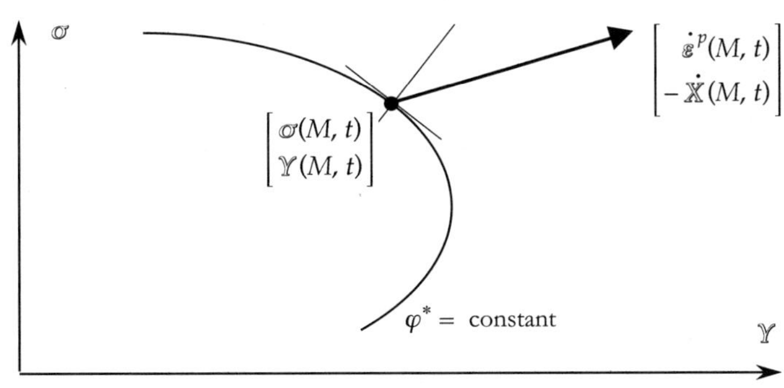

Figure 6.8. *Admissible approximation at* $(M, t) \in \Omega \times [0, T]$.

To simplify the notations, let

$$\eta(\dot{\varepsilon}^p, \dot{\mathbb{X}}, \sigma, \mathbb{Y}) = \varphi(\dot{\varepsilon}^p, -\dot{\mathbb{X}}) + \varphi^*(\sigma, \mathbb{Y}) - Tr(\sigma\, \dot{\varepsilon}^p) + \mathbb{Y} \circ \dot{\mathbb{X}} \qquad (6.71)$$

Taking into account the properties (6.7) and (6.8), one can define a measure of the error in constitutive relation as follows:

Definition

One calls error in constitutive relation associated with an admissible solution s_{Ad} *the real quantity*

$$\mathbf{e}_{RdC}(s_{Ad}) = \frac{1}{T}\int_0^T \int_\Omega \eta(\dot{\varepsilon}^p_{Ad}, \dot{\mathbb{X}}_{Ad}, \sigma_{Ad}, \mathbb{Y}_{Ad})\, d\Omega\, dt \qquad (6.72)$$

By construction, one has

$$\mathbf{e}_{RdC}(s_{Ad}) \geq 0 \quad \text{and} \quad \left\{\mathbf{e}_{RdC}(s_{Ad}) = 0 \Leftrightarrow s_{Ad} = s_{ex}\right\}$$

where s_{ex} denotes the exact solution of the reference problem.

REMARKS

1. An alternative consists in defining the error measure by

$$\tilde{\mathbf{e}}_{RdC}(s_{Ad}) = \operatorname*{Sup}_{t \in [0, T]} \int_\Omega \eta(\dot{\varepsilon}^p_{Ad}, \dot{\mathbb{X}}_{Ad}, \sigma_{Ad}, \mathbb{Y}_{Ad})\, d\Omega$$

2. For many constitutive models, particularly P1, VP1, P2 and VP2, the potentials φ and φ^* can become infinite. The error measure $\mathbf{e}_{RdC}(s_{Ad})$ is useless unless these potentials remain finite. We will return to this point in detail in Section 6.4.5.

6.4.3 Error in the solution

The objective of this section is to define a measure of the error in the solution which enables one to evaluate the distance between an admissible solution s_{Ad} and the exact solution of the reference model s_{ex}. Having done this, we will give a theorem [LADEVEZE, 1996] enabling one to connect this error in the solution to the error in constitutive relation introduced in (6.72).

For the problems considered here, this theorem may be viewed as the counterpart of the PRAGER - SYNGE theorem for linear problems.

Definition of the error in the solution

Definition

One calls error in the solution over $[0, t]$ *between* s_{Ad} *and the exact solution* s_{ex} *the real quantity*

$$\delta(s_{Ad}, s_{ex}; t) = \frac{1}{2}\int_\Omega \mathrm{Tr}\left[(\sigma_{Ad} - \sigma_{ex})\mathbf{K}^{-1}(\sigma_{Ad} - \sigma_{ex})\right]_{|t} d\Omega$$
$$+ \frac{1}{2}\int_\Omega \left[(\mathbb{Y}_{Ad} - \mathbb{Y}_{ex}) \circ \Lambda^{-1} (\mathbb{Y}_{Ad} - \mathbb{Y}_{ex})\right]_{|t} d\Omega$$
$$+ \int_0^t \int_\Omega \tilde{\eta}(s_{Ad}, s_{ex})\, d\Omega\, d\tau \tag{6.73}$$

where

$$\tilde{\eta}(s_{Ad}, s_{ex}) = \eta(\dot{\varepsilon}^p_{Ad}, \dot{\mathbb{X}}_{Ad}, \sigma_{ex}, \mathbb{Y}_{ex}) + \eta(\dot{\varepsilon}^p_{ex}, \dot{\mathbb{X}}_{ex}, \sigma_{Ad}, \mathbb{Y}_{Ad}) \tag{6.74}$$

Due to the positivity of the constitutive operators $\mathbf{K}$ and Λ and to the definition of $\tilde{\eta}$, the quantity δ is always positive or zero.

The connection we are about to establish between the error in the solution $\delta(s_{Ad}, s_{ex}; t)$ and the error in constitutive relation will lead to the following important result:

Property

$$\text{If } \delta(s_{Ad}, s_{ex}; t) = 0, \text{ then } s_{Ad} = s_{ex} \textit{ in } [0, t]. \tag{6.75}$$

Connection with the error in constitutive relation

Theorem

For any admissible solution s_{Ad}, *one has*

$$\mathbf{e}_{RdC}(s_{Ad}) = \frac{1}{T}\,\delta(s_{Ad}, s_{ex}; T) \tag{6.76}$$

Proof

Observing that $\eta(s_{ex}) = 0$, one gets

$$\begin{aligned}\eta(s_{Ad}) &= \eta(\dot{\varepsilon}^p_{Ad}, \dot{\mathbb{X}}_{Ad}, \sigma_{Ad}, \mathbb{Y}_{Ad}) + \eta(\dot{\varepsilon}^p_{ex}, \dot{\mathbb{X}}_{ex}, \sigma_{ex}, \mathbb{Y}_{ex})\\ &= \varphi(\dot{\varepsilon}^p_{Ad}, -\dot{\mathbb{X}}_{Ad}) + \varphi^*(\sigma_{Ad}, \mathbb{Y}_{Ad}) - \mathrm{Tr}(\sigma_{Ad}\,\dot{\varepsilon}^p_{Ad}) + \mathbb{Y}_{Ad} \circ \dot{\mathbb{X}}_{Ad}\\ &\quad + \varphi(\dot{\varepsilon}^p_{ex}, -\dot{\mathbb{X}}_{ex}) + \varphi^*(\sigma_{ex}, \mathbb{Y}_{ex}) - \mathrm{Tr}(\sigma_{ex}\,\dot{\varepsilon}^p_{ex}) + \mathbb{Y}_{ex} \circ \dot{\mathbb{X}}_{ex}\\ &= \tilde{\eta}(s_{Ad}, s_{ex})\\ &\quad - \mathrm{Tr}[(\sigma_{Ad} - \sigma_{ex})(\dot{\varepsilon}^p_{Ad} - \dot{\varepsilon}^p_{ex})] + (\mathbb{Y}_{Ad} - \mathbb{Y}_{ex}) \circ (\dot{\mathbb{X}}_{Ad} - \dot{\mathbb{X}}_{ex})\end{aligned}$$

The admissibility conditions satisfied by s_{Ad} and s_{ex} yield

$$\int_\Omega \mathrm{Tr}\big[(\sigma_{Ad} - \sigma_{ex})(\varepsilon_{Ad} - \varepsilon_{ex})\big]\, d\Omega = 0$$

$$\dot{\varepsilon}^e_{Ad} - \dot{\varepsilon}^e_{ex} = \mathbf{K}^{-1}(\dot{\sigma}_{Ad} - \dot{\sigma}_{ex})$$

$$\dot{\mathbb{X}}_{Ad} - \dot{\mathbb{X}}_{ex} = \Lambda^{-1}(\dot{\mathbb{Y}}_{Ad} - \dot{\mathbb{Y}}_{ex})$$

Thus, by integration on Ω, one gets

$$\begin{aligned}\int_\Omega \eta(s_{Ad})\, d\Omega &= \int_\Omega \tilde{\eta}(s_{Ad}, s_{ex})\, d\Omega\\ &\quad - \int_\Omega \mathrm{Tr}[(\sigma_{Ad} - \sigma_{ex})(\dot{\varepsilon}^p_{Ad} - \dot{\varepsilon}^p_{ex})]\, d\Omega\\ &\quad + \int_\Omega (\mathbb{Y}_{Ad} - \mathbb{Y}_{ex}) \circ (\dot{\mathbb{X}}_{Ad} - \dot{\mathbb{X}}_{ex})\, d\Omega\\ &= \int_\Omega \tilde{\eta}(s_{Ad}, s_{ex})\, d\Omega\\ &\quad + \frac{1}{2}\frac{d}{dt}\int_\Omega Tr\big[(\sigma_{Ad} - \sigma_{ex})\mathbf{K}^{-1}(\sigma_{Ad} - \sigma_{ex})\big]\, d\Omega\\ &\quad + \frac{1}{2}\frac{d}{dt}\int_\Omega \big[(\mathbb{Y}_{Ad} - \mathbb{Y}_{ex}) \circ \Lambda^{-1}(\mathbb{Y}_{Ad} - \mathbb{Y}_{ex})\big]\, d\Omega\end{aligned}$$

The announced result is obtained simply by integrating the last relation over the time interval $[0, T]$ and dividing all the terms by T.

Thus, Property (6.75) is a consequence of Theorem (6.76). This proves that the real quantity

$$\delta(s_{Ad}, s_{ex}; T)/T$$

can indeed be viewed as an error in the solution which enables one to evaluate the distance between the exact solution and the admissible solution.

6.4.4 Relative error

One associates with the absolute error the relative error

$$\varepsilon = \mathbf{e}_{RdC}/D$$

where the denominator D is defined by

$$D = \frac{2}{T} \operatorname*{Sup}_{t \in [0,T]} \left[d_1(t) + d_2(t) \right]$$

with

$$d_1(t) = \int_0^t \int_\Omega \operatorname{Sup}\left[\varphi(\dot{\varepsilon}^p, -\dot{\mathbb{X}}) + \varphi^*(\sigma, \mathbb{Y}), \frac{R_0}{\|\sigma_D\|} \operatorname{Tr}[\dot{\sigma} \mathbf{K}^{-1} \sigma] \right] d\Omega \, d\tau$$

$$d_2(t) = \frac{1}{2} \int_\Omega \left[\operatorname{Tr}[\sigma \mathbf{K}^{-1} \sigma] + \mathbb{Y} \circ \Lambda^{-1} \mathbb{Y} \right]_{|t} d\Omega$$

REMARK. If one omits the term $\frac{R_0}{\|\sigma_D\|} \operatorname{Tr}[\dot{\sigma} \mathbf{K}^{-1} \sigma]$ from the definition of $d_1(t)$, one has

$$d_1(t) + d_2(t) = \delta(s_{Ad}, 0, t)$$

The term $\frac{R_0}{\|\sigma_D\|} \operatorname{Tr}[\dot{\sigma} \mathbf{K}^{-1} \sigma]$ is introduced solely to improve the behavior of the relative error in the case of a purely elastic problem.

6.4.5 Construction of the admissible fields

Let us assume that the reference problem has been solved by the incremental method and reuse the notations of Section 6.3.3.

Now, we are seeking to reconstruct, based on the results of the numerical model, an admissible solution $s_{Ad} = (\underline{U}_{Ad}, \sigma_{Ad}, \varepsilon^p_{Ad}, \mathbb{X}_{Ad}, \mathbb{Y}_{Ad})$.

For the error to remain finite, the admissible solution must verify the two constraints

$$\varphi(\dot{\varepsilon}^p_{Ad}, -\dot{\mathbb{X}}_{Ad}) < +\infty \quad \text{and} \quad \varphi^*(\sigma_{Ad}, \mathbb{Y}_{Ad}) < +\infty \tag{6.77}$$

Construction of σ_{Ad}

The construction of the equilibrated stress field is identical to that presented in 6.3.3 for the error measure in DRUCKER's sense.

Construction of $\underline{U}_{Ad}$

A priori, the finite element field $\underline{U}_h(t_n) = \mathbf{N}(M)\mathbf{q}(t_n)$ verifies the kinematic constraints. Therefore, one could choose this field as $\underline{U}_{Ad}(t_n)$.

However, taking into account the state laws, one would then have

$$\varepsilon^p_{Ad} = \varepsilon(\underline{U}_h) - \mathbf{K}^{-1}\sigma_{Ad} \tag{6.78}$$

If plasticity (or viscoplasticity) takes place at constant volume, for φ to be finite one should have $\mathrm{Tr}(\varepsilon^p_{Ad}) = 0$. In general, the plastic strain defined by (6.78) does not satisfy this condition. Thus, it is necessary to modify the construction of the displacement field.

More precisely, it is necessary to construct the field $\underline{U}_{Ad}(t_n)$ such that

$\underline{U}_{Ad}(t_n)$ verifies the kinematic constraints and

$\mathrm{Tr}[\varepsilon(\underline{U}_{Ad}) - \mathbf{K}^{-1}\sigma_{Ad}] = 0$

Such construction is possible if one adapts the method proposed in [GASTINE - LADEVEZE - MARIN - PELLE, 1992] for incompressible elasticity. This adaptation was described in [MOËS, 1996]. Having constructed the field $\underline{U}_{Ad}(t_n)$ at each nodal instant, $\underline{U}_{Ad}$ is obtained by linear interpolation. Assuming the time variation of the given displacements between two nodal instants to be linear, one obtains a displacement which verifies the kinematic constraints at all times. For large strains, if plasticity or viscoplasticity takes place at constant volume, the medium is quasi-incompressible. This situation requires a customized numerical treatment and a specific evaluation of the discretization errors which can be achieved according to [GASTINE - LADEVEZE - MARIN - PELLE, 1992].

Construction of ε^p_{Ad}

Referring to the construction of $\underline{U}_{Ad}$, it suffices to take

$$\varepsilon^p_{Ad} = \varepsilon(\underline{U}_{Ad}) - \mathbf{K}^{-1}\sigma_{Ad}$$

Construction of the internal variables $\mathbb{X}_{Ad}$ ***and*** $\mathbb{Y}_{Ad}$

The constitutive relations being used are local in the space variables. Therefore, the determination of the internal variables is carried out independently at each point of the structure (in practice, at a certain number of integration points as necessary for the correct evaluation of the error).

A very natural method consists in determining $\mathbb{X}_{Ad}$ and $\mathbb{Y}_{Ad}$ such that they verify $\mathbb{Y}_{Ad} = \Lambda\,\mathbb{X}_{Ad}$ and minimize the error at each point M,

$$\operatorname*{Inf}_{\mathbb{X} \in \boldsymbol{X}} \int_0^T \eta(\dot{\varepsilon}^p_{Ad}, \dot{\mathbb{X}}, \sigma_{Ad}, \Lambda\,\mathbb{X})\, dt \tag{6.79}$$

where $\boldsymbol{X} = \{\, \mathbb{X} \setminus \forall t \in [0, T],\ \varphi(\dot{\varepsilon}^p_{Ad}, -\dot{\mathbb{X}}) < +\infty \text{ and } \varphi^*(\sigma_{Ad}, \Lambda\,\mathbb{X}) < +\infty \}$

Example of implementation

Let us consider, for example, the plasticity model P1 with linear strain hardening. One has

$$\mathbb{X} = p \quad \text{and} \quad \mathbb{Y} = R = \lambda\, p \quad (\lambda > 0, \text{constant})$$

and

$$\boldsymbol{X} = \{\, t \mapsto p(M, t) \setminus \forall t \in [0, T],\ \lambda\, p \geq R_0 + \left\| (\sigma_{Ad})_D \right\| \text{ and } \dot{p} \geq \left\| \dot{\varepsilon}^p_{Ad} \right\| \}$$

Taking (6.17) and (6.18) into account, (6.79) is equivalent to

$$\operatorname*{Inf}_{p \in \boldsymbol{X}} \left[R_0\, p(T) + \frac{\lambda}{2}\, [p(T)]^2 \right]$$

REMARK. To make the cost of the minimization (6.79) reasonable, one can proceed incrementally. One assumes that $\mathbb{X}_n = \mathbb{X}_{Ad}(t_n)$ is known and seeks $\mathbb{X}_{n+1} = \mathbb{X}_{Ad}(t_{n+1})$ by solving on $[t_n, t_{n+1}]$ the minimization problem

$$\operatorname*{Inf}_{\hat{\mathbb{X}}} \int_{t_n}^{t_{n+1}} \eta(\dot{\varepsilon}^p_{Ad}, \dot{\hat{\mathbb{X}}}, \sigma_{Ad}, \Lambda \hat{\mathbb{X}})\, dt$$

with

$$\hat{\mathbb{X}}(t) = \frac{\mathbb{X}_{n+1} - \mathbb{X}_n}{t_{n+1} - t_n}(t - t_n) + \mathbb{X}_n \tag{6.80}$$

Of course, $\mathbb{X}_{n+1}$ should be chosen such that it belongs to $\boldsymbol{\mathcal{X}}$.

In practice, it is often sufficient to evaluate the integral in (6.80) by the trapezoidal rule. Thus, one obtains a convex minimization problem whose only unknown is $\mathbb{X}_{n+1}$,

$$\operatorname*{Inf}_{\mathbb{X}_{n+1}} \left[\eta\left(\dot{\varepsilon}^p_{[t_n, t_{n+1}]}, \frac{\mathbb{X}_{n+1} - \mathbb{X}_n}{t_{n+1} - t_n}, \sigma_{Ad}(t_n), \Lambda \mathbb{X}_n \right) + \eta\left(\dot{\varepsilon}^p_{[t_n, t_{n+1}]}, \frac{\mathbb{X}_{n+1} - \mathbb{X}_n}{t_{n+1} - t_n}, \sigma_{Ad}(t_{n+1}), \Lambda \mathbb{X}_{n+1} \right) \right] \tag{6.81}$$

In (6.81), $\dot{\varepsilon}^p_{[t_n, t_{n+1}]}$ designates the constant value of $\dot{\varepsilon}^p_{Ad}$ over $[t_n, t_{n+1}]$.

6.4.6 Error indicators in time and in space

The construction of time and space indicators is based on the same principles as for the error in DRUCKER's sense and it relies on the use of the same "intermediate" problems.

Error indicator in time

Let us consider Problem [(6.61) - (6.63)], this time using internal variables to describe the constitutive law. Equation (6.63) is replaced by:

- state laws: $\sigma = \mathbf{K}(\varepsilon(\underline{U}) - \varepsilon^p)$ and $\mathbb{Y} = \Lambda \mathbb{X}$;
- evolution laws: $\dot{\varepsilon}^p \in \partial_{\sigma}\varphi^*$ and $-\dot{\mathbb{X}} \in \partial_{\mathbb{Y}}\varphi^*$.

Definition

Fields $\underline{U}_{h,Ad}$, $\sigma_{h,Ad}$, $\varepsilon^p_{h,Ad}$, $\mathbb{X}_{h,Ad}$, $\mathbb{Y}_{h,Ad}$ *defined on* $\Omega \times [0, T]$ *are said to be admissible if:*

- $\underline{U}_{h,Ad}$ *verifies the kinematic constraints and the initial condition;*
- $\sigma_{h,Ad}$ *verifies the equilibrium equations in the weak finite element sense and is zero at* $t = 0$;
- $\sigma_{h,Ad}$, $\varepsilon^p_{h,Ad}$, $\mathbb{X}_{h,Ad}$, $\mathbb{Y}_{h,Ad}$ *verify the state laws*

$$\sigma_{h,Ad} = \mathbf{K}(\varepsilon(\underline{U}_{h,Ad}) - \varepsilon^p_{h,Ad}) \quad \textit{and} \quad \mathbb{Y}_{h,Ad} = \Lambda \mathbb{X}_{h,Ad}$$

The quality of $s_{h,Ad} = (\underline{U}_{h,Ad}, \sigma_{h,Ad}, \varepsilon^p_{h,Ad}, \mathbb{X}_{h,Ad}, \mathbb{Y}_{h,Ad})$ as an approximate solution of Problem [(6.61) - (6.63)] can be evaluated by the error in constitutive relation based on the dissipation

$$\mathbf{e}_{RdC}(s_{h,Ad}) = \frac{1}{T}\int_0^T \int_\Omega \eta(\dot{\varepsilon}^p_{h,Ad}, \dot{\mathbb{X}}_{h,Ad}, \sigma_{h,Ad}, \mathbb{Y}_{h,Ad})\, d\Omega\, dt$$

For a given space discretization, this error measure enables one to evaluate the errors due to the use of the incremental method and to the use of a NEWTON-type iterative method over each time increment. Its implementation can be easily carried out using the techniques described in Section 6.4.5. Here, however, one uses directly the field $\tilde{\sigma}_{h,n}$ as the stress field $\sigma_{h,Ad}(t_n)$.

Then, one lets

$$\mathbf{i}_{time} = \mathbf{e}_{RdC}(s_{h,Ad})$$

A zero value of $\mathbf{i}_{time}$ means that $s_{h,Ad}$ is the exact solution of Problem [(6.61) - (6.63)]. This error measure constitutes an indicator of the time error between the numerical model and reference model [(6.1) - (6.3)].

Error indicator in space

Let us consider Problem [(6.65) - (6.67)], this time rewriting the constitutive relation in the form:

- *State laws*

$$\forall t_n \in \Delta, \quad \sigma^n = \mathbf{K}(\varepsilon(\underline{U}^n) - \varepsilon^{p,n}) \quad \textit{and} \quad \mathbb{Y}^n = \Lambda \mathbb{X}^n \tag{6.82}$$

- *Evolution laws*

$$\forall t_n \in \Delta, \quad \dot{\varepsilon}^{p,n} \in \partial_{\sigma}\varphi^*(\sigma^n, \mathbb{Y}^n) \quad \textit{and} \quad -\dot{\mathbb{X}}^n \in \partial_{\mathbb{Y}}\varphi^*(\sigma^n, \mathbb{Y}^n) \tag{6.83}$$

where the superscript n indicates that a value is taken at t_n and, for any quantity z,

$$\dot{z}^n = \frac{(z^n - z^{n-1})}{(t_n - t_{n-1})}$$

Definition

$s^{\Delta}_{Ad} = (\underline{U}^{\Delta}_{Ad}, \sigma^{\Delta}_{Ad}, \varepsilon^{p,\Delta}_{Ad}, \mathbb{X}^{\Delta}_{Ad}, \mathbb{Y}^{\Delta}_{Ad})$ is said to be admissible with respect to Problem [(6.65) - (6.67)] *if Equations* (6.65), (6.66) *and* (6.82) *are verified.*

Here again, the quality of s^{Δ}_{Ad} as an approximate solution of [(6.65) - (6.67)] can be evaluated by the error in constitutive relation

$$\mathbf{e}_{RdC}(s^{\Delta}_{Ad}) = \frac{1}{T}\sum_{n=1}^{p}\left[(t_n - t_{n-1})\int_{\Omega}\eta(\dot{\varepsilon}^{p,n}_{Ad}, \dot{\mathbb{X}}^n_{Ad}, \sigma^n_{Ad}, \mathbb{Y}^n_{Ad})\, d\Omega\right]$$

For a given time discretization, this error measure enables one to evaluate the errors due to the use of a discretization of the finite element type in space. Its implementation is easy: to construct s^{Δ}_{Ad} it suffices to take the values of the fields s_{Ad} at the nodal instants.

Then, one lets

$$\mathbf{i}_{space} = \mathbf{e}_{RdC}(s^{\Delta}_{Ad})$$

Indicator of the convergence error of the iterations

Besides the errors due to the time discretization and those due to the space discretization, the incremental method also introduces errors due to the approximate resolution over each increment of the nonlinear problem (6.57). With respect to the convergence thresholds typically used in finite element codes, these errors are often negligible compared to the time and space errors. Nevertheless, the concept of error in constitutive relation also enables the construction of a specific indicator to evaluate these errors. Here, this indicator is defined in the framework of the dissipation error following the approach by [MOES, 1996], also described in [LADEVEZE - MOES, 1998b]; it would be defined in a similar way for the error in DRUCKER's sense.

The reference problem is the problem discretized in time and space. Let

$$\underline{U}_h^\Delta = \{\underline{U}_h^0, \underline{U}_h^1, \dots, \underline{U}_h^n, \dots, \underline{U}_h^p\}$$
$$\varepsilon_h^\Delta = \{\varepsilon(\underline{U}_h^0), \varepsilon(\underline{U}_h^1), \dots, \varepsilon(\underline{U}_h^n), \dots, \varepsilon(\underline{U}_h^p)\}$$
$$\sigma_h^\Delta = \{\sigma_h^0, \sigma_h^1, \dots, \sigma_h^n, \dots, \sigma_h^p\}$$

where $\underline{U}_h^n$ denotes a displacement field of the finite element type and σ_h^n a stress field defined on Ω at time t_n.

The problem discretized in space and time can be formulated as follows:

Find $\underline{U}_h^\Delta$ and σ_h^Δ such that:

$$\forall t_n \in \Delta, \quad \underline{U}_h^n \text{ is KA and of the finite element type} \tag{6.84}$$

$$\forall t_n \in \Delta \quad and \quad \forall \underline{U}_h^* \in \boldsymbol{\mathcal{U}}_{h0},$$
$$\int_\Omega \mathrm{Tr}\left[\sigma_h^n\, \varepsilon(\underline{U}_h^*)\right] d\Omega = \int_\Omega \underline{f}_d \cdot \underline{U}_h^*\, d\Omega + \int_{\partial_2\Omega} \underline{F}_d \cdot \underline{U}_h^*\, dS \tag{6.85}$$

$$\forall t_n \in \Delta, \quad \sigma_h^n = \mathbf{K}(\varepsilon(\underline{U}_h^n) - \varepsilon_h^{p,n}) \quad \text{and} \quad \mathbb{Y}_h^n = \Lambda\, \mathbb{X}_h^n, \tag{6.86}$$

$$\forall t_n \in \Delta, \quad \dot{\varepsilon}_h^{p,n} \in \partial_\sigma \varphi^*(\sigma_h^n, \mathbb{Y}_h^n) \quad \text{and} \quad -\dot{\mathbb{X}}_h^n \in \partial_{\mathbb{Y}} \varphi^*(\sigma_h^n, \mathbb{Y}_h^n) \tag{6.87}$$

Then, for $s_{h,Ad}^\Delta = (\underline{U}_{h,Ad}^\Delta, \sigma_{h,Ad}^\Delta, \varepsilon_{h,Ad}^{p,\Delta}, \mathbb{X}_{h,Ad}^\Delta, \mathbb{Y}_{h,Ad}^\Delta)$ admissible with respect to Problem [6.84 - 6.87], i.e. for $s_{h,Ad}^\Delta$ verifying 6.84, 6.85 and 6.86, one defines an error built on the dissipation by

$$\mathbf{e}_{RdC}(s_{h,Ad}^\Delta) = \frac{1}{T} \sum_{n=1}^{p} \left[(t_n - t_{n-1}) \int_\Omega \eta(\dot{\varepsilon}_{h,Ad}^{p,n}, \dot{\mathbb{X}}_{h,Ad}^n, \sigma_{h,Ad}^n, \mathbb{Y}_{h,Ad}^n)\, d\Omega\right]$$

This error measure enables one to evaluate the quality of the solution of the nonlinear problem over each increment. Thus, let

$$\mathbf{i}_{iter} = \mathbf{e}_{RdC}(s_{h,Ad}^\Delta)$$

REMARK. It is easy to construct relative versions of the above error indicators. For example, for $\mathbf{i}_{time} = \mathbf{e}_{RdC}(s_{h,Ad})$, the relative indicator is defined by

$$\mathbf{I}_{time} = \mathbf{i}_{time} / D_{time} \quad \text{with} \quad D_{time} = \frac{2}{T} \operatorname*{Sup}_{t \in [0,T]} \left| d_{h,1}(t) + d_{h,2}(t) \right|$$

and

$$d_{h,1}(t) = \int_0^t \int_\Omega \operatorname{Sup} \left| \varphi(\dot{\varepsilon}^p_h, -\dot{\mathbb{X}}_h) + \varphi^*(\sigma_h, \Upsilon_h), \frac{R_0}{\| \sigma_{h,D} \|} \operatorname{Tr}[\dot{\sigma}_h \mathbf{K}^{-1} \sigma_h] \right| d\Omega \, d\tau$$

$$d_{h,2}(t) = \frac{1}{2} \int_\Omega \left[\operatorname{Tr}[\sigma_h \mathbf{K}^{-1} \sigma_h] + \Upsilon_h \circ \Lambda^{-1} \Upsilon_h \right]_{|t} d\Omega$$

Similar definitions are easily constructed for the space indicator and for the indicator of convergence of the iterations.

6.5 Examples

Here we present some simple examples treated by MOES. The structure being considered is a portal frame (Figure 6.9) subjected to two consecutive solicitations: first, a load distribution $\underline{F}_1(t)$ is applied on the upper part over the time interval $[0, 2]$; then, a second load distribution $\underline{F}_2(t)$ is applied on the right part over the time interval $[2, 4]$ (Figure 6.10).

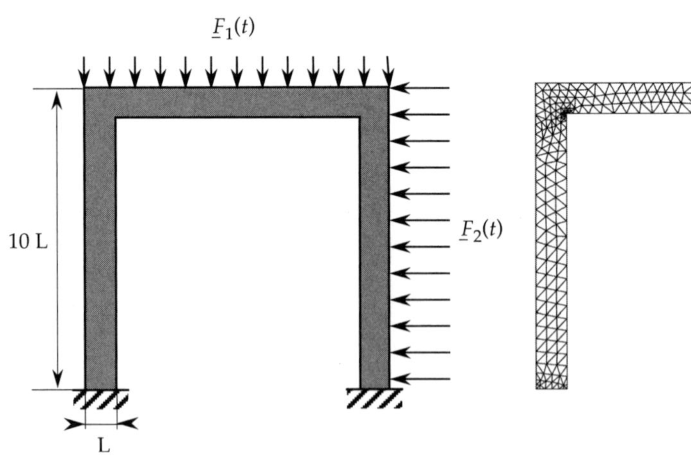

Figure 6.9. *Portal frame and sample mesh.*

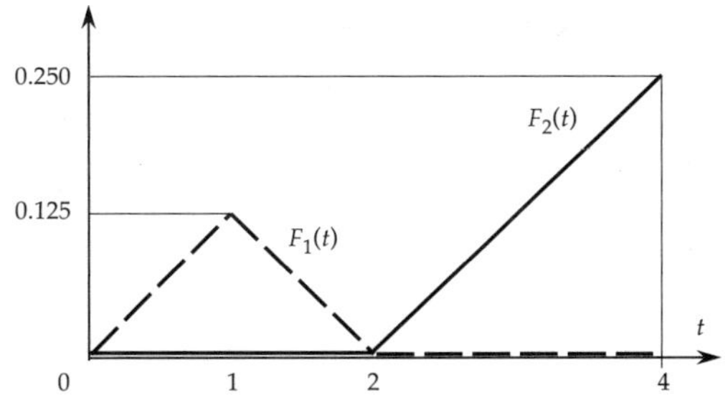

Figure 6.10. *Evolution of the loads versus time.*

The constitutive relation is a PRANDTL - REUSS law with linear strain hardening $g'(p) = k\,p$ (k constant) characterized by the adimensionalized parameters $E = 244.95$, $\nu = 0.3$, $R_0 = 1$, $k = 8.16$
All calculations were performed with 3-node triangles.

6.5.1 Comparison of the two errors

This example is taken from [LADEVEZE - MOES, 1998b]. The mesh was created using 3-node triangles and is shown in Figure 6.9, and Figure 6.11 shows the evolution of the two errors as functions of time. Qualitatively, the evolutions of the error in DRUCKER's sense and of the dissipation error are very similar. Over the time interval $[1, 3]$, the behavior is elastic and, therefore, the error over this interval is constant.

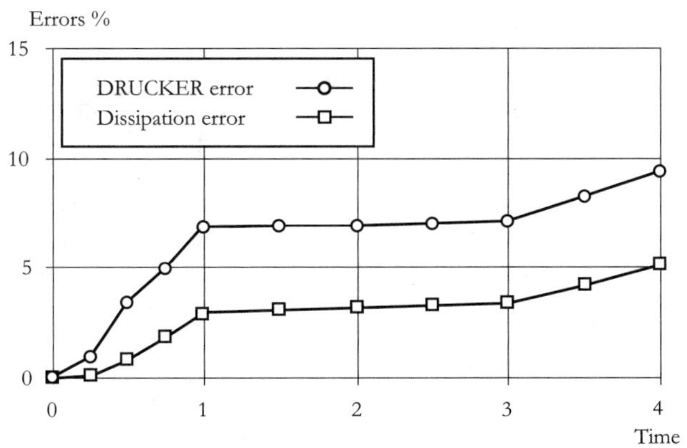

Figure 6.11. *Comparison of the error in DRUCKER's sense and the dissipation error.*

6.5.2 Behavior of the errors and of their specific indicators

Initially, for a given time discretization, the calculation was carried out on five increasingly refined meshes. The results obtained are shown in Figure 6.12 and Figure 6.13.

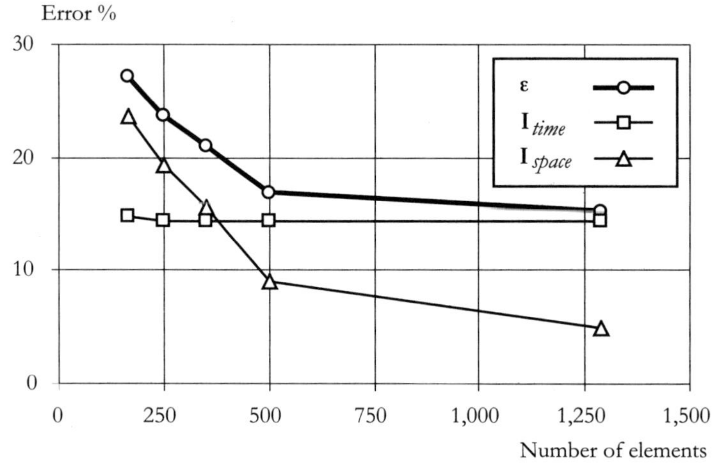

Figure 6.12. *Influence of the mesh - Error in DRUCKER's sense.*

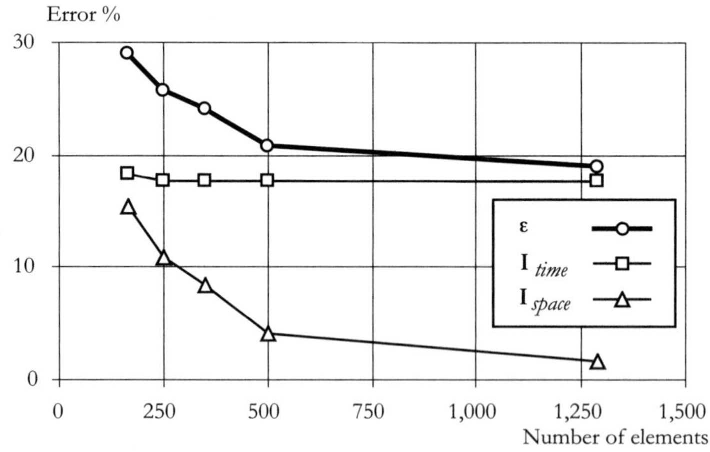

Figure 6.13. *Influence of the mesh - Dissipation error.*

We note that for the two types of errors the time error indicator is relatively insensitive to the space discretization.

Figure 6.14 and Figure 6.15 show the evolution of the errors for a fixed mesh and for increasingly refined time discretizations.

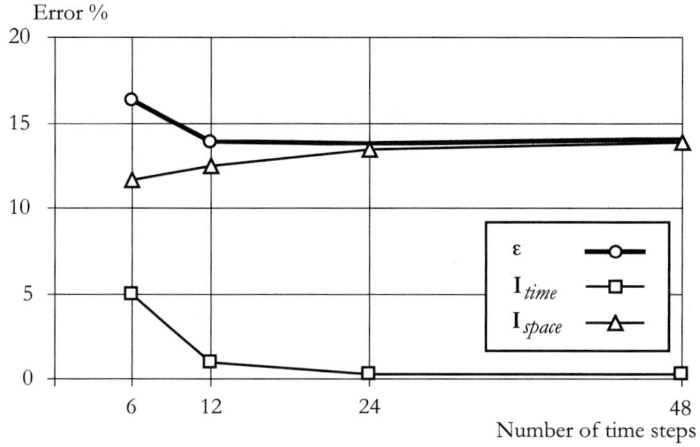

Figure 6.14. *Influence of the number of time steps - Error in DRUCKER's sense.*

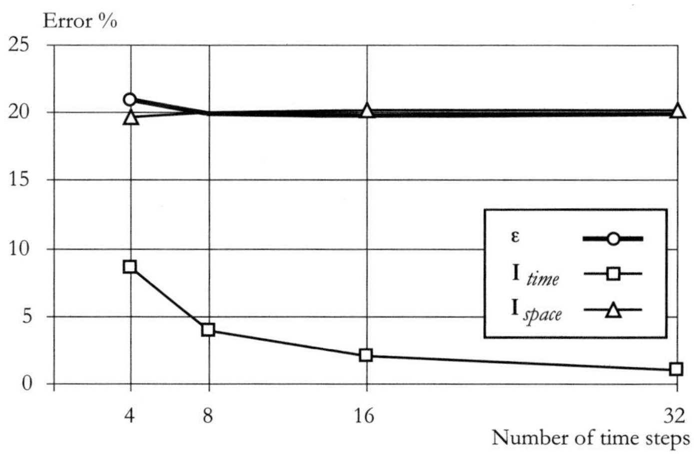

Figure 6.15. *Influence of the number of time steps - Dissipation error.*

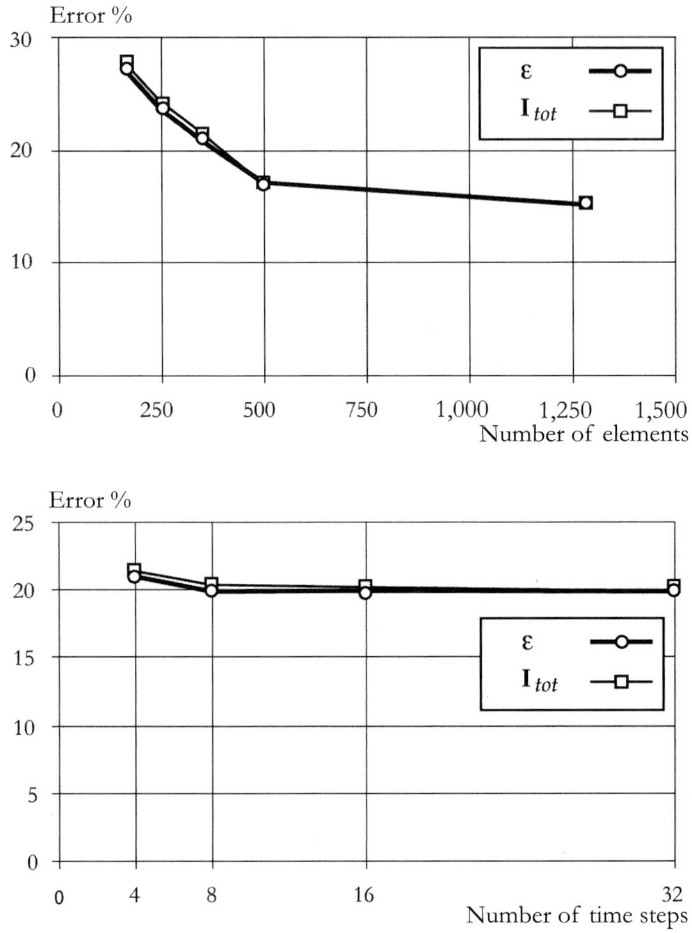

Figure 6.16. *Comparison of* $\boldsymbol{\varepsilon}$ *and* $\mathbf{I}_{tot}$ *- Error in* DRUCKER*'s sense.*

We note that the space error indicator behaves well: it is relatively insensitive to the time discretization provided it is not too coarse.

In Figure 6.16, DRUCKER's error is compared with the quantity

$$\mathbf{I}_{tot} = [\mathbf{I}_{time}^2 + \mathbf{I}_{space}^2]^{1/2}$$

Finally, in Figure 6.17, the dissipation error is compared with the quantity

$$\mathbf{I}_{tot} = \mathbf{I}_{time} + \mathbf{I}_{space}$$

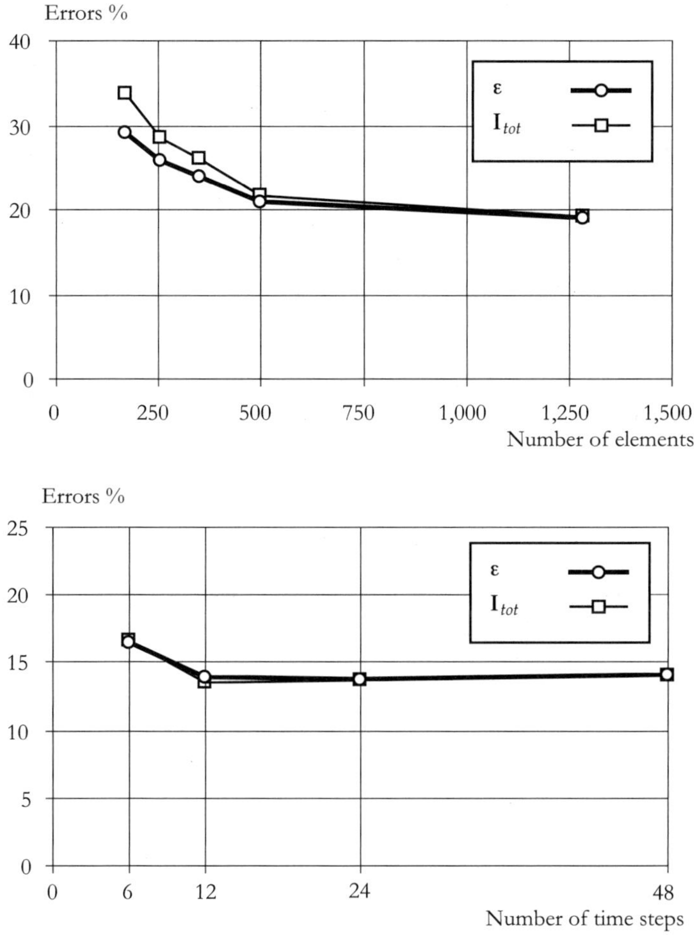

Figure 6.17. *Comparison of* $\boldsymbol{\varepsilon}$ *and* $\mathbf{I}_{tot}$ *- Dissipation error.*

These results are presented for various meshes and various time discretizations. In all situations, one can see that the quantity $\mathbf{I}_{tot}$ constitutes a good approximation of the error $\boldsymbol{\varepsilon}$.

This is due to the fact that the convergence thresholds of the nonlinear calculations over an increment were chosen sufficiently small so that the convergence errors of the iterations would be negligible. Thus, on all the examples treated, one has $\mathbf{I}_{iter} < 0.2\%$.

6.6 Adaptive control of the calculations

The objective of adaptive control is to reach an error ε_0 specified by the user at minimum numeric cost.

As in the linear case, our proposed adaptation principle consists in performing a relatively coarse initial calculation leading to an error ε, then modifying the calculation's parameters (essentially the mesh and the size of the time steps) in such a way that the new calculation yields the desired precision ε_0.

In the framework of plasticity and error in DRUCKER's sense, this approach was developed for the first time in [COFFIGNAL - LADEVEZE, 1983] for the modification of the mesh alone, then in [GALLIMARD - LADEVEZE - PELLE, 1995] for the simultaneous modification of the mesh and time steps. More recently, it was used in the framework of the dissipation error to control the discretization parameters as well as the iterative procedure [LADEVEZE - MOES, 1998b], [PELLE - RYCKELYNCK, 1998].

The determination of the parameters of the new calculation is obviously the crucial point of the procedure.

In Section 6.5.2, we observed that:

- *for DRUCKER's error*

$$\varepsilon^2 \approx \mathbf{I}^2_{space} + \mathbf{I}^2_{time} \tag{6.88}$$

- *for the dissipation error*

$$\varepsilon \approx \mathbf{I}_{space} + \mathbf{I}_{time} \tag{6.89}$$

as long as the convergence thresholds of the nonlinear calculation over an increment are chosen sufficiently small.

Let us assume that

$$\mathbf{I}_{space} = O(h^p) \quad \text{and} \quad \mathbf{I}_{time} = O((\Delta t)^q) \quad (p > 0, q > 0) \tag{6.90}$$

and that the numerical cost of an analysis is evaluated by the number of elements N of the mesh raised to a power α times the number of time steps P raised to a power β, i.e.

$$\text{numerical cost} = N^{\alpha} P^{\beta} \tag{6.91}$$

Then, by a reasoning similar to that in Chapter 5, one obtains the problem 6.92.

Minimize the cost function

$$\left[\sum_{E} \frac{1}{r_E^d}\right]^{\alpha} \left[\sum_{\Delta t} \frac{1}{r_{\Delta t}}\right]^{\beta} \tag{6.92}$$

under the constraints

$$(\varepsilon_0)^{\gamma} = (\mathbf{I}_{space,0})^{\gamma} + (\mathbf{I}_{time,0})^{\gamma}$$

$$\sum_{E} (r_E^p)^{\gamma} (\mathbf{I}_{space,E})^{\gamma} = (\mathbf{I}_{space,0})^{\gamma} \quad \textit{and} \quad \sum_{\Delta t} (r_{\Delta t}^q)^{\gamma} (\mathbf{I}_{time,\Delta t})^{\gamma} = (\mathbf{I}_{time,0})^{\gamma}$$

In the above expressions:

- d is the spatial dimension (1, 2 or 3);
- $\gamma = 1$ for the dissipation error and $\gamma = 2$ for DRUCKER's error ;
- the coefficients r_E are defined as in (4.3) and the coefficients $r_{\Delta t}$ of the time discretization are defined in the same way

$$r_{\Delta t} = \frac{\Delta t^*}{\Delta t}$$

where Δt is the current time step size and Δt^* the (unknown) size which corresponds to the optimum;

- $\mathbf{I}_{space,E}$ is the contribution of an element E of the current mesh to the space error indicator;
- $\mathbf{I}_{time,\Delta t}$ is the contribution of the time step size Δt of the current time discretization to the time error indicator;
- $\mathbf{I}_{space,0}$ and $\mathbf{I}_{time,0}$ are (unknown) target values of the space and time indicators.

One has the following property:

Property

The target values $\mathbf{I}_{space,0}$ *and* $\mathbf{I}_{time,0}$ *depend only on* $\varepsilon_0, \alpha, \beta, \gamma, \delta, p$ *and* q*, and minimization problem* (6.92) *decouples into two minimization problems: one which enables the determination of the* r_E *and the other the determination of the* $r_{\Delta t}$.

Proof

Introducing a LAGRANGE multiplier A, the resolution of (6.92) is the same as seeking the extrema of

$$\left[\sum_E \frac{1}{r_E^d}\right]^\alpha \left[\sum_{\Delta t} \frac{1}{r_{\Delta t}}\right]^\beta + A\left[\sum_E (r_E^p)^\gamma (\mathbf{I}_{space,E})^\gamma + \sum_{\Delta t} (r_{\Delta t}^q)^\gamma (\mathbf{I}_{time,\Delta t})^\gamma - (\varepsilon_0)^\gamma\right]$$

The extremality conditions in r_E and in $r_{\Delta t}$ yield

$$N^{\alpha-1} P^\beta \frac{d\,\alpha}{r_E^d} = A\,p\,\gamma\, r_E^{p\gamma} (\mathbf{I}_{space,E})^\gamma$$

$$N^\alpha P^{\beta-1} \frac{\beta}{r_{\Delta t}} = A\,q\,\gamma\, r_{\Delta t}^{q\gamma} (\mathbf{I}_{time,\Delta t})^\gamma$$

Summing up the first equations over all the elements E and the second equations over all the time steps Δt, one gets

$$(\varepsilon_0)^\gamma = \frac{N^\alpha P^\beta}{A\,\gamma}\left[\frac{d\,\alpha}{p} + \frac{\beta}{q}\right]$$

then

$$(\mathbf{I}_{space,0})^\gamma = d\,q\,\alpha \frac{(\varepsilon_0)^\gamma}{d\,q\,\alpha + p\,\beta} \quad \text{and} \quad (\mathbf{I}_{time,0})^\gamma = p\,\beta \frac{(\varepsilon_0)^\gamma}{d\,q\,\alpha + p\,\beta}$$

which proves the first part of the property.

Then, decoupling of the minimization problem is immediate:

- *Find the* r_E *solution of*

$$\textit{Minimize} \quad \left[\sum_E \frac{1}{r_E^d}\right]^\alpha \quad \textit{with} \quad \sum_E (r_E^p)^\gamma (\mathbf{I}_{space,E})^\gamma = (\mathbf{I}_{space,0})^\gamma$$

• *Find the $r_{\Delta t}$ solutions of*

$$\text{Minimize} \quad \left[\sum_{\Delta t} \frac{1}{r_{\Delta t}} \right]^{\beta} \quad \text{with} \quad \sum_{\Delta t} (r_{\Delta t}^{q})^{\gamma} (\mathbf{I}_{time,\Delta t})^{\gamma} = (\mathbf{I}_{time,0})^{\gamma}$$

These two problems are similar to those already studied in Chapter 5.

After the optimum parameters have been determined, one reiterates the whole calculation over $[0, T]$ with these new discretization parameters.

REMARKS AND COMMENTS

1. The adaptation principle which we propose here consists, overall, in subdividing the interval being studied into subintervals and performing the optimization successively over each subinterval.

An alternative approach consists in modifying the calculation's parameters (mesh, time step...) once the error reaches a critical value set by the user, then continuing the calculations with these new parameters. Examples of this procedure can be found in [COFFIGNAL - LADEVEZE, 1983]. The difficulty resides in the prediction of good parameters for continuing the calculation. Indeed, a change of parameters at a given instant influences only the "instantaneous" part of the error. It is essentially for this reason that our preference goes towards the adaptation principle that we detailed previously.

2. The above method assumes that the numerical cost is evaluated correctly by (6.91). Acceptable values of α and β should be determined by numerical tests.

3. For the convergence rate (6.90), one can use theoretical results if they exist; otherwise, one uses results of numerical tests.

4. As in the elastic case, if the error obtained at the end of the first analysis is very far from the desired value, several iterations of the adaptive procedure may be needed.

5. For nonlinear problems, in which history is involved, solved by the incremental method, if at the end of the analysis the desired accuracy is not achieved, one must reiterate the analysis over $[0, T]$. If several iterations are necessary, this can result in significant computational costs.

6. In [LADEVEZE - MOES, 1998b], the error indicators $\mathbf{i}_{space}$, $\mathbf{i}_{time}$ and $\mathbf{i}_{tot}$

were shown to be relatively insensitive to the value of $\mathbf{i}_{iter}$, provided that it is sufficiently small: $\mathbf{i}_{iter} \leq 0.1 \times \mathrm{Min}(\mathbf{i}_{space}, \mathbf{i}_{time})$.

7. The dissipation error and the associated specific error indicators constitute the basis of a truly adaptive nonlinear calculation strategy which exploits the advantages of the LATIN-method. This nonincremental method provides at each iteration a displacement field and a stress field (and, more generally, fields of internal variables) defined on $\Omega \times [0, T]$. At the end of an iteration, if the error measure obtained is greater than the desired accuracy, one modifies the parameters (mesh, time step...) of the algorithm and continues the calculation by performing an additional iteration. Thus, the parameters related to the final calculation, whose accuracy is prescribed, are determined progressively in the course of the iterations. In the early iterations, the problems to be solved, which correspond to coarse discretizations (in time, in space...), are numerically very inexpensive. Gradually, these various discretizations are adjusted according to the error levels obtained until the desired accuracy is achieved [PELLE - RYCKELYNCK, 1998]. Thus, it is possible to automate nonlinear analyses completely.

6.7 Generalizations

The concept of dissipation error developed above in the framework of standard and normal formulations can be extended to other types of constitutive relation formulations, especially bistandard formulations [HALPHEN - NGUYEN, 1975] and formulations described using a bipotential [DE SAXCE, 1992], [BERGA - DE SAXCE, 1994], [DE SAXCE - FENG, 1998], [HJIAI - DE SAXCE - ABRIAK, 1996], [HJIAI, 1999]. These extensions are still based on the notion of admissibility defined in Section 6.4.1, but require some adjustments of the definition of the error in constitutive relation [LADEVEZE, 1996].

6.7.1 Extension to bistandard formulations

Let $s_{Ad} = (\dot{\varepsilon}^p_{Ad}, \dot{\mathbb{X}}_{Ad}, \sigma_{Ad}, \mathbb{Y}_{Ad})$ be an admissible approximate solution and let us reuse the notations of Section 6.3.2.

Definition

One calls error in constitutive relation associated with an admissible approximate solution s_{Ad} *the real quantity*

$$\mathbf{e}_{RdC}(s_{Ad}) = \frac{1}{T}\int_0^T \int_\Omega \eta(\dot{\varepsilon}^p_{Ad}\,, \dot{\mathbb{X}}_{Ad}\,, \sigma_{Ad}\,, \mathbb{Y}_{Ad})\, d\Omega\, dt \tag{6.93}$$

where

$$\begin{aligned} &2\,\eta(\dot{\varepsilon}^p_{Ad}, \dot{\mathbb{X}}_{Ad}, \sigma_{Ad}, \mathbb{Y}_{Ad}) \\ &\quad = \varphi(\dot{\varepsilon}^p_{Ad}, -\dot{\mathbb{X}}_{Ad}) + \varphi^*(\sigma_{Ad}, \mathbb{Y}_{Ad}) - \mathrm{Tr}(\sigma_{Ad}\dot{\varepsilon}^p_{Ad}) + \mathbb{Y}_{Ad} \circ \dot{\mathbb{X}}_{Ad} \\ &\quad\quad + \Phi(\dot{\varepsilon}^p_{Ad}, -\dot{\mathbb{Y}}_{Ad}) + \Phi^*(\sigma_{Ad}, \mathbb{X}_{Ad}) - \mathrm{Tr}(\sigma_{Ad}\dot{\varepsilon}^p_{Ad}) + \mathbb{X}_{Ad} \circ \dot{\mathbb{Y}}_{Ad} \end{aligned}$$

By construction, one also has

$$\mathbf{e}_{RdC}(s_{Ad}) \geq 0 \quad \text{and} \quad \left\{\mathbf{e}_{RdC}(s_{Ad}) = 0 \Leftrightarrow s_{Ad} = s_{ex}\right\}$$

As before, one can define an error in the solution between an admissible solution and the exact solution by letting

$$\begin{aligned} \delta\,(s_{Ad}\,, s_{ex}\,; t) = &\frac{1}{2}\int_\Omega \mathrm{Tr}\Big[(\sigma_{Ad} - \sigma_{ex})\mathbf{K}^{-1}(\sigma_{Ad} - \sigma_{ex})\Big]_{|t}\, d\Omega \\ &+ \frac{1}{2}\int_\Omega \Big[(\mathbb{Y}_{Ad} - \mathbb{Y}_{ex}) \circ \Lambda^{-1}\,(\mathbb{Y}_{Ad} - \mathbb{Y}_{ex})\Big]_{|t}\, d\Omega \\ &+ \int_0^t \int_\Omega \tilde{\eta}(s_{Ad}, s_{ex})\, d\Omega\, d\tau \end{aligned}$$

with

$$\tilde{\eta}(s_{Ad}\,, s_{ex}) = \eta(\dot{\varepsilon}^p_{Ad}\,, \dot{\mathbb{X}}_{Ad}\,, \sigma_{ex}\,, \mathbb{Y}_{ex}) + \eta(\dot{\varepsilon}^p_{ex}\,, \dot{\mathbb{X}}_{ex}\,, \sigma_{Ad}\,, \mathbb{Y}_{Ad})$$

and one gets the following result:

Theorem

For any admissible solution s_{Ad}*, one has*

$$\mathbf{e}_{RdC}(s_{Ad}) = \frac{1}{T}\,\delta\,(s_{Ad}\,, s_{ex}\,; T)$$

The proof is similar to that given in Section 6.4.3.

6.7.2 Extension to bipotential formulations

Now, let us consider the case of a constitutive relation described by a bipotential according to the terminology introduced by DE SAXCE.

Let

$$x \bullet y = \mathrm{Tr}(\sigma \dot{\varepsilon}^p) - \Upsilon \circ \dot{\mathbb{X}}$$

$$x = \begin{bmatrix} \dot{\varepsilon}^p \\ -\dot{\mathbb{X}} \end{bmatrix} \quad \text{and} \quad y = \begin{bmatrix} \sigma \\ \Upsilon \end{bmatrix}$$

A bipotential is a function $\boldsymbol{b} : \boldsymbol{e} \times \boldsymbol{f} \to \mathbb{R}, (x, y) \mapsto \boldsymbol{b}(x, y)$ convex in $\boldsymbol{x}$ for a given $\boldsymbol{y}$, convex in $\boldsymbol{y}$ for given $\boldsymbol{x}$, and such that

$$\forall (x, y) \in \boldsymbol{e} \times \boldsymbol{f}, \qquad b(x, y) - x \bullet y \geq 0$$

Definition

A material with bipotential is a material whose constitutive relation takes the form

$$\boldsymbol{b}\left[(\dot{\varepsilon}^p, -\dot{\mathbb{X}}), (\sigma, \Upsilon)\right] - \mathrm{Tr}(\sigma \dot{\varepsilon}^p) + \Upsilon \circ \dot{\mathbb{X}} = 0 \tag{6.94}$$

where $\boldsymbol{b}$ *is a bipotential.*

Property (6.94) is equivalent to either of the following properties:

- *Property* 1

$$\begin{aligned} \dot{\varepsilon}^p &\in \partial_\sigma \boldsymbol{b}\left[(\dot{\varepsilon}^p, -\dot{\mathbb{X}}), (\sigma, \Upsilon)\right] \\ -\dot{\mathbb{X}} &\in \partial_\Upsilon \boldsymbol{b}\left[(\dot{\varepsilon}^p, -\dot{\mathbb{X}}), (\sigma, \Upsilon)\right] \end{aligned} \tag{6.95}$$

- *Property* 2

$$\begin{aligned} \sigma &\in \partial_{\dot{\varepsilon}^p} \boldsymbol{b}\left[(\dot{\varepsilon}^p, -\dot{\mathbb{X}}), (\sigma, \Upsilon)\right] \\ \Upsilon &\in \partial_{(-\dot{\mathbb{X}})} \boldsymbol{b}\left[(\dot{\varepsilon}^p, -\dot{\mathbb{X}}), (\sigma, \Upsilon)\right] \end{aligned} \tag{6.96}$$

In this context, the construction of a measure of the error in constitutive relation is easy.

For

$$s_{Ad} = (\dot{\varepsilon}^p_{Ad}, \dot{\mathbb{X}}_{Ad}, \sigma_{Ad}, \mathbb{Y}_{Ad})$$

it suffices to let

$$\eta(s_{Ad}) = \boldsymbol{b}\left((\dot{\varepsilon}^p_{Ad}, -\dot{\mathbb{X}}_{Ad}), (\sigma_{Ad}, \mathbb{Y}_{Ad})\right) - \mathrm{Tr}(\sigma_{Ad}\dot{\varepsilon}^p_{Ad}) + \mathbb{Y}_{Ad} \circ \dot{\mathbb{X}}_{Ad}$$

then to define the error by

$$\mathbf{e}_{RdC}(s_{Ad}) = \frac{1}{T}\int_0^T \int_\Omega \eta(s_{Ad})\, d\Omega\, dt \tag{6.97}$$

6.7.3 Application to unilateral contact problems

Here, we are considering the problem of two elastic solids Ω^1 and Ω^2 in unilateral contact with each other at an interface Γ^c (Figure 6.18). We assume that the contact is of the dry friction type (COULOMB law).

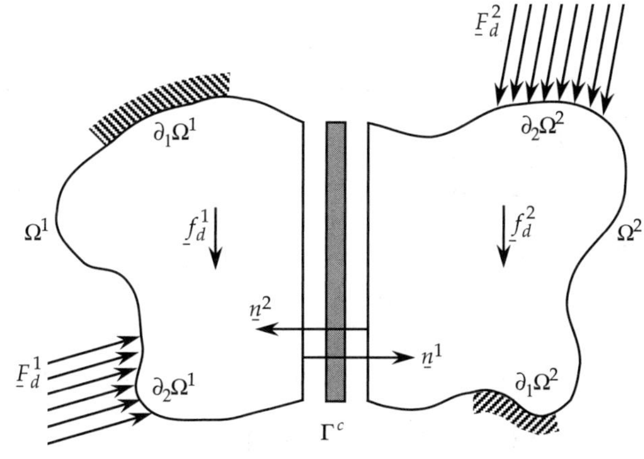

Figure 6.18. *Notations.*

In order to facilitate the definition of an error in constitutive relation, it is useful to consider the interface as an entirely separate mechanical entity. This entity provides for the transmission of displacements and forces across Γ^c and possesses its own constitutive relation.

Formulation of the problem

Let us orient Γ^c by choosing the unit normal vector, e.g. $\underline{n}^c = \underline{n}^1$.

Let us introduce at the interface the mechanical quantities $\underline{W}^1, \underline{W}^2, \underline{F}^1, \underline{F}^2$ and $\underline{F}^c$ representing two displacement fields (the displacements on both sides of the interface), two fields of surface force densities $\underline{F}^1$ and $\underline{F}^2$ (the forces transmitted to Ω^1 and Ω^2), and a field of surface density of "interior" forces $\underline{F}^c$ defined on Γ^c. Equilibrium at the interface is expressed by

$$\underline{F}^c = -\,\underline{F}^1 \quad \text{and} \quad \underline{F}^c = \underline{F}^2$$

Let us introduce the displacement jump, which for the interface plays a role similar to a strain

$$\underline{W}^c = \underline{W}^2 - \underline{W}^1$$

For any vector $\underline{V}$, let

$$V_n = \underline{V} \cdot \underline{n}^c \quad \text{and} \quad \underline{V}_T = \underline{V} - V_n\,\underline{n}^c$$

The constitutive relation of the dry friction type can be formulated as follows [DUVAUT - LIONS, 1972]:

$$\begin{aligned} &W_n^c \geq 0, \qquad F_n^c \geq 0\,, \qquad \left\| \underline{F}_T^c \right\| \leq \mu\, F_n^c \\ &W_n^c \cdot F_n^c = 0\,,\ \underline{W}_T^c \wedge \underline{F}_T^c = 0\,,\ \underline{W}_T^c \cdot \underline{F}_T^c \leq 0 \end{aligned} \tag{6.98}$$

where μ denotes the friction coefficient.

Using the approach by DE SAXCE, let us introduce the function

$$\boldsymbol{b}(\underline{V}, \underline{F}) = \chi_K(\underline{V}) + \chi_{C(\mu)}(\underline{F}) + \mu\, F_n \left\| \underline{V}_T \right\| \tag{6.99}$$

where χ_K and $\chi_{C(\mu)}$ are the indicator functions of the convex sets

$$K = \{\, \underline{V} \backslash V_n \leq 0 \,\}$$

$$C(\mu) = \{\, \underline{F} \backslash F_n \geq 0 \text{ and } \left\| \underline{F}_T \right\| \leq \mu\, F_n \,\}$$

One can find in [DE SAXCE, 1992] the detailed proof of the properties of $\boldsymbol{b}$, in particular, that $\boldsymbol{b}(\underline{V}, \underline{F})$ is a bipotential.

Then, the constitutive relation (6.98) is equivalent to the conditions

$$\boldsymbol{b}(-\underline{W}^c, \underline{F}^c) + \underline{W}^c \cdot \underline{F}^c = 0 \tag{6.100}$$

The unilateral contact problem can therefore be formulated as follows:

Find $(\underline{U}^1, \sigma^1)$ *defined on* Ω^1, $(\underline{U}^2, \sigma^2)$ *defined on* Ω^2, *and* $(\underline{W}^1, \underline{W}^2, \underline{F}^c)$ *defined on* Γ^c *such that:*

- *Kinematic constraints*

$$\underline{U}^1_{|\partial_1\Omega^1} = 0, \quad \underline{U}^2_{|\partial_1\Omega^2} = 0, \quad \underline{U}^1_{|\Gamma^c} = \underline{W}^1, \quad \underline{U}^2_{|\Gamma^c} = \underline{W}^2 \tag{6.101}$$

- *Equilibrium equations*

For $\ell = 1, 2, \quad \forall \underline{U}^*$ *zero on* $\partial_1\Omega^\ell$,

$$\int_{\Omega^\ell} \mathrm{Tr}\left[\sigma^\ell \varepsilon(\underline{U}^*)\right] d\Omega$$

$$= \int_{\Omega^\ell} \underline{f}^\ell_d \cdot \underline{U}^* \, d\Omega + \int_{\partial_2\Omega^\ell} \underline{F}^\ell_d \cdot \underline{U}^* \, dS + \int_{\Gamma^c} \underline{F}^\ell \cdot \underline{U}^* \, dS \tag{6.102}$$

$$on \quad \Gamma^c: \ \underline{F}^c = \underline{F}^2 = -\underline{F}^1 \tag{6.103}$$

- *Constitutive relations*

$$For\, \ell = 1, 2, \quad \sigma^\ell = \mathbf{K}^\ell \varepsilon(\underline{U}^\ell) \tag{6.104}$$

$$\boldsymbol{b}(-\underline{W}^c, \underline{F}^c) + \underline{W}^c \cdot \underline{F}^c = 0 \tag{6.105}$$

Error in constitutive relation

Definition

The fields

$$s_{Ad} = (\underline{U}^1_{Ad}, \sigma^1_{Ad}, \underline{U}^2_{Ad}, \sigma^2_{Ad}, \underline{W}^1_{Ad}, \underline{W}^2_{Ad}, \underline{F}^c_{Ad}, \underline{F}^1_{Ad}, \underline{F}^2_{Ad})$$

are said to be admissible for the contact problem if they verify Equations (6.101), (6.102) *and* (6.103).

Then, one can define an error in constitutive relation by

$$\mathbf{e}^2_{RdC}(s_{Ad}) = \sum_{\ell=1}^{2} \left\| \sigma^{\ell}_{Ad} - K\,\varepsilon(\underline{U}^{\ell}_{Ad}) \right\|^2_{\Omega^\ell} + 2\int_{\Gamma^c} \left[b(-\underline{W}^c_{Ad}, \underline{F}^c_{Ad}) + \underline{W}^c_{Ad} \cdot \underline{F}^c_{Ad} \right] d\Gamma \qquad (6.106)$$

Taking into account the properties of the bipotential, $\mathbf{e}_{RdC}(s_{Ad})$ is zero if and only if s_{Ad} verifies the constitutive relations (6.104) and (6.105), i.e. if and only if s_{Ad} is the solution of the unilateral contact problem. An implementation of this error measure is detailed in [COOREVITS - HILD - PELLE, 1999] for the frictionless case ($\mu = 0$). In fact, this particular case involves a classical potential and one can do without the notion of bipotential.

REMARKS

1. The first implementation of constitutive relation error estimators created from a bipotential was given in [HJIAI, 1999].

2. The presence of a coefficient 2 before the integral on Γ^c enables one, in the frictionless case, to establish a relation between the error in constitutive relation and the errors in the solution

$$\mathbf{e}^2_{RdC}(s_{Ad}) \geq \sum_{\ell=1}^{2} \left\| \sigma^{\ell}_{Ad} - \sigma^{\ell}_{ex} \right\|^2_{\Omega^\ell}$$

3. The plasticity and viscoplasticity models used in soil mechanics are not standard, but they can be formulated using bipotentials. Therefore, dissipation errors can be introduced to verify and control these types of finite element calculations.

4. Few of the works on error estimators have focused on problems involving unilateral contact. Nevertheless, let us mention the error estimator described in [WRIGGERS - SCHERF, 1995], which is of the same type as ZZ2.

5. Among the many works dealing with the numerical treatment of contact, one can mention [ODEN - MARTINS, 1985], [RAOUS - CHABRAND - LEBON, 1988], [FREMOND, 1988], [LICHT - PRATT - RAOUS, 1991], [KIKUCHI - ODEN, 1988], [ALART, 1997], [COCU - PRATT - RAOUS, 1995], [HILD, 2000].

6.7.4 Extension to problems with large transformations

Many problems involving large transformations (large displacements and large strains) contain critical points, especially instability points. This is the case of elasticity or plasticity problems with large strains, in which phenomena such as buckling or necking can appear. For buckling, besides [KOITER, 1967], [HUCHINSON, 1974], the reader could refer to [POTIER FERRY, 1987], [BAZANT - CEDOLIN, 1991], [NGUYEN, 2000]. For plasticity or viscoplasticity with large transformations, among many references, one can mention [MANDEL, 1972], [DAFALIAS, 1983], [STOLZ, 1987], [ROUGEE, 1991], [LADEVEZE, 1996], and for the computational aspects [SIMO, 1988], [IBRAHIMBEGOVIC - GHARZEDDINE, 1999]. The specific numerical strategies used to pass through the critical points are variants of [RIKS, 1991] and [CRISFIELD, 1991]. There are alternative approaches to the classical incremental numerical methods, for example, asymptotic numerical techniques [COCHELIN - DAMIL - POTIER FERRY, 1994] and the "LATIN" method introduced in [LADEVEZE, 1996], which was developed for buckling in [BOUCARD - LADEVEZE - POSS - ROUGEE, 1997].

Using the material formulation of problems with large transformations proposed in [LADEVEZE, 1996], it is easy to construct constitutive relation error measures. Indeed, in this formulation, the constitutive relations are expressed in a way equivalent to that used with small perturbations. An early implementation was done in [BUSSY - COFFIGNAL - LADEVEZE, 1985]. There are also error indicators built on the equilibrium residuals or on stress smoothing, particularly for metal forming problems [FOURMENT - CHENOT, 1994, 1995], [ANTUNEZ, 1998], [PERIC - DUTKO - OWEN, 1998]. However, these indicators are limited to the characterization of the quality of the spatial discretization. Let us also mention the family of indicators proposed by [BRINK - STEIN, 1998], which uses our technique of construction of admissible fields.

6.7.5 Extension to damage-prone materials

Structural calculations involving damage-prone materials constitute a class of problems which are more difficult to grasp than the one before. These studies, which started in the 1980s, deal with the modeling and simulation of structural deterioration phenomena leading to the occurrence of cracks which grow until they become unstable [PREDELEANU, 1987], [BAZANT - CEDOLIN, 1991]. In damage theory, whose bases can be found in [LEMAITRE - CHABOCHE, 1985], a crack is treated as a completely deteriorated zone.

The main difficulties appear beyond the first critical point which, with some adjustments, can be predicted by HILL's theory derived in 1958 [BENALAL - BILLARDON - GEYMONAT, 1989]. Beyond that point, classical damage models present serious deficiencies which mainly take the form of an abnormal dependence of the calculated solution on the mesh. In mathematical terms, the problem ceases to be elliptic.

Today, much research effort is being spent on the construction of consistent damage models.

One approach consists in introducing nonlocal damage mechanics [BAZANT - PIJAUDIER CABOT, 1988], [PIJAUDIER-CABOT - DUBE - LABORDERIE - BODE, 1992]. A more mathematical route leads to the construction of "localization limiters" [BELYTSCHKO - LASRY, 1989], [DE BORST, 1988]. Another approach, suitable for composite materials, is to use a "damage mesomodel", optionally with a delay effect [LADEVEZE, 1992]. Applications in the area of delamination are given in [ALLIX - LADEVEZE, 1992, 1996], [LADEVEZE, 1995], [LADEVEZE - ALLIX - DEU - LEVEQUE, 2000].

Several works have proposed error estimators for these simulations [AUBRY - LUCAS - TIE, 1998], [ORTIZ - QUIGLEY, 1991], [PERIC - YU - OWEN, 1994], [WUNDERLICH - CRAMER - STEINL, 1998], [HUERTA - DIEZ - RODRIGUEZ - PIJAUDIER-CABOT, 1998]. These estimators, which are often based on stress smoothing, are suitable only to assess the quality of the mesh.

The extension of the concept of error in constitutive relation to problems involving damage-prone materials was proposed in [LADEVEZE - MOES - DOUCHIN, 1999], [LADEVEZE, 2001]. It is based on the observation that after a reformulation of the problem the state laws and the evolution laws stem from convex potentials, but for different dualities. Therefore, the error in constitutive relation is constructed as the sum of two contributions. The first contribution is the classical error associated with the dissipation, and the second is an error associated with the satisfaction of the state laws, i.e. with the free energy. Obviously, the latter requires one to return to the classical notion of admissibility, which consists in satisfying only the kinematic constraints and equilibrium conditions. One difficulty worth mentioning is that, for problems involving instability phenomena, a small value of the error does not necessarily mean that the calculated solution is very close to the exact solution.

Chapter 7

The constitutive relation error method in dynamics

7.1 Introduction

For dynamic problems – whether transient dynamics or vibrations – few works propose error estimators.

For vibration problems, the issue of the determination of true bounds of the eigenfrequencies was solved in [LADEVEZE - PELLE, 1983, 1989]. As an application, upper bounds of the errors due to finite element approximations were given in [LADEVEZE - PELLE, 1983, 1984, 1989]. This problem, which is essentially the same as determining lower bounds of the eigenfrequencies, had previously been solved only in very specific situations [WEINSTEIN, 1963], [FICHERA, 1965]. Our approach, which is presented in the first part of this chapter (Section 7.2), uses an extension of the concept of error in constitutive relation to vibration problems. Based on this technique, COFFIGNAL developed error estimators which are simple to use. These estimators enable one to evaluate the quality of the eigenmodes calculated by finite element methods;

they were applied first to beam problems [COFFIGNAL, 1993], then to plate problems [BOISSE - COFFIGNAL, 1997]. More recently, another family of works has appeared [MADAY – PATERA - PERAIRE, 1999] using as the reference a vibration problem discretized on a "very fine" mesh, i.e. a mesh much finer than that used to calculate the approximate eigenfrequencies and eigenmodes. This approach and ours have many points in common. Another approach based on residuals was proposed in [WIBERG - BANSYS - HAGER, 1999]. Other works deal with HELMOTZ's problem, i.e. forced vibrations. Their main finding is that the classical error and the interpolation error are not of the same order: there is a "pollution" error which can become very significant as frequencies increase [IHLENBURG - BABUSKA, 1995], [BOUILLARD - IHLENBURG, 1999].

For transient dynamic problems, stress smoothing techniques or equilibrium residuals have been used in [ZIENKIEWICZ - XIE, 1991], [EWING - LAZAROV - VASSILEV, 1992], [WIBERG - LI, 1993, 1994, 1998], [BELYTSCHKO - TABARRA, 1993]. Of course, as was the case in nonlinear evolution problems, these approaches are not capable of taking into account all sources of error: the time discretization, the stopping of the iterations....

For these problems, the concept of error in constitutive relation opens the way to significant advances in the difficult problem of the estimation of discretization errors both in space and in time.

The initial works focused on a direct extension to dynamics of the error measure used in linear statics [LADEVEZE - PELLE, 1989a]. The first application for implicit NEWMARK schemes can be found in [COOREVITS - LADEVEZE - PELLE, 1992].

Another more general extension was proposed in [LADEVEZE, 1993]. The corresponding error measure, which can be viewed as an extension to dynamics of the error in DRUCKER's sense (Chapter 6), enables one to account for some nonlinearities.

This last version is the one we are presenting in the second part of this chapter. Let us point out that this error measure is also the basis of model updating methods developed at LMT-Cachan [LADEVEZE - REYNIER - MAÏA, 1994], [LADEVEZE, 1998], [CHOUAKI - LADEVEZE - PROSLIER, 1998],

[LADEVEZE - CHOUAKI, 1999].

Various implementations are possible; these depend on the extension conditions imposed on the fields being constructed. The difficulty is to construct an error estimator with reasonable effectivity index values. The implementation proposed in [COMBE - PELLE, 1998], [COMBE - LADEVEZE - PELLE, 1998, 1999] in the context of explicit schemes used in fast dynamics has already led to very interesting results. In particular, these works provide a comparison between constitutive relation error estimators and the estimators proposed in [ZIENKIEWICZ - WOOD - HINE - TAYLOR, 1984], [ZIENKIEWICZ - XIE, 1991], [WIBERG - LI, 1993].

As was the case for nonlinear problems, the concept of error in constitutive relation is an effective means of constructing error indicators enabling the separation of the various discretization errors and, thus, the development of adaptive procedures.

7.2 Linear vibrations

7.2.1 Formulation of the problem

A classical presentation of the formulation of a vibration problem for an elastic structure Ω consists in using RAYLEIGH's quotient [WEINSTEIN - STENGER, 1972], [GIBERT, 1988], [GERADIN - RIXEN, 1991]; one can refer to [MORAND - OHAYON, 1992] for fluid-structure problems and to [BRUNEAU, 1999] for acoustic problems. In 3D elasticity, for example, RAYLEIGH's quotient is defined for any nonzero displacement field $\underline{U}$ by

$$R(\underline{U}) = \frac{\frac{1}{2}\int_\Omega \mathrm{Tr}\left[\varepsilon(\underline{U})\mathbf{K}\varepsilon(\underline{U})\right] d\Omega}{\frac{1}{2}\int_\Omega \rho\, \underline{U}\cdot\underline{U}\, d\Omega}$$

Thus, $R(\underline{U})$ represents the quotient of the strain energy $E_D(\underline{U})$ associated with $\underline{U}$ over the "kinetic" energy $E_C(\underline{U})$ calculated using the displacement $\underline{U}$ in place of the velocity.

Subsequently, we will denote $\langle \underline{U}_1, \underline{U}_2 \rangle$ the inner product associated with the strain energy and $(\underline{U}_1, \underline{U}_2)$ the inner product associated with the kinetic energy. Thus, one has

$$R(\underline{U}) = \frac{\langle \underline{U}, \underline{U} \rangle}{(\underline{U}, \underline{U})} \quad (\underline{U} \neq 0) \tag{7.1}$$

and the following very classical property:

Property

The extrema of RAYLEIGH's quotient on the space $\boldsymbol{\mathcal{U}}_0$ of the fields which are kinematically admissible to zero are the squares of the eigenpulsations and are reached at the eigenmodes.

Thus, seeking the eigenpulsations and eigenmodes is equivalent to solving the problem

Find the real numbers λ for which there exists $\underline{U} \in \boldsymbol{\mathcal{U}}_0$ with $\underline{U} \neq 0$ such that

$$\forall \underline{U}^* \in \boldsymbol{\mathcal{U}}_0, \quad \langle \underline{U}, \underline{U}^* \rangle = \lambda (\underline{U}, \underline{U}^*) \tag{7.2}$$

Under very general conditions which are always verified for classical structural calculation problems, Problem (7.2) has a denumerable sequence of solutions such that

$$0 < \lambda_1 \leq \lambda_1 \leq \ldots \leq \lambda_n \leq \ldots \quad \text{and} \quad \lim_{n \to \infty} \lambda_n = +\infty \tag{7.3}$$

The eigenpulsations are defined by $\omega_n = \sqrt{\lambda_n}$ and the eigenfrequencies by $f_n = \omega_n / 2\pi$.

The eigenmodes associated with ω_n are the fields $\underline{U}_n \in \boldsymbol{\mathcal{U}}_0$ such that

$$\forall \underline{U}^* \in \boldsymbol{\mathcal{U}}_0, \quad \langle \underline{U}_n, \underline{U}^* \rangle = \omega_n^2 (\underline{U}_n, \underline{U}^*) \tag{7.4}$$

These modes always form a finite-dimension subspace of $\boldsymbol{\mathcal{U}}_0$.

REMARK. Below, for the sake of simplicity, we will always assume that the displacement boundary conditions are such that

$$\{ \underline{U} \in \boldsymbol{\mathcal{U}}_0 \text{ and } \varepsilon(\underline{U}) = 0 \} \Rightarrow \underline{U} = 0 \tag{7.5}$$

Thus, $\langle \underline{U}_1, \underline{U}_2 \rangle$ is a genuine inner product on $\boldsymbol{\mathcal{U}}_0$.

If the displacement boundary conditions are insufficient for Property (7.5) to be verified, the first eigenpulsation is $\omega_1 = 0$ and the associated eigenmodes are the rigid body modes. All the methods we are about to present can be extended, with minor modifications, to this type of situation.

7.2.2 Finite element discretization

The discretized problem

Let us again use the notations of Section 1.3. Discretizing Problem (7.2) is equivalent to seeking the extrema of RAYLEIGH's quotient on $\boldsymbol{\mathcal{U}}_{h0}$:

Find the real numbers Λ *for which there exists* $\underline{U}_h \in \boldsymbol{\mathcal{U}}_{h0}$ *nonzero such that*

$$\forall \underline{U}_h^* \in \boldsymbol{\mathcal{U}}_{h0}, \quad \langle \underline{U}_h , \underline{U}_h^* \rangle = \Lambda \, (\underline{U}_h , \underline{U}_h^*) \tag{7.6}$$

which leads to the eigenvalue problem of finite dimension $N = \dim \boldsymbol{\mathcal{U}}_{h0}$:

Find the real numbers Λ *for which there exists* $\mathbf{q} \neq 0$ *such that*

$$\mathbb{K}\mathbf{q} = \Lambda \mathbb{M}\mathbf{q} \tag{7.7}$$

where $\mathbb{K}$ is the stiffness matrix and $\mathbb{M}$ the mass matrix defined by

$$\mathbb{K} = \int_\Omega \mathbf{B}^T\mathbf{K}\mathbf{B}\, d\Omega \quad \text{and} \quad \mathbb{M} = \int_\Omega \mathbf{N}^T \rho \mathbf{N}\, d\Omega \tag{7.8}$$

Problem (7.7) has N eigenvalues

$$0 < \Lambda_1 \leq \Lambda_2 \leq \ldots \leq \Lambda_N \tag{7.9}$$

associated with the eigenvectors

$$\mathbb{K}\mathbf{q}_n = \Lambda_n \mathbb{M}\mathbf{q}_n \tag{7.10}$$

Thus, one obtains approximations of the eigenpulsations and eigenmodes

$$n = 1, 2, \ldots, N: \quad \Omega_n = \sqrt{\Lambda_n} \quad \text{and} \quad \underline{U}_n(M) = \mathbf{N}(M)\mathbf{q}_n \tag{7.11}$$

The approximate modes $\underline{U}_n$ form an orthogonal basis of $\boldsymbol{\mathcal{U}}_0$ in the sense of each of the inner products $(\,\cdot\,,\,\cdot\,)$ and $\langle \cdot\,, \cdot \rangle$. This basis is usually normalized by setting $(\underline{U}_n , \underline{U}_m) = \delta_{nm}$ and, thus, $\langle \underline{U}_n , \underline{U}_m \rangle = \Lambda_n \, \delta_{nm}$.

Upper bounds of the exact eigenpulsations

Using the classical properties of RAYLEIGH's quotient, one can prove [COURANT - HILBERT, 1953], [SANCHEZ-HUBERT - SANCHEZ-PALENCIA, 1989] the following comparison relation

$$n = 1, 2, \dots, N: \quad \lambda_n \leq \Lambda_n \tag{7.12}$$

Lower bounds of the exact eigenpulsations

Additionally, in order to verify the quality of the approximate eigenpulsations, it is necessary to determine lower bounds of the eigenpulsations. This is a more difficult problem. In the framework of structural calculations, the intermediate problem method [WEINSTEIN, 1963] and the orthogonal invariant method [FICHERA, 1965] enable one to achieve lower bounds in specific situations: simple geometric shapes and isotropic materials. However, these methods are not suitable for finite element analysis and cannot be used with complex geometries. In the next section, we present a numerical method which enables one to bound the exact eigenpulsations and, thus, to verify the quality of the approximations obtained.

7.2.3 The new formulation

The method we are proposing relies on a new "static" formulation of a vibration problem derived using a quotient similar to RAYLEIGH's quotient and the classical properties of this quotient. Here, we briefly present this new formulation and the corresponding bounding method. Then, we give the main lines of its implementation. For further details, the reader may refer to [LADEVEZE - PELLE, 1983, 1984, 1988, 1989a].

Static formulation

This formulation gives a special role to the mass densities of a set of loads $\underline{\Gamma}$ defined on Ω. The corresponding space is denoted $\boldsymbol{\mathcal{A}}$. In particular, $\boldsymbol{\mathcal{A}}$ contains vector fields which are only piecewise continuous. The space $\boldsymbol{\mathcal{A}}$ does not coincide with $\boldsymbol{\mathcal{U}}$. One should note that the elements $\underline{\Gamma} \in \boldsymbol{\mathcal{A}}$ can also be interpreted as acceleration amplitudes, which explains the notations used.

Let us introduce the static problem associated with a mass density of loads $\underline{\Gamma} \in \boldsymbol{\mathcal{A}}$:

Find $\underline{U} \in \boldsymbol{\mathcal{U}}_0$ *such that*

$$\forall \underline{U}^* \in \boldsymbol{\mathcal{U}}_0, \quad \int_\Omega \mathrm{Tr}\left[\varepsilon(\underline{U})\mathbf{K}\varepsilon(\underline{U}^*)\right] d\Omega = \int_\Omega \rho \underline{\Gamma} \cdot \underline{U}^* d\Omega$$

i.e., with the notations introduced above

$$\forall \underline{U}^* \in \boldsymbol{\mathcal{U}}_0, \quad \langle \underline{U}, \underline{U}^* \rangle = (\underline{\Gamma}, \underline{U}^*) \tag{7.13}$$

Equation (7.13) has a unique solution which will be denoted $\underline{U} = \mathbf{G}(\underline{\Gamma})$, where $\mathbf{G}$ is GREEN's classical operator.

Now, one can define a new quotient as

$$\forall \underline{\Gamma} \in \boldsymbol{\mathcal{A}}, \quad \tilde{R}(\underline{\Gamma}) = \frac{(\underline{\Gamma}, \underline{\Gamma})}{\langle \mathbf{G}(\underline{\Gamma}), \mathbf{G}(\underline{\Gamma}) \rangle} \qquad (\underline{\Gamma} \neq 0) \tag{7.14}$$

Seeking the extrema of $\tilde{R}$ on $\boldsymbol{\mathcal{A}}$ leads to the following problem:

Find the real numbers μ *for which there exists* $\underline{\Gamma} \in \boldsymbol{\mathcal{A}}$ *nonzero such that*

$$\forall \underline{\Gamma}^* \in \boldsymbol{\mathcal{A}}, \quad (\underline{\Gamma}, \underline{\Gamma}^*) = \mu \langle \mathbf{G}(\underline{\Gamma}), \mathbf{G}(\underline{\Gamma}^*) \rangle \tag{7.15}$$

The usefulness of this new quotient resides in the following fundamental property, whose proof is given in [LADEVEZE - PELLE, 1989a]:

Theorem

Problems (7.2) *and* (7.15) *have the same solutions.*

REMARKS AND COMMENTS

1. A priori, a mass density of quantities of acceleration is a much less regular field than a displacement field. Thus, one advantage of the quotient $\tilde{R}$ is that it is defined on a space $\boldsymbol{\mathcal{A}}$ of fields much less regular than that on which the RAYLEIGH classical quotient is defined, which, from a numerical point of view, is a definite plus.

Typically, for example in 3D elasticity, one has

$$\boldsymbol{\mathcal{U}}_0 \subset \boldsymbol{\mathcal{U}} = \left[H^1(\Omega)\right]^3 \quad \text{and} \quad \boldsymbol{\mathcal{A}} = \left[L^2(\Omega)\right]^3$$

Thus, it is possible to perform the discretization using discontinuous fields while preserving conformity.

2. However, one difficulty is the presence in the denominator of GREEN's operator, which is generally not known explicitly. We will show below how to overcome this problem.

7.2.4 Method for finding bounds of the exact eigenfrequencies

Let $\boldsymbol{\mathcal{A}}_h$ be a subspace of $\boldsymbol{\mathcal{A}}$ of finite dimension N. The discretization of Problem (7.15) leads to the following eigenvalue problem:

Find the real numbers Π for which there exists $\underline{\Gamma}_h \in \boldsymbol{\mathcal{A}}_h$ nonzero such that

$$\forall \underline{\Gamma}^* \in \boldsymbol{\mathcal{A}}_h, \quad (\underline{\Gamma}_h, \underline{\Gamma}_h^*) = \Pi \left\langle \mathbf{G}(\underline{\Gamma}_h), \mathbf{G}(\underline{\Gamma}_h^*) \right\rangle \tag{7.16}$$

We will designate the solutions of (7.16) by

$$0 < \Pi_1 \leq \Pi_1 \leq \ldots \leq \Pi_N \tag{7.17}$$

with which are associated the N eigenmodes $\underline{\Gamma}_n$ such that

$$\forall \underline{\Gamma}_h^* \in \boldsymbol{\mathcal{A}}_h, \quad (\underline{\Gamma}_n, \underline{\Gamma}_h^*) = \Pi_n \left\langle \mathbf{G}(\underline{\Gamma}_n), \mathbf{G}(\underline{\Gamma}_h^*) \right\rangle \tag{7.18}$$

and one can set $(\underline{\Gamma}_n, \underline{\Gamma}_m) = \delta_{nm}$.

One can prove [LADEVEZE - PELLE, 1989] the following property:

Property

$$\forall n = 1, 2, \ldots, N: \quad \Pi_n \left[1 + \frac{\Pi_n}{k_S}\right]^{-1} \leq \lambda_n \leq \Pi_n \tag{7.19}$$

where k_S is a constant, characteristic of the choice of $\boldsymbol{\mathcal{A}}_h$, defined by

$$k_S = \operatorname*{Inf}_{\underline{\Gamma} \in \boldsymbol{\mathcal{A}}_h^{(\perp)}} \tilde{R}(\underline{\Gamma}) \tag{7.20}$$

In (7.20), $\mathcal{A}_h^{(\perp)}$ denotes the orthogonal complement of $\mathcal{A}_h$ in $\mathcal{A}$ for the inner product $(\bullet, \bullet)$.

An initial step in the implementation of the bounding method consists in obtaining an estimate – more precisely, a satisfactory lower bound – of the constant k_S.

This point being set aside, there is another major difficulty to overcome. Indeed, let us consider a basis $(\tilde{\underline{N}}_1(M), \tilde{\underline{N}}_2(M), \dots, \tilde{\underline{N}}_N(M))$ of subspace $\mathcal{A}_h$, and write

$$\underline{\Gamma}_h(M) = \sum_{n=1}^{N} \tilde{\underline{N}}_n(M)\, \tilde{q}_n = \tilde{\mathbf{N}}(M)\, \tilde{\mathbf{q}} \tag{7.21}$$

Then, Problem (7.16) becomes:

Find the real numbers Π *for which there exists* $\tilde{\mathbf{q}}$ *nonzero such that*

$$\tilde{\mathbb{M}}\tilde{\mathbf{q}} = \Pi \tilde{\mathbb{K}}\, \tilde{\mathbf{q}} \tag{7.22}$$

where the matrices $\tilde{\mathbb{M}}$ and $\tilde{\mathbb{K}}$ are defined by

$$\tilde{\mathbb{M}}_{nm} = (\tilde{\underline{N}}_n, \tilde{\underline{N}}_m) \quad \text{and} \quad \tilde{\mathbb{K}}_{nm} = \left\langle \mathbf{G}(\tilde{\underline{N}}_n), \mathbf{G}(\tilde{\underline{N}}_m) \right\rangle \tag{7.23}$$

Equation (7.22) is formally the same as (7.10). With the exception of very simple one-dimensional problems, operator $\mathbf{G}$ is not known explicitly. Therefore, it is impossible to construct matrix $\tilde{\mathbb{K}}$ directly.

REMARK. A similar bounding method based on RAYLEIGH's classical quotient in displacement was proposed in [LADEVEZE - PELLE, 1983, 1989], relying on the following property:

Property

Let $\mathcal{U}_h$ *be a subspace of* $\mathcal{U}$ *of finite dimension* N*; one has*

$$\forall n = 1, 2, \dots, N: \quad \Lambda_n \left[1 + \frac{\Lambda_n}{k_C}\right]^{-1} \leq \lambda_n \leq \Lambda_n$$

where Λ_n *are the solutions of* (7.10) *and* k_C *is a constant, dependent on the choice of* $\boldsymbol{\mathcal{U}}_h$, *defined by*

$$k_C = \operatorname*{Inf}_{\underline{U} \in \boldsymbol{\mathcal{U}}_h^{\langle\perp\rangle}} R(\underline{U})$$

Here $\boldsymbol{\mathcal{U}}_h^{<\perp>}$ denotes the orthogonal complement of $\boldsymbol{\mathcal{U}}_h$ in $\boldsymbol{\mathcal{U}}$ for the inner product $\langle \cdot , \cdot \rangle$.

In the framework of standard displacement methods, obtaining bounds requires only the determination of a lower bound for the constant k_C. Nevertheless, this determination uses techniques whose principles are similar to those which we are going to present below, except that they are more difficult to implement.

7.2.5 Implementation of the bounding method

The basic idea consists in carrying out a new approximation for Problem (7.16) by replacing the exact stress field

$$\sigma(\underline{\Gamma}) = \mathbf{K}\varepsilon(\mathbf{G}(\underline{\Gamma}))$$

by a stress field $\hat{\sigma}(\underline{\Gamma})$ statically admissible for load $\underline{\Gamma}$. Initially, we assume that we know how to construct an operator $\hat{\sigma}$ which associates with any field $\underline{\Gamma} \in \boldsymbol{\mathcal{A}}_h$ a stress field $\hat{\sigma}(\underline{\Gamma})$ in equilibrium with $\underline{\Gamma}$.

Obtaining lower bounds

Let us now consider the following discretized problem:

Find the real numbers $\hat{\Pi}$ *for which there exists* $\hat{\underline{\Gamma}}_h \in \boldsymbol{\mathcal{A}}_h$ *nonzero such that*

$$\forall \underline{\Gamma}^* \in \boldsymbol{\mathcal{A}}_h, \quad (\hat{\underline{\Gamma}}_h, \underline{\Gamma}_h^*) = \hat{\Pi} \langle\!\langle \hat{\sigma}(\hat{\underline{\Gamma}}_h), \hat{\sigma}(\underline{\Gamma}_h^*) \rangle\!\rangle \tag{7.24}$$

where $\langle\!\langle \cdot , \cdot \rangle\!\rangle$ denotes the inner product

$$\langle\!\langle \sigma_1, \sigma_2 \rangle\!\rangle = \int_\Omega \operatorname{Tr}\left[\sigma_1 \mathbf{K}^{-1} \sigma_2\right] d\Omega$$

Problem (7.24) is an eigenvalue problem of dimension N, which has N solutions such that

$$0 < \hat{\Pi}_1 \leq \hat{\Pi}_2 \leq \ldots \leq \hat{\Pi}_N$$

The resolution of that problem enables one to obtain lower bounds of the exact eigenpulsations:

Property

$$\forall n = 1, 2, \ldots, N: \quad \hat{\Pi}_n \left[1 + \frac{\hat{\Pi}_n}{k_S} \right]^{-1} \leq \lambda_n \tag{7.25}$$

Proof

Problem (7.24) is equivalent to seeking the extrema on $\mathcal{A}_h$ of the quotient

$$\hat{R}(\underline{\Gamma}) = \frac{(\underline{\Gamma}, \underline{\Gamma})}{(\!(\hat{\sigma}(\underline{\Gamma}), \hat{\sigma}(\underline{\Gamma}))\!)}$$

Besides, the complementary energy is minimum for the exact solution

$$\langle \mathbf{G}(\underline{\Gamma}), \mathbf{G}(\underline{\Gamma}) \rangle = (\!(\mathbf{K}\, \varepsilon[\mathbf{G}(\underline{\Gamma})], \mathbf{K}\varepsilon[\mathbf{G}(\underline{\Gamma})])\!) \leq (\!(\hat{\sigma}(\underline{\Gamma}), \hat{\sigma}(\underline{\Gamma}))\!)$$

Therefore,

$$\forall \underline{\Gamma} \in \mathcal{A}_h, \quad \hat{R}(\underline{\Gamma}) \leq \tilde{R}(\underline{\Gamma})$$

The monotonicity principle [WEINSTEIN - STENGER, 1972] applied to both quotients $\hat{R}$ and $\tilde{R}$ yields

$$\forall n = 1, 2, \ldots, N: \quad \hat{\Pi}_n \leq \Pi_n$$

Using (7.19), one obtains the desired result.

Obtaining upper bounds

Using the results of the resolution of Problem (7.24) and performing some inexpensive additional calculations, one can also obtain upper bounds of the exact eigenpulsations.

Let $\hat{\underline{\Gamma}}_1, \hat{\underline{\Gamma}}_2, \ldots, \hat{\underline{\Gamma}}_N$ be the eigenmodes of (7.24) associated with the eigenpulsations $\hat{\Pi}_1, \hat{\Pi}_2, \ldots, \hat{\Pi}_N$, and let $\mathcal{U}_r$ be a subspace of $\mathcal{U}_0$ of finite dimension r. Let us associate with each mode $\hat{\underline{\Gamma}}_n$ the displacement field $\mathbf{d}(\hat{\underline{\Gamma}}_n)$ solution of the problem

$$\forall \underline{U}^* \in \mathcal{U}_r, \quad \langle \mathbf{d}(\hat{\underline{\Gamma}}_n), \underline{U}^* \rangle = (\mathbf{P}(\hat{\underline{\Gamma}}_n), \underline{U}^*) \tag{7.26}$$

where $\mathbf{P}$ denotes the orthogonal projection operator from $\boldsymbol{\mathcal{A}}$ onto $\boldsymbol{\mathcal{U}}_r$ for the inner product $(\bullet,\bullet)$.

For $q \leq r$, let us denote by $\boldsymbol{\mathcal{W}}_p$ the subspace of $\boldsymbol{\mathcal{U}}_r$ of dimension p ($p \leq q$) generated by $\mathbf{P}(\hat{\underline{\Gamma}}_1), \mathbf{P}(\hat{\underline{\Gamma}}_2), \dots, \mathbf{P}(\hat{\underline{\Gamma}}_q)$.

Then, let us consider the eigenvalue problem of dimension p:

Find the real numbers $\hat{\Lambda}$ *for which there exists* $\underline{W} \in \boldsymbol{\mathcal{W}}_p$ *nonzero such that*

$$\forall \underline{W}^* \in \boldsymbol{\mathcal{W}}_p, \quad (\underline{W}, \underline{W}^*) = \hat{\Lambda} \langle \mathbf{d}(\underline{W}), \mathbf{d}(\underline{W}^*) \rangle \tag{7.27}$$

The resolution of (7.27) provides upper bounds of the first N exact eigenpulsations.

Property

$$\forall n = 1, 2, \dots, p: \quad \lambda_n \leq \hat{\Lambda}_n \tag{7.28}$$

Proof

Problem (7.28) is equivalent to seeking the extrema on $\boldsymbol{\mathcal{W}}_p$ of the quotient

$$\underset{\sim}{R}(\underline{W}) = \frac{(\underline{W}, \underline{W})}{\langle \mathbf{d}(\underline{W}), \mathbf{d}(\underline{W}) \rangle}$$

Let us consider the potential energy associated with Problem (7.13)

$$E_p(\underline{U}) = \frac{1}{2} \langle \underline{U}, \underline{U} \rangle - (\underline{\Gamma}, \underline{U})$$

This quantity is minimum for the field $\mathbf{G}(\underline{\Gamma})$, from which one easily deduces

$$\forall \underline{\Gamma} \in \boldsymbol{\mathcal{A}}, \quad \forall \underline{U} \in \boldsymbol{\mathcal{U}}_0, \quad 2(\underline{\Gamma}, \underline{U}) - \langle \underline{U}, \underline{U} \rangle \leq \langle \mathbf{G}(\underline{\Gamma}), \mathbf{G}(\underline{\Gamma}) \rangle \tag{7.29}$$

Applying (7.29) for $\underline{\Gamma} = \underline{W}$ and $\underline{U} = \mathbf{d}(\underline{W})$, one obtains

$$2(\underline{W}, \mathbf{d}(\underline{W})) - \langle \mathbf{d}(\underline{W}), \mathbf{d}(\underline{W}) \rangle \leq \langle \mathbf{G}(\underline{W}), \mathbf{G}(\underline{W}) \rangle$$

and because of

$$(\underline{W}, \mathbf{d}(\underline{W})) = \langle \mathbf{d}(\underline{W}), \mathbf{d}(\underline{W}) \rangle$$

one has

$$\langle \mathbf{d}(\underline{W}), \mathbf{d}(\underline{W})\rangle \le \langle \mathbf{G}(\underline{W}), \mathbf{G}(\underline{W})\rangle$$

From which one deduces

$$\forall \underline{W} \in \boldsymbol{W}_p, \quad \tilde{R}(\underline{W}) \le \underset{\sim}{R}(\underline{W})$$

Finally, using the monotonicity principle, one obtains the desired result.

REMARK. In practice, the number p of eigenpulsations for which bounds are being sought is small compared to N. Therefore, this procedure is very inexpensive.

Construction of an operator $\hat{\sigma}$

The main purpose of constructing an operator $\hat{\sigma}$ is to make the construction and the resolution of Problem (7.24) possible.

This is achieved, in principle, by:

- a "mixed" discretization of quotient $\tilde{R}(\underline{\Gamma})$;
- a transformation of Problem (7.24), whose unknowns are of the "load" type, into a problem whose unknowns are of the "stress" type.

Mixed discretization of $\tilde{R}(\underline{\Gamma})$

For $\underline{\Gamma} \in \boldsymbol{A}$, let us denote by $\boldsymbol{S}(\underline{\Gamma})$ the set of the stress fields which are in equilibrium with the load $\underline{\Gamma}$,

$$\sigma \in \boldsymbol{S}(\underline{\Gamma}) \Leftrightarrow \{ \forall \underline{U}^* \in \boldsymbol{U}_0, (\!(\sigma, \sigma(\underline{U}^*))\!) = (\underline{\Gamma}, \underline{U}^*) \} \tag{7.30}$$

One has the classical property

$$\langle \mathbf{G}(\underline{\Gamma}), \mathbf{G}(\underline{\Gamma})\rangle = \inf_{\sigma \in \boldsymbol{S}(\underline{\Gamma})} (\!(\sigma, \sigma)\!) \tag{7.31}$$

In order to discretize $\tilde{R}(\underline{\Gamma})$, one chooses a subspace $\boldsymbol{A}_N$ of $\boldsymbol{A}$ of finite dimension N and a subspace $\boldsymbol{S}_P$ of $\boldsymbol{S}$ of finite dimension P.

For a given $\underline{\Gamma}$, one defines $\hat{\sigma}(\underline{\Gamma})$ as the field in $\boldsymbol{S}_P$ which is closest to $\mathbf{K}\varepsilon[\mathbf{G}(\underline{\Gamma})]$ in the sense of (7.31).

The field $\hat{\sigma}(\underline{\Gamma})$ is therefore the solution of the following problem:

Find $\sigma \in \boldsymbol{S}_P$ *such that*

$$\begin{aligned} &\forall \underline{U}^* \in \boldsymbol{U}_0, && \langle\!\langle \sigma , \sigma(\underline{U}^*) \rangle\!\rangle = (\underline{\Gamma}, \underline{U}^*) \\ &\forall \sigma^* \in \boldsymbol{S}_{P0} = \boldsymbol{S}_P \cap \boldsymbol{S}(0), && \langle\!\langle \sigma , \sigma^* \rangle\!\rangle = 0 \end{aligned} \tag{7.32}$$

where $\boldsymbol{S}(0)$ *denotes the space of the self-equilibrated stress fields (i.e. the stress fields which are in equilibrium with zero load).*

One can verify that (7.32) has a unique solution if and only if

$$\forall \underline{\Gamma} \in \boldsymbol{A}_N, \quad \boldsymbol{S}_P \cap \boldsymbol{S}(\underline{\Gamma}) \neq \varnothing . \tag{7.33}$$

Therefore, with this condition, the operator $\hat{\sigma}$ is completely determined.

Transformation of the problem

Let us introduce the linear operator $\mathbf{D}$ such that

$$\forall \underline{U}^* \in \boldsymbol{U}_0, \quad \langle\!\langle \sigma , \sigma(\underline{U}^*) \rangle\!\rangle = (\mathbf{D}(\sigma), \underline{U}^*) \tag{7.34}$$

For example, in 2D or 3D elasticity, one has

$$\mathbf{D}(\sigma) = -\frac{1}{\rho} \underline{\text{div}}\, \sigma$$

The operator $\mathbf{D}$ is defined on a subspace $\boldsymbol{D}$ of $\boldsymbol{S}$. This subspace consists of the stress fields σ which are sufficiently regular so that $\mathbf{D}(\sigma) \in \boldsymbol{A}$.

If the vibration problem being considered includes, on a portion $\partial_2\Omega$ of the boundary $\partial\Omega$, boundary conditions of the type $\sigma\underline{n} = 0$, these conditions must be imposed to the fields of $\boldsymbol{D}$ in order for (7.34) to be verified.

Let us consider a given subspace $\boldsymbol{S}_P$ of $\boldsymbol{S}$ of finite dimension P and the eigenvalue problem:

Find the real numbers $\overline{\Pi}$ *for which there exists* $\bar{\sigma} \in \boldsymbol{S}_P$ *nonzero such that*

$$\forall \sigma^* \in \boldsymbol{S}_P, \quad (\mathbf{D}(\bar{\sigma}), \mathbf{D}(\sigma^*)) = \overline{\Pi} \langle\!\langle \bar{\sigma} , \sigma^* \rangle\!\rangle \tag{7.35}$$

Clearly, $\boldsymbol{S}_{P0}$ is the kernel of the restriction of $\mathbf{D}$ to $\boldsymbol{S}_P$. If $\boldsymbol{S}_{P0} \neq \varnothing$, 0 is an eigenvalue of (7.35) and the associated eigenmodes are elements of the subspace $\boldsymbol{S}_{P0}$.

Let $\boldsymbol{\mathcal{A}}_N = \mathbf{D}(\boldsymbol{\mathcal{S}}_P)$ be the subspace of $\boldsymbol{\mathcal{A}}$ the image of $\boldsymbol{\mathcal{S}}_P$ through $\mathbf{D}$. The main interest of Problem (7.35) resides in the following property:

Property

- *The nonzero eigenvalues of* (7.35) *are exactly the* N *eigenvalues of Problem* (7.24) *associated with the subspace* $\boldsymbol{\mathcal{A}}_N$.
- *If* $\bar{\sigma}_n$ *is an eigenmode of Problem* (7.35) *associated with* $\bar{\Pi}_n \neq 0$, *then* $\underline{\Gamma}_n = \mathbf{D}(\bar{\sigma}_n)$ *is an eigenmode of* (7.24) *associated with* $\hat{\Pi}_n = \bar{\Pi}_n$ *and one has* $\bar{\sigma}_n = \hat{\sigma}(\underline{\Gamma}_n)$.

This property was proven in [LADEVEZE - PELLE, 1989a].

Implementation

Thus, in order to solve Problem (7.24), one uses the following method:

- one chooses a subspace $\boldsymbol{\mathcal{S}}_P$ of $\boldsymbol{\mathcal{S}}$ of finite dimension P;
- one sets the subspace $\boldsymbol{\mathcal{A}}_N$ as $\boldsymbol{\mathcal{A}}_N = \mathbf{D}(\boldsymbol{\mathcal{S}}_P)$, which ensures that condition (7.33) is verified;
- if $(\mathbb{s}_1, \mathbb{s}_2, \ldots, \mathbb{s}_P)$ is a basis of $\boldsymbol{\mathcal{S}}_P$ and if one lets

$$\sigma = \sum_{n=1}^{P} \alpha_n \mathbb{s}_n \quad \text{and} \quad \boldsymbol{\alpha} = [\alpha_1, \alpha_2, \ldots, \alpha_P]^T$$

Problem (7.35) becomes:

Find the real numbers $\bar{\Pi}$ *for which there exists* $\boldsymbol{\alpha}$ *nonzero such that*

$$\overline{\mathbb{M}}\boldsymbol{\alpha} = \bar{\Pi}\,\overline{\mathbb{K}}\boldsymbol{\alpha} \tag{7.36}$$

where matrices $\overline{\mathbb{M}}$ *and* $\overline{\mathbb{K}}$ *are defined by*

$$\overline{\mathbb{M}}_{nm} = (\mathbf{D}(\mathbb{s}_n), \mathbf{D}(\mathbb{s}_m)) \qquad \text{and} \qquad \overline{\mathbb{K}}_{nm} = (\!(\mathbb{s}_n, \mathbb{s}_m)\!)$$

REMARKS

1. In the context of the finite element method, the fields $\mathbb{s}_n$ are chosen with local support; the fields $\mathbf{D}(\mathbb{s}_n)$ are too.

The finite elements obtained in this manner are part of the general family of "equilibrium" finite elements introduced in [FRAEIJS DE VEUBEKE, 1965].

Examples of such elements for elastic membranes are given in [LADEVEZE - PELLE, 1989a].

One should note that the construction of these elements is based on techniques which are very similar to those used in constructing statically admissible fields in order to estimate discretization errors in linear elasticity.

2. The matrix $\overline{\mathbb{M}}$ often has a large kernel. Nevertheless, it was shown in [PELLE, 1985] that if (7.36) is being solved by the subspace iteration method [BATHE - WILSON, 1976], [IMBERT, 1984] it is very easy to initialize the algorithm by a subspace orthogonal to that kernel, thus excluding the kernel modes.

3. Problem (7.35) can be viewed as a discretized form of RAYLEIGH's dual quotient, also called TOUPIN's quotient.

7.2.6 Evaluation of the constant k_S

Taking into account (7.25), in order to obtain lower bounds of the exact pulsations one must determine a lower bound of k_S. Let us recall that

$$k_S = \operatorname*{Inf}_{\underline{\Gamma} \in \mathcal{A}_h^{(\perp)}} \tilde{R}(\underline{\Gamma})$$

In order to obtain the desired lower bound, one must first establish some preliminary results.

Let us consider a mesh of the structure consisting of a collection of elements E.

One has the following property:

Property

Any load $\underline{\Gamma} \in \mathcal{A}_h^{(\perp)}$ *can be decomposed into* $\underline{\Gamma} = \underline{\Gamma}_0 + \underline{\Gamma}_1$, *with* ($G_E$ *denoting the center of inertia of an element* E),

$$\forall E, \quad \underline{\Gamma}_{0\,|E} = \underline{a}_E + \underline{\omega}_E \wedge \underline{X} \quad (\underline{a}_E \text{ and } \underline{\omega}_E \text{ constant on } E,\ \underline{X} = \underline{G_E M})$$

$$(\underline{\Gamma}_1, \underline{\Gamma}_0^*) = 0, \quad \forall \underline{\Gamma}_0^* \text{ such that for any } E : \underline{\Gamma}_{0\,|E}^* = \underline{a}_E^* + \underline{\omega}_E^* \wedge \underline{X}$$

Proof

It suffices to construct $\underline{\Gamma}_0$ element by element, letting

$$\underline{a}_E = \frac{\int_E \rho \, \underline{\Gamma} \, dE}{\int_E \rho \, dE} \quad \text{and} \quad \underline{\omega}_E = \boldsymbol{\mathcal{I}}_E^{-1}\left[\int_E \rho \, \underline{X} \wedge \underline{\Gamma} \, dE\right]$$

where $\boldsymbol{\mathcal{I}}_E$ is the inertia operator of element E at G_E.

This is equivalent to decomposing $\boldsymbol{\mathcal{A}}_h^{(\perp)}$ into two orthogonal subspaces $\boldsymbol{\mathcal{E}}_0$ and $\boldsymbol{\mathcal{E}}_1$ such that:

- $\boldsymbol{\mathcal{E}}_0$ is of finite dimension;
- $\forall \underline{\Gamma}_1 \in \boldsymbol{\mathcal{E}}_1$, the load $\underline{\Gamma}_{1|E}$ is in equilibrium on E,

$$\int_E \underline{\Gamma}_{1|E} \, \rho \, dE = 0 \quad \text{and} \quad \int_E \underline{X} \wedge \underline{\Gamma}_{1|E} \, \rho \, dE = 0 \tag{7.37}$$

We will also use the following property:

Property

For $\underline{\Gamma} = \underline{\Gamma}_0 + \underline{\Gamma}_1$, *one has*

$$\frac{1}{\tilde{R}(\underline{\Gamma})} \leq \frac{1}{\tilde{R}(\underline{\Gamma}_0)} + \frac{1}{\tilde{R}(\underline{\Gamma}_1)} \tag{7.38}$$

Proof

Let $m_0 = (\underline{\Gamma}_0, \underline{\Gamma}_0)$ and $m_1 = (\underline{\Gamma}_1, \underline{\Gamma}_1)$.

One has

$$\frac{1}{\tilde{R}(\underline{\Gamma})} = \frac{\langle \mathbf{G}(\underline{\Gamma}_0), \mathbf{G}(\underline{\Gamma}_0) \rangle + \langle \mathbf{G}(\underline{\Gamma}_1), \mathbf{G}(\underline{\Gamma}_1) \rangle + 2 \langle \mathbf{G}(\underline{\Gamma}_0), \mathbf{G}(\underline{\Gamma}_1) \rangle}{m_0 + m_1}$$

$$\frac{1}{\tilde{R}(\underline{\Gamma}_0)} + \frac{1}{\tilde{R}(\underline{\Gamma}_1)} = \frac{m_1 \langle \mathbf{G}(\underline{\Gamma}_0), \mathbf{G}(\underline{\Gamma}_0) \rangle + m_0 \langle \mathbf{G}(\underline{\Gamma}_1), \mathbf{G}(\underline{\Gamma}_1) \rangle}{m_0 \, m_1}$$

then

$$\frac{1}{\tilde{R}(\underline{\Gamma}_0)} + \frac{1}{\tilde{R}(\underline{\Gamma}_1)} - \frac{1}{\tilde{R}(\underline{\Gamma})}$$

$$= \frac{m_1^2 \langle \mathbf{G}(\underline{\Gamma}_0), \mathbf{G}(\underline{\Gamma}_0)\rangle + m_0^2 \langle \mathbf{G}(\underline{\Gamma}_1), \mathbf{G}(\underline{\Gamma}_1)\rangle - 2m_0 m_1 \langle \mathbf{G}(\underline{\Gamma}_0), \mathbf{G}(\underline{\Gamma}_1)\rangle}{m_0\, m_1(m_0 + m_1)}$$

$$= \frac{< m_1\mathbf{G}(\underline{\Gamma}_0) - m_0\mathbf{G}(\underline{\Gamma}_1), m_1\mathbf{G}(\underline{\Gamma}_0) - m_0\mathbf{G}(\underline{\Gamma}_1) >}{m_0\, m_1(m_0 + m_1)} \geq 0$$

The lower bound of k_S is given by the following property:

Property

$$k_S \geq \frac{1}{\operatorname*{Sup}_E \beta_E^2 + \alpha_S^2} \tag{7.39}$$

with:

- $1/\beta_E$, *first nonzero eigenpulsation of the free element* E, *defined by*

$$\frac{1}{\beta_E^2} = \operatorname*{Inf}_{\underline{U}_E \in \boldsymbol{\mathcal{U}}_E} \frac{\int_E \mathrm{Tr}[\varepsilon(\underline{U}_E)\mathbf{K}\varepsilon(\underline{U}_E)]\, dE}{\int_E \rho \underline{U}_E \cdot \underline{U}_E\, dE} \tag{7.40}$$

where $\boldsymbol{\mathcal{U}}_E$ *is the set of the displacement fields defined on* E *which verify*

$$\int_E \underline{U}_E\, dE = 0 \quad \text{and} \quad \int_E \underline{X} \wedge \underline{U}_E\, dE = 0 \tag{7.41}$$

- $1/\alpha_S$ *defined by*

$$\frac{1}{\alpha_S^2} = \operatorname*{Inf}_{\underline{\Gamma}_0 \in \boldsymbol{\mathcal{E}}_0} \frac{(\underline{\Gamma}_0, \underline{\Gamma}_0)}{(\!(\hat{\sigma}(\underline{\Gamma}_0), \hat{\sigma}(\underline{\Gamma}_0))\!)}$$

where $\hat{\sigma}$ *is an operator similar to that constructed in* (7.32).

Proof

Let $\underline{\Gamma} = \underline{\Gamma}_0 + \underline{\Gamma}_1 \in \boldsymbol{\mathcal{A}}_h^{(\perp)}$. From Property (7.38), one deduces

$$\frac{1}{\tilde{R}(\underline{\Gamma})} \leq \frac{\langle \mathbf{G}(\underline{\Gamma}_0), \mathbf{G}(\underline{\Gamma}_0)\rangle}{(\underline{\Gamma}_0, \underline{\Gamma}_0)} + \frac{\langle \mathbf{G}(\underline{\Gamma}_1), \mathbf{G}(\underline{\Gamma}_1)\rangle}{(\underline{\Gamma}_1, \underline{\Gamma}_1)} \tag{7.42}$$

Let $\hat{\sigma}$ be an operator which associates with any field $\underline{\Gamma}_0 \in \boldsymbol{\mathcal{E}}_0$ a stress field $\hat{\sigma}(\underline{\Gamma}_0)$ in equilibrium with the load $\underline{\Gamma}_0$. One obtains

$$\langle \mathbf{G}(\underline{\Gamma}_0), \mathbf{G}(\underline{\Gamma}_0) \rangle = (\!(\mathbf{K}\varepsilon[\mathbf{G}(\underline{\Gamma}_0)], \mathbf{K}\varepsilon[\mathbf{G}(\underline{\Gamma}_0)])\!) \leq (\!(\hat{\sigma}(\underline{\Gamma}_0), \hat{\sigma}(\underline{\Gamma}_0))\!)$$

and therefore

$$\frac{\langle \mathbf{G}(\underline{\Gamma}_0), \mathbf{G}(\underline{\Gamma}_0) \rangle}{(\underline{\Gamma}_0, \underline{\Gamma}_0)} \leq \frac{(\!(\hat{\sigma}(\underline{\Gamma}_0), \hat{\sigma}(\underline{\Gamma}_0))\!)}{(\underline{\Gamma}_0, \underline{\Gamma}_0)} \leq \alpha_S^2 \tag{7.43}$$

Denoting $\underline{\Gamma}_{1|E}$ the restriction of $\underline{\Gamma}_1 \in \boldsymbol{\mathcal{E}}_1$ to E, let us consider the problem:

Find $\underline{U}_E \in \boldsymbol{\mathcal{U}}_E$ *such that*

$$\forall \underline{U}^*, \quad \int_E \mathrm{Tr}\left[\varepsilon(\underline{U}_E)\mathbf{K}\varepsilon(\underline{U}^*)\right] dE = \int_E \rho\, \underline{\Gamma}_{1|E} \cdot \underline{U}^* dE \tag{7.44}$$

Due to the equilibrium conditions verified by $\underline{\Gamma}_{1|E}$, this problem has at least one solution; this solution is unique because of the conditions (7.41).

Then, the stress field σ defined by

$$\sigma_{|E} = \mathbf{K}\varepsilon(\underline{U}_E)$$

is statically admissible for the load $\underline{\Gamma}_1$: $\sigma \in \boldsymbol{\mathcal{S}}(\underline{\Gamma}_1)$.

Thus, one also has

$$\langle \mathbf{G}(\underline{\Gamma}_1), \mathbf{G}(\underline{\Gamma}_1) \rangle \leq (\!(\sigma, \sigma)\!)$$

as well as

$$\langle \mathbf{G}(\underline{\Gamma}_1), \mathbf{G}(\underline{\Gamma}_1) \rangle \leq \sum_E \int_E \mathrm{Tr}\left[\sigma_{|E}\mathbf{K}^{-1}\sigma_{|E}\right] dE$$

and, finally,

$$\langle \mathbf{G}(\underline{\Gamma}_1), \mathbf{G}(\underline{\Gamma}_1) \rangle = \sum_E \int_E \mathrm{Tr}\left[\varepsilon(\underline{U}_E)\mathbf{K}\varepsilon(\underline{U}_E)\right] dE \tag{7.45}$$

Using (7.44) for $\underline{U}^* = \underline{U}_E$, one has

$$\int_E \mathrm{Tr}\left[\varepsilon(\underline{U}_E)\mathbf{K}\varepsilon(\underline{U}_E)\right] dE = \int_E \rho\, \underline{\Gamma}_1 \cdot \underline{U}_E \, dE \tag{7.46}$$

Applying the CAUCHY - SCHWARZ inequality, one obtains

$$\left[\int_E \rho\, \underline{\Gamma}_1 \cdot \underline{U}_E \, dE\right]^2 \leq \left[\int_E \rho\, \underline{\Gamma}_1 \cdot \underline{\Gamma}_1 \, dE\right] . \left[\int_E \rho\, \underline{U}_E \cdot \underline{U}_E \, dE\right]$$

$$\leq \left[\int_E \rho\, \underline{\Gamma}_1 \cdot \underline{\Gamma}_1 \, dE\right] . \beta_E^2 \left[\int_E \mathrm{Tr}\left[\varepsilon(\underline{U}_E)\mathbf{K}\varepsilon(\underline{U}_E)\right] dE\right]$$

hence

$$\int_E \mathrm{Tr}\left[\varepsilon(\underline{U}_E)\mathbf{K}\varepsilon(\underline{U}_E)\right] dE \leq \beta_E^2 \int_E \rho\, \underline{\Gamma}_1 \cdot \underline{\Gamma}_1 \, dE$$

which leads successively to

$$\left\langle \mathbf{G}(\underline{\Gamma}_1), \mathbf{G}(\underline{\Gamma}_1)\right\rangle \leq \sum_E \int_E \mathrm{Tr}\left[\sigma_{|E}\mathbf{K}^{-1}\sigma_{|E}\right] dE$$

$$\left\langle \mathbf{G}(\underline{\Gamma}_1), \mathbf{G}(\underline{\Gamma}_1)\right\rangle \leq \sum_E \beta_E^2 \int_E \rho\, \underline{\Gamma}_1 \cdot \underline{\Gamma}_1 \, dE$$

$$\left\langle \mathbf{G}(\underline{\Gamma}_1), \mathbf{G}(\underline{\Gamma}_1)\right\rangle \leq \mathrm{Sup}_E\, \beta_E^2 \,(\underline{\Gamma}_1, \underline{\Gamma}_1) \tag{7.47}$$

Combining the properties (7.42), (7.43) and (7.47), one obtains

$$\frac{1}{\tilde{R}(\underline{\Gamma})} \leq \mathrm{Sup}_E\, \beta_E^2 + \alpha_S^2$$

hence the announced result.

REMARKS

1. In many common situations, it is possible to construct the discretization subspaces $\boldsymbol{\mathcal{A}}_h$ in such a way that

$$\forall \underline{\Gamma} \in \boldsymbol{\mathcal{A}}_h^{(\perp)}, \quad \int_E \underline{\Gamma}\, \rho \, dE = 0 \quad \text{and} \quad \int_E \underline{X} \wedge \underline{\Gamma}\, \rho\, dE = 0$$

Under these conditions, one has $\boldsymbol{\mathcal{E}}_0 = \{0\}$ and therefore

$$k_S \geq \frac{1}{\mathrm{Sup}_E\, \beta_E^2} \tag{7.48}$$

2. In order to evaluate β_E, one can use classical upper bounds from the literature which can be found in [PAYNE, 1967], [LADEVEZE J - LADEVEZE P, 1978, 1979] and in the references cited in these articles.

3. The calculation of α_S requires the resolution of a problem similar to (7.36); here, however, only the smallest nonzero eigenvalue needs to be calculated.

4. The works by [MADAY - PATERA - PERAIRE, 1999] on the bounding of eigenvalues are related to the previous approach and its kinematic version. The difference is that in these works the authors take as the reference a finite element model constructed using a very fine mesh, as opposed to the "continuous medium" reference considered here. Consequently, the numerical evaluation of the constants corresponding to our characteristic constants k_C and k_S is carried out by solving local problems on the fine mesh.

7.2.7 Practical method for constructing bounds

The procedure which is to be followed in practice, to determine bounds of the first p eigenfrequencies, may be summarized as follows:

• Choose a subspace $\boldsymbol{S}_P$ of $\boldsymbol{S}$ of finite dimension P; then $\boldsymbol{A}_N = \mathbf{D}(\boldsymbol{S}_P)$. For $\sigma \in \boldsymbol{S}_P$, let $\sigma = \mathbb{S}\alpha$, with $\mathbb{S} = (s_1, s_2, \dots, s_P)$ a basis of $\boldsymbol{S}_P$.

• Construct the matrices $\overline{\mathbb{M}}$ and $\overline{\mathbb{K}}$ and calculate the p smallest eigenvalues of the problem

$$\overline{\mathbb{M}}\alpha = \bar{\Pi}\,\overline{\mathbb{K}}\alpha$$

Thus, one obtains $\bar{\Pi}_1, \bar{\Pi}_2, \dots, \bar{\Pi}_p$, the associated modes $\alpha_1, \alpha_2, \dots, \alpha_p$, and an approximation of the first eigenfrequencies

$$\bar{f}_n = \frac{1}{2\pi}\sqrt{\bar{\Pi}_n} \tag{7.49}$$

• Choose a subspace $\boldsymbol{U}_r$ of $\boldsymbol{U}_0$ of finite dimension r and solve the $2 \times p$ linear problems ($n = 1, 2, \dots, p$):

1. *Find the projections* $\underline{P}_n = \mathbf{P}(\underline{\Gamma}_n)$ *such that*

$$\forall \underline{U}^* \in \boldsymbol{U}_r, \quad (\underline{P}_n, \underline{U}^*) = (\underline{\Gamma}_n, \underline{U}^*)$$

2. *Find the fields* $\underline{D}_n = \mathbf{d}(\underline{P}_n)$ *such that*

$$\forall \underline{U}^* \in \boldsymbol{\mathcal{U}}_r, \quad \langle \underline{D}_n , \underline{U}^* \rangle = (\underline{P}_n , \underline{U}^*)$$

where $\underline{\Gamma}_n = \mathbf{D}(\mathbb{S}\alpha_n)$.

- Construct the $p \times p$ matrices

$$\mathbb{A}_{nm} = (\underline{P}_n , \underline{P}_m) \quad \text{and} \quad \mathbb{B}_{nm} = \langle \underline{D}_n , \underline{D}_m \rangle$$

and solve the small eigenvalue problem

$$\mathbb{A} X = \hat{\Lambda} \mathbb{B} X$$

whose solutions yield the upper bounds of the exact eigenfrequencies

$$M_n = \hat{\Lambda}_n^{1|2} / 2\pi$$

- Determine a lower bound $k_S \geq 1/B$ of the constant k_S and calculate the lower bounds of the exact eigenfrequencies

$$m_n = \frac{1}{2\pi} \left[\frac{\bar{\Pi}_n}{1 + \bar{\Pi}_n B} \right]^{1|2}$$

which yield an approximate value of the exact frequency

$$F_n = \frac{m_n + M_n}{2} \tag{7.50}$$

and an upper bound of the error made

$$e_n = \frac{M_n - m_n}{M_n + m_n} \tag{7.51}$$

7.2.8 Example

The implementation of the bounding method was carried out by PELLE for the case of the transverse vibrations of an elastic membrane fixed along its boundary. The results presented here are taken from [LADEVEZE - PELLE, 1989a].

The exact frequencies of this problem are known and consist of the real numbers of the form $\pi^2(p^2 + q^2)$, where p and q are strictly positive integers.

Initially, the calculation of the error bounds was performed on three meshes; the coarsest mesh (192 elements) is shown in Figure 7.1.

The two other meshes, constructed following the same principle, consist of 400 and 576 elements, respectively. Figure 7.2 shows the error bounds (7.51) associated with these three meshes, calculated for the first 24 eigenfrequencies.

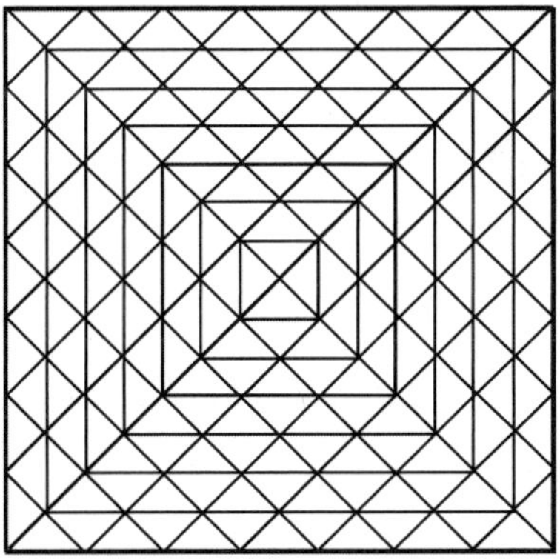

Figure 7.1. *Regular mesh with 192 elements.*

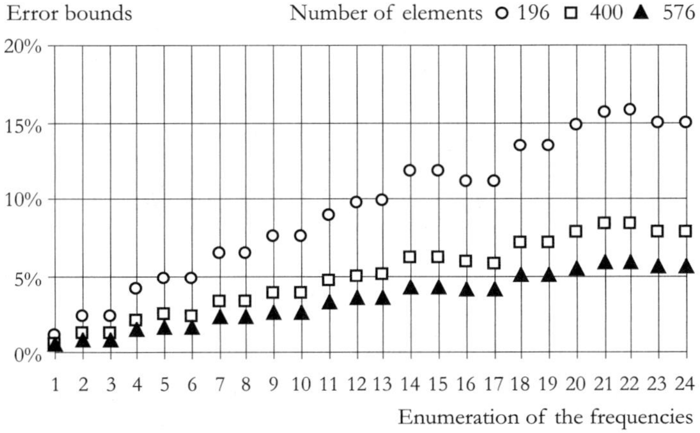

Figure 7.2. *Evolution of the error bounds with the mesh's fineness.*

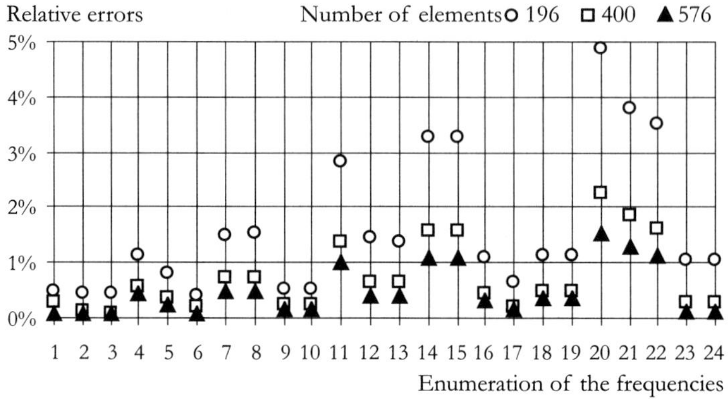

Figure 7.3. *Quality of the* $\bar{f}_n$ *approximations.*

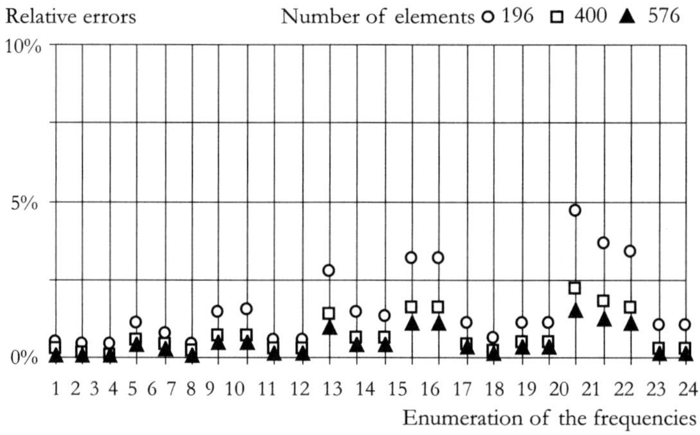

Figure 7.4. *Quality of the* $\bar{\Pi}_n$ *approximations.*

REMARKS

1. One should note that as far as the resolution of the approximate problem (7.35) is concerned the method proposed above yields excellent approximations $\bar{f}_n$, as shown by Figure 7.3 which gives the relative errors with respect to the exact eigenfrequencies for the three meshes. These approximations are much better than those obtained by a standard kinematic method.

2. Besides bounds, the resolution of Problem (7.36) also provides very good approximations of the exact eigenfrequencies, as shown by Figure 7.4 which gives the relative errors between the $\overline{\Pi}_n$ and the exact eigenfrequencies

$$|\lambda_n - \overline{\Pi}_n| / \lambda_n$$

7.2.9 Adaptive vibration calculations

Using the calculation of the error bounds presented above, PELLE [PELLE, 1985] developed an adaptive method for the calculation of the first p eigenfrequencies with a predetermined accuracy.

Optimum mesh

Let p be the number of frequencies that one wishes to calculate. As the accuracy criterion, let us introduce the average error

$$\bar{e} = \frac{1}{p} \sum_{n=1}^{p} e_n$$

Definition

A mesh $\mathcal{T}^*$ *is said to be optimum in relation to a given accuracy* e_0 *if:*

$\bar{e}^* = e_0$ *(desired accuracy)*

N^* *(number of elements of* $\mathcal{T}^*$*) is minimum*

Error indicators

By itself, a knowledge of the average accuracy, which is by nature a global quantity, is not sufficient to construct an optimum mesh. To obtain local information, we introduce error indicators, based on the error in constitutive relation, whose local contributions are easy to obtain.

Let us use the notations of Section 7.2.7 again. By construction, the field $\hat{\sigma}_n = \mathbb{S}\alpha_n$ is statically admissible for the load $\underline{\Gamma}_n$, and the field $\underline{D}_n$ is kinematically admissible. These fields $\underline{D}_n$ and $\hat{\sigma}_n$ constitute displacement and stress approximations of the exact modes corresponding to the eigenfrequency f_n.

Then, one defines the following indicators:

- for one mode

$$\varepsilon_n = \frac{\left\| \hat{\sigma}_n - \mathbf{K}\varepsilon\,(\underline{D}_n) \right\|_{\sigma,\Omega}}{\left\| \hat{\sigma}_n + \mathbf{K}\varepsilon\,(\underline{D}_n) \right\|_{\sigma,\Omega}}$$

- for the first p modes

$$\varepsilon = \left[\frac{1}{p} \sum_{n=1}^{p} \varepsilon_n^2 \right]^{1/2}$$

Similarly, the local contributions on each element E of the mesh are defined by:

- for one mode

$$\varepsilon_{n,E} = \frac{\left\| \hat{\sigma}_n - \mathbf{K}\varepsilon\,(\underline{D}_n) \right\|_{\sigma,E}}{\left\| \hat{\sigma}_n + \mathbf{K}\varepsilon\,(\underline{D}_n) \right\|_{\sigma,\Omega}}$$

- for the first p modes

$$\varepsilon_E = \left[\frac{1}{p} \sum_{n=1}^{p} \varepsilon_{n,E}^2 \right]^{1/2}$$

Thus, one has

$$\varepsilon_n^2 = \sum_E \varepsilon_{n,E}^2 \quad \text{and} \quad \varepsilon^2 = \sum_E \varepsilon_E^2$$

REMARKS

1. If the subspaces $\mathcal{U}_r$ and $\mathcal{A}_N$ have been constructed such that $\mathcal{U}_r \subset \mathcal{A}_N$, one can show [LADEVEZE - PELLE, 1989a] that if ε is zero then

$$\forall n = 1, 2, \ldots, p, \quad \underline{D}_n \text{ is an exact mode and } \hat{\sigma}_n = \mathbf{K}\varepsilon(\underline{D}_n)$$

Therefore, the p calculated eigenfrequencies are exact.

2. Let us consider the solutions of the discretized kinematic formulation (7.6) or (7.10):

- approximate eigenpulsations Ω_n;
- associated eigenmodes $\underline{U}_n = \mathbf{N}\mathbf{q}_n$.

Let $\underline{\Gamma}_n = \Omega_n \rho \underline{U}_n$. Then, if $\hat{\sigma}_n$ is a stress field in equilibrium with the volume load $\underline{\Gamma}_n$, one can build a very simple indicator by letting

$$\varepsilon_n = \frac{\| \hat{\sigma}_n - \mathbf{K}\varepsilon(\underline{U}_n) \|_{\sigma,\Omega}}{\| \hat{\sigma}_n + \mathbf{K}\varepsilon(\underline{U}_n) \|_{\sigma,\Omega}} \tag{7.52}$$

and

$$\varepsilon = \left[\frac{1}{p} \sum_{n=1}^{p} \varepsilon_n^2 \right]^{1/2} \tag{7.53}$$

If this indicator is zero, then for any $n = 1, 2, \ldots, p$, Ω_n is an exact eigenvalue and $\underline{U}_n$ is an exact eigenmode. Of course, in order to construct $\hat{\sigma}_n$, one can use the construction of statically admissible fields developed in statics. This type of indicator was recently applied to the case of beams and plates [COFFIGNAL, 1993], [BOISSE - COFFIGNAL, 1998].

Structure of the error bounds

To make things clear, let us consider the case where the lower bound of k_S is given by (7.48),

$$B = \operatorname*{Sup}_{E} \beta_E^2$$

Let $\hat{\Lambda}_n = \overline{\Pi}_n (1 + \theta_n)^2$. Under the assumption, which is valid if the mesh is sufficiently fine, that θ_n and $B\,\overline{\Pi}_n$ are small compared to 1, one has

$$e_n \approx \frac{1}{2}\theta_n + \frac{1}{4} B \overline{\Pi}_n$$

$$\bar{e} \approx \delta + B\xi \tag{7.54}$$

with

$$\delta = \frac{1}{2p} \sum_{n=1}^{p} \theta_n \quad \text{and} \quad \xi = \frac{1}{4p} \sum_{n=1}^{p} \bar{\Pi}_n$$

The benefit of (7.54) is that it shows that the error is composed of two terms of different natures.

The first term δ involves only quantities which are global on the structure, while the second term $B\,\xi$ depends (through the constant B) on the maximum size of the elements used in the mesh. If, for example, starting from a given mesh, one were to increase the sizes of some of the elements, the order of magnitude of δ would remain virtually unchanged, whereas B would increase significantly. Similarly, if one were to refine the mesh for all but the largest elements, one could hope to reduce the value of δ, but B would remain unchanged.

This leads us to connect the evolution of the error indicator ε, which is a global quantity, with the evolution of the term δ alone, and not with that of $\bar{e}$ directly.

Determination of an optimum mesh

Starting from an initial calculation with a mesh $\mathcal{T}$, one calculates the quantities $\bar{e}$, B, δ, ξ, ε and the contribution ε_E of each element E.

The problem consists in determining an optimum mesh $\mathcal{T}^*$. One assumes that

$$\varepsilon = O(h^q), \quad B = O(h^\beta) \quad \text{and} \quad \delta = O(h^\gamma)$$

Let d be the space dimension ($d = 1, 2$ or 3) and, as in Chapter 5, let

$$r_E = h_E^* / h_E$$

where h_E is the current size of element E and h_E^* is the size being sought in order to achieve optimality.

By the same reasoning as in Chapter 5, the value of the indicator (7.53) on the optimum mesh can be obtained by

$$(\varepsilon^*)^2 = \sum_E r_E^{2q}\, \varepsilon_E^2$$

Here, however, the value of ε^* is a priori unknown. In order to estimate this quantity, it suffices to couple the evolutions of $\boldsymbol{\varepsilon}$ and e by letting

$$\varepsilon^* = \varepsilon \left[\frac{\delta^*}{\delta}\right]^{q/\gamma}$$

Besides, one has $e_0 \approx \delta^* + B^* \xi^*$ and

$$B^* = B\left[\frac{h^*}{h}\right]^{\beta} \quad \text{with} \quad h = \operatorname*{Sup}_{E \in \boldsymbol{\tau}} h_E \quad \text{and} \quad h^* = \operatorname*{Sup}_{E^* \in \boldsymbol{\tau}^*} h_E^*$$

Taking on first approximation $\xi^* \approx \xi$, one gets

$$(\varepsilon^*)^2 = \frac{\varepsilon^2}{\delta^{2q/\gamma}}\left[e_0 - \xi B\left[\frac{h^*}{h}\right]^{\beta}\right]^{2q/\gamma}$$

Thus, the objective is to determine the r_E which are solutions of the problem

$$\text{Minimize} \qquad N^* = \sum_E \frac{1}{r_E^d} \tag{7.55}$$

with the constraints

$$\sum_E r_E^{2q}\, \varepsilon_E^2 = \frac{\varepsilon^2}{\delta^{2q/\gamma}}\left[e_0 - \xi B\left[\frac{h^*}{h}\right]^{\beta}\right]^{2q/\gamma}$$

$$\text{and} \quad \forall E, \quad r_E h_E \le h^*$$

Clearly, the resolution of (7.55) is more complex than that of Problem (5.9) for the static case.

Below, in order to simplify the expressions and the calculations, we will assume that $\beta = \gamma$.

Property

If the elements of the mesh are ordered in such a way that

$$\frac{\varepsilon_{E_1}^{2/(2q+d)}}{h_{E_1}} \le \frac{\varepsilon_{E_2}^{2/(2q+d)}}{h_{E_2}} \le \ldots \le \frac{\varepsilon_{E_N}^{2/(2q+d)}}{h_{E_N}}$$

and if m^* is the largest integer which minimizes the function

$$m \mapsto \frac{P(m)^d}{e_0^{d/\beta}} \left[\sum_{k=1}^{m} h_{E_k}^2 + \frac{h_{E_m}^d}{\varepsilon_{E_m}^{2d/(2q+d)}} \sum_{i=m+1}^{N} \varepsilon_{E_i}^{2d/(2q+d)} \right]$$

in which

$$P(m) = \left[\frac{\delta}{\varepsilon^{\beta/q}} \left[\psi(m) \right]^{\beta/2q} + \frac{\xi B}{h^\beta} \right]^{1/\beta}$$

$$\textit{and } \psi(m) = \sum_{n=1}^{j} \frac{\varepsilon_{E_n}^2}{h_{E_n}^{2q}} + \frac{\varepsilon_{E_m}^{4q/(2q+d)}}{h_{E_m}^{2q}} \sum_{n=j+1}^{N} \varepsilon_{E_n}^{2d/(2q+d)}$$

then

$$\textit{for } 0 \le k \le m^*, \quad r_{E_k} = \frac{e_0^{1/\beta}}{P(m^*)} \frac{1}{h_{E_k}}$$

$$\textit{for } m^* < i \le N, \quad r_{E_i} = \frac{e_0^{1/\beta}}{P(m^*)} \frac{1}{h_{E_{m^*}}} \left[\frac{\varepsilon_{E_{m^*}}}{\varepsilon_{E_i}} \right]^{2/(2q+d)}$$

Proof

Initially, let us assume that the maximum size h^* of the optimum mesh is known and introduce the Lagrangian

$$J = \sum_E \frac{1}{r_E^d} + A \left[\sum_E r_E^{2q} \varepsilon_E^2 - G(h^*) \right] + \sum_E b_E \left[r_E h_E - h^* \right]$$

where

$$G(h^*) = \frac{\varepsilon^2}{\delta^{2q/\beta}} \left[e_0 - \xi B \left[\frac{h^*}{h} \right]^\beta \right]^{2q/\beta} \tag{7.56}$$

Using the technique of [KUHN - TUCKER, 1950], [EKELAND - TEMAM, 1974], one obtains the three conditions

$$-\frac{d}{r_E^{d+1}} + 2qA\, r_E^{2q-1} \varepsilon_E^2 + b_E h_E = 0$$

$$b_E \geq 0,$$
$$\sum_E b_E \left[r_E h_E - h^* \right] = 0$$

The set of the elements of the mesh $\mathcal{T}$ can be divided into two parts

$$\mathcal{T} = \mathcal{T}_- \cup \mathcal{T}_+$$

with:

- if $E_- \in \mathcal{T}_-$, $r_{E_-} h_{E_-} = h^*$ and $b_{E_-} \geq 0$;
- if $E_+ \in \mathcal{T}_+$, $r_{E_+} h_{E_+} < h^*$ and $b_{E_+} = 0$.

For $E_+ \in \mathcal{T}_+$, one has $b_{E_+} = 0$, which yields

$$r_{E_+} = \left[\frac{d}{2q} \frac{1}{A\, \varepsilon_{E_+}^2} \right]^{1/(2q+d)}$$

hence

$$\left[\frac{d}{2q\, A} \right]^{2q/(2q+d)} = \frac{1}{\sum\limits_{E \in \mathcal{T}_+} \varepsilon_E^{2q/(2q+d)}} \left[G(h^*) - \sum_{E \in \mathcal{T}_-} r_E^{2q}\, \varepsilon_E^2 \right]$$

Let $E_- \in \mathcal{T}_-$; one has $b_{E_-} \geq 0$ and therefore

$$\left[\frac{d}{2q\, A} \right]^{2q/(2q+d)} \geq r_{E_-}^{2q}\ \varepsilon_{E_-}^{4q/(2q+d)}$$

This yields

$$\forall E_- \in \mathcal{T}_-, \quad \frac{G(h^*)}{(h^*)^{2q}} - \sum_{E \in \mathcal{T}_-} \frac{\varepsilon_E^2}{h_E^{2q}} \geq \frac{\varepsilon_{E_-}^{4q/(2q+d)}}{h_{E_-}^{2q}} \sum_{E \in \mathcal{T}_+} \varepsilon_E^{2d/(2q+d)} \tag{7.57}$$

Similarly, for $E_+ \in \mathcal{T}_+$, one has $r_{E_+} h_{E_+} < h^*$, which yields

$$\left[\frac{d}{2q} \frac{1}{A\, \varepsilon_{E_+}^2} \right]^{2q/(2q+d)} = r_{E_+}^{2q} < \left[\frac{h^*}{h_{E_+}} \right]^{2q}$$

then

$$\forall E_+ \in \mathcal{T}_+, \quad \frac{G(h^*)}{(h^*)^{2q}} - \sum_{E \in \mathcal{T}_-} \frac{\varepsilon_E^2}{h_E^{2q}} \leq \frac{\varepsilon_{E_+}^{4q/(2q+d)}}{h_{E_+}^{2q}} \sum_{E \in \mathcal{T}_+} \varepsilon_E^{2d/(2q+d)} \tag{7.58}$$

Let us number the elements of the mesh from 1 to N in such a way that

$$\frac{\varepsilon_1^{2/(2q+d)}}{h_1} \leq \frac{\varepsilon_2^{2/(2q+d)}}{h_2} \leq \ldots \leq \frac{\varepsilon_N^{2/(2q+d)}}{h_N} \tag{7.59}$$

in which, to keep it simple, we wrote j for E_j.

If an element numbered k is in $\mathcal{T}_-$ and an element numbered i in $\mathcal{T}_+$, one has

$$\frac{h_i}{\varepsilon_i^{2/(2q+d)}} \left[\frac{d}{2\,qA}\right]^{1/(2q+d)} = h_i^* < h^* = h_k^* \leq \frac{h_k}{\varepsilon_k^{2/(2q+d)}} \left[\frac{d}{2\,qA}\right]^{1/(2q+d)} \tag{7.60}$$

Taking (7.59) into account, this yields $k < i$.

Therefore, these exists an integer m $(1 \leq m \leq N)$ such that

$$\mathcal{T}_- = \{E_1, E_2, \ldots, E_m\} \quad \text{and} \quad \mathcal{T}_+ = \{E_{m+1}, E_{m+2}, \ldots, E_N\}$$

Let

$$a(j) = \sum_{n=1}^{j} \frac{\varepsilon_n^2}{h_n^{2q}}; \quad b(j) = \sum_{n=j+1}^{N} \varepsilon_n^{2d/(2q+d)} \quad \text{and} \quad c(j) = \frac{\varepsilon_j^{4q/(2q+d)}}{h_j^{2q}}$$

The inequalities (7.59), (7.60) and (7.59) yield

$$a(m) + c(m)\,b(m) \leq \frac{G(h^*)}{(h^*)^{2q}} < a(m) + c(m+1)\,b(m)$$

Finally, noting that

$$a(m+1) + c(m+1)\,b(m+1) = a(m) + c(m+1)\,b(m)$$

one gets

$$\psi(m) \leq \frac{G(h^*)}{(h^*)^{2q}} < \psi(m+1)$$

with $\psi(m) = a(m) + c(m)\,b(m)$.

Since function ψ is increasing (i.e. $n < j \Rightarrow \psi(n) \leq \psi(j)$), for a fixed h^*, there exists only one integer m such that

$$\psi(m) \leq \frac{G(h^*)}{(h^*)^{2q}} < \psi(m+1)$$

Since function ψ is defined only for integer values, one chooses

$$\psi(m) = \frac{G(h^*)}{(h^*)^{2q}}$$

which corresponds to

$$h^* = \frac{e_0^{1/\beta}}{\left[\dfrac{\delta}{\varepsilon^{\beta/q}}\,[\psi(m)]^{\beta/2q} + \dfrac{\xi B}{h^{\beta}}\right]^{1/\beta}} \tag{7.61}$$

Therefore, again for a fixed h^*, one has:

- for $E_k \in \mathcal{T}_-$ $(1 \leq k \leq m)$

$$r_k = \frac{h^*}{h_k} \tag{7.62}$$

- for $E_i \in \mathcal{T}_+$ $(m+1 \leq i \leq N)$

$$r_i = \frac{h^*}{h_m}\left[\frac{\varepsilon_m}{\varepsilon_i}\right]^{2/(2q+d)} \tag{7.63}$$

Thus, the number of elements in the mesh is

$$N^*(m) = \frac{1}{e_0^{d/\beta}}\left[\frac{\delta}{\varepsilon^{\beta/q}}\,[\psi(m)]^{\beta/2q} + \frac{\xi B}{h^{\beta}}\right]^{d/\beta} \times \left[\sum_{k=1}^{m} h_k^2 + \frac{h_m^d}{\varepsilon_m^{2d/(2q+d)}} \sum_{i=m+1}^{N} \varepsilon_i^{2d/(2q+d)}\right] \tag{7.64}$$

which is a function of the unknown m alone.

The optimum mesh is determined in two steps:

- One seeks the largest integer m^* which minimizes $N^*(m)$.
- Having obtained m^*, one calculates h^* using (7.61), then determines

the size modification coefficients using (7.62) and (7.63).

REMARKS AND COMMENTS

1. Examples of the implementation of the previous procedure are given in [LADEVEZE - PELLE, 1989a].

2. One should note that on the optimum mesh two zones can be distinguished

$$\Omega_- = \bigcup_{k=1}^{m^*} E_k \quad \text{and} \quad \Omega_+ = \bigcup_{i=m^*+1}^{N} E^*$$

In Ω_-, the optimum sizes are uniform and equal to h^*.

In Ω_+, the sizes are not uniform. Indeed, an easy calculation shows that for an element E^* located in this zone one has

$$\varepsilon_{E^*}^2 = \left[\frac{h^*}{h_{m^*}}\right]^{(2q+d)} \varepsilon_{m^*}^2$$

In Ω_+, it is the contributions to the error which are uniform.

7.3 Transient dynamics

7.3.1 New formulation of the reference problem

Let us again consider the reference problem [(1.1) - (1.4)].

If one introduces the quantity of acceleration $\rho \ddot{\underline{U}}$ as an unknown in its own right denoted $\underline{\Gamma}$, this problem becomes:

Find $\underline{U}(M,t)$, $\sigma(M,t)$ *and* $\underline{\Gamma}(M, t)$ *defined on* $\Omega \times [0, T]$ *such that:*

- *Kinematic constraint equations and initial conditions*

$$\underline{U} \in \boldsymbol{\mathcal{U}}^{[0,T]},$$

$$\forall t \in [0,T], \quad \underline{U}_{|\partial_1\Omega} = \underline{U}_d \tag{7.65}$$

$$\forall M \in \Omega, \quad \underline{U}_{|t=0} = \underline{U}_0, \quad \dot{\underline{U}}_{|t=0} = \underline{V}_0 \tag{7.66}$$

• *Equilibrium equations*

$$\sigma \in \boldsymbol{S}^{[0,T]} \quad \text{and} \quad \underline{\Gamma} \in \boldsymbol{A}^{[0,T]}$$

$$\forall t \in [0, T], \quad \forall \underline{U}^* \in \boldsymbol{U}_0,$$

$$-\int_\Omega \mathrm{Tr}\left[\sigma \varepsilon(\underline{U}^*)\right] d\Omega + \int_\Omega \underline{f}_d \cdot \underline{U}^* \, d\Omega + \int_{\partial_2\Omega} \underline{F}_d \cdot \underline{U}^* \, dS$$

$$= \int_\Omega \underline{\Gamma} \cdot \underline{U}^* \, d\Omega \qquad (7.67)$$

• *Constitutive relations*

$$\forall t \in [0, T], \quad \forall M \in \Omega, \quad \sigma_{|t} = \mathbf{A}\left(\varepsilon(\dot{\underline{U}})_{|\tau}\,;\, \tau \in [0, t]\right) \qquad (7.68)$$

$$\forall t \in [0, T], \quad \forall M \in \Omega, \quad \underline{\Gamma} = \rho \, \ddot{\underline{U}} \qquad (7.69)$$

In (7.67), $\boldsymbol{A}^{[0,T]}$ denotes the space in which the field $\underline{\Gamma}(M, t)$ is being sought.

The use of $\underline{\Gamma} = \rho \ddot{\underline{U}}$ as the *constitutive relation* enables one to consider the quantity of acceleration as an independent variable in addition to the displacement $\underline{U}$. The connection between these two quantities is reestablished by the constitutive relation (7.69).

The advantage of this approach is that it clearly separates the kinematic quantity $\underline{U}$, which verifies the kinematic constraints and the initial conditions, from the dynamic quantities σ and $\underline{\Gamma}$, which verify the equilibrium.

REMARKS

1. The introduction of $\underline{\Gamma} = \rho \ddot{\underline{U}}$ as the constitutive relation is natural in model updating problems in dynamics [LADEVEZE, 1993], in which the modeling errors affect not only the stiffness, but also the mass and the damping. One can find in [LADEVEZE, 1998], [LADEVEZE - CHOUAKI, 1999] an approach derived from the concept of error in constitutive relation which enables one to evaluate the quality of a model with respect to a reference experiment (static response, free or forced vibrations), thus leading to methods for updating numerical models: this approach, contrary to other updating methods, consists in deriving a true a posteriori error estimate.

2. The initial conditions concern the values of the displacement and of the velocity at the initial time $t = 0$. We will always assume that these initial conditions allow, through the constitutive relation, the determination of the initial value of the stress. (This is obviously the case in elasticity.) In the nonlinear case, the anelastic strain ε^p and the internal variables are always assumed to be zero at $t = 0$. If necessary, the initial value of $\underline{\Gamma}$ can be found using the equilibrium at the initial time

$$\underline{\Gamma}_0 = \underline{\Gamma}_{|t=0} = (\underline{\text{div}}\, \sigma + \underline{f}_d)_{|t=0} \tag{7.70}$$

A common situation is when the displacement field and the stress field are zero at the initial time.

3. For the quantities of acceleration, we use notations similar to those we used for the mass densities of loads in vibration problems. There is no risk of confusion in doing so. Besides, these quantities evolve in spaces which differ only by their physical dimension, and their introduction follows from the same reasoning.

4. One can also replace the constitutive relation (7.69) by a relation of a more general form [LADEVEZE, 1993]

$$\underline{\Gamma} = \rho \ddot{\underline{U}} + \mathbf{a}\left(\dot{\underline{U}}_{|\tau};\, \tau \in [0, t]\right) \tag{7.71}$$

where $\mathbf{a}$ is a given operator.

In this way, it is possible to introduce certain types of damping.

5. For problems of slow dynamics, it is often advantageous to use a truncated modal basis. In [DUTTA - RAMAKRISHNAN, 1997], an error estimator was developed for the control of such calculations.

7.3.2 Extension of DRUCKER's stability condition

We assume that the classical "material" constitutive relation (7.68) is strictly stable in DRUCKER's sense (Chapter 6).

For the constitutive relation (7.71), we introduce a condition of the same type. Thus, let us consider two pairs of "acceleration-displacement quantities" which verify (7.71) and the initial conditions

$$\forall t\in[0,T],\quad \begin{cases} \underline{\Gamma}_{|t}=\rho\,\ddot{\underline{U}}+\mathbf{a}\left(\dot{\underline{U}}_{|\tau};\tau\in[0,t]\right) \\ \underset{\approx}{\Gamma}_{|t}=\rho\,\ddot{\underset{\approx}{U}}+\mathbf{a}\left(\dot{\underset{\approx}{U}}_{|\tau};\tau\in[0,t]\right)\end{cases} \tag{7.72}$$

$$\underline{U}_{|t=0}=\underset{\approx}{U}_{|t=0}=\underline{U}_0;\quad \dot{\underline{U}}_{|t=0}=\dot{\underset{\approx}{U}}_{|t=0}=\underline{V}_0$$

Definition

The relation (7.71) *is said to be stable if:*

$\forall(\underline{\Gamma},\underline{U})$ *and* $(\underset{\approx}{\Gamma},\underset{\approx}{U})$ *verifying* (7.72), *one has*

$$\forall t\in[0,T],\quad \int_0^t(\underline{\Gamma}-\underset{\approx}{\Gamma})\cdot(\dot{\underline{U}}-\dot{\underset{\approx}{U}})\,dt\geq\frac{1}{2}\rho\left[\dot{\underline{U}}-\dot{\underset{\approx}{U}}\right]^2_{|t} \tag{7.73}$$

Overall, if the material relation is strictly stable in DRUCKER's sense and if (7.73) is verified, one says that the constitutive relations are strictly stable in DRUCKER's sense.

The interest of the definition (7.73) resides in the following property:

Property

If the constitutive relations (7.68) *and* (7.71) *are strictly stable in* DRUCKER*'s sense, the solution of the reference problem is unique.*

Proof

Let $(\underline{U},\sigma,\underline{\Gamma})$ and $(\underset{\approx}{U},\underset{\sim}{\sigma},\underset{\approx}{\Gamma})$ be two solutions of the reference problem. Letting $\underline{U}^*=\dot{\underline{U}}-\dot{\underset{\approx}{U}}$, one deduces from the equilibrium equation (7.67) that

$$\forall t\in[0,T],\quad \int_\Omega \mathrm{Tr}\left[(\sigma-\underset{\sim}{\sigma})\varepsilon(\dot{\underline{U}}-\dot{\underset{\approx}{U}})\right]d\Omega+\int_\Omega(\underline{\Gamma}-\underset{\approx}{\Gamma})\cdot(\dot{\underline{U}}-\dot{\underset{\approx}{U}})\,d\Omega=0$$

Integrating over $[0,t]$ and using the stability inequalities, one obtains

$\forall t\in[0,T]$,

$$0=\int_0^t\int_\Omega\left[\mathrm{Tr}[(\sigma-\underset{\sim}{\sigma})\varepsilon(\dot{\underline{U}}-\dot{\underset{\approx}{U}})]+(\underline{\Gamma}-\underset{\approx}{\Gamma})\cdot(\dot{\underline{U}}-\dot{\underset{\approx}{U}})\right]d\Omega\,dt$$

$$\geq\frac{1}{2}\int_\Omega\left[\rho\left[\dot{\underline{U}}-\dot{\underset{\approx}{U}}\right]^2_{|t}\right]d\Omega$$

and thus, taking into account the initial conditions,

$$\underline{U} = \underset{\approx}{U} \quad \text{on } \Omega \times [0, T]$$

Then, from the constitutive relations, one obtains

$$\sigma = \underset{\sim}{\sigma} \quad \text{and} \quad \underline{\Gamma} = \underset{\approx}{\Gamma} \quad \text{over } \Omega \times [0, T]$$

REMARK. In the common case where $\mathbf{a} = 0$, the property (7.73) is obviously verified; thus, the previous result applies.

7.3.3 Error in constitutive relation

To simplify the presentation, we will assume that the constitutive relations are stable in DRUCKER's sense and given by (7.68) and (7.69). The following definitions and properties can be easily extended to the case where (7.69) is replaced by (7.71) [LADEVEZE, 1993].

Admissible fields

Definition

A triplet $s_{Ad} = (\underline{U}_{KA}, \sigma_{DA}, \underline{\Gamma}_{DA})$ *is said to be admissible for Problem* [(7.65) - (7.69)] *if:*

- $\underline{U}_{KA}$ *verifies the kinematic constraints and the initial conditions;*
- σ_{DA} *and* $\underline{\Gamma}_{DA}$ *verify the dynamic equilibrium* (7.67) *;*
- σ_{DA} *and* $\underline{\Gamma}_{DA}$ *are zero at the initial time.*

The quality of s_{Ad} as an approximate solution is evaluated from the way in which the constitutive relations (7.68) and (7.69) are verified.

Measure of the error in constitutive relation

Let us consider an admissible triplet $s_{Ad} = (\underline{U}_{KA}, \sigma_{DA}, \underline{\Gamma}_{DA})$.

Through the constitutive relations, one associates the field $\underline{U}_{KA}$ with the fields

$$\sigma_{KA} = \mathbf{A}\left(\varepsilon(\underline{\dot{U}}_{KA})\,|_{\tau}\,;\, \tau \in [0, t]\right) \quad \text{and} \quad \underline{\Gamma}_{KA} = \rho\, \underline{\ddot{U}}_{KA}$$

Similarly, through the inverse constitutive relation, one associates the field σ_{DA} with a strain field ε_{DA} and, after integrating over time and taking into account the initial conditions, one associates the field $\underline{\Gamma}_{DA}$ with a displacement field $\underline{U}_{DA}$. Be careful, though, that a priori $\varepsilon_{DA} \neq \varepsilon(\underline{U}_{DA})$. Moreover, in general, the strain ε_{DA} is not integrable.

For any point M and for any $t \in [0, T]$, let

$$\eta(M, t) = a \int_0^t (\underline{\Gamma}_{KA} - \underline{\Gamma}_{DA}) \cdot (\underline{\dot{U}}_{KA} - \underline{\dot{U}}_{DA})\, d\tau$$
$$+ (1-a) \int_0^t \mathrm{Tr}[(\sigma_{KA} - \sigma_{DA})(\varepsilon(\underline{\dot{U}}_{KA}) - \dot{\varepsilon}_{DA})]\, d\tau \qquad (7.74)$$

where a is a weighting parameter such that $0 < a < 1$. A common value is $a = \frac{1}{2}$. This is an extension of the definition (6.41).

Property

For a constitutive behavior strictly stable in DRUCKER*'s sense, one has*

$$\forall t \in [0, T], \quad \eta(M, t) \geq 0, \quad \textit{and}$$
$$[\forall M \in \Omega, \quad \forall t \in [0, T], \quad \eta(M, t) = 0]$$
$$\Leftrightarrow [s_{Ad} \textit{ is the solution of the reference problem}] \qquad (7.75)$$

Proof

Using the strict stability of the constitutive behavior and noting that

$$\int_0^t (\underline{\Gamma}_{KA} - \underline{\Gamma}_{DA}) \cdot (\underline{\dot{U}}_{KA} - \underline{\dot{U}}_{DA})\, d\tau = \frac{\rho}{2} \left[\underline{\dot{U}}_{KA} - \underline{\dot{U}}_{DA}\right]^2_{|t}$$
$$a > 0 \quad \text{and} \quad 1 - a > 0$$

it follows that $\forall t \in [0, T],\ \eta(M, t) \geq 0$.

Now, let us assume that $\forall M \in \Omega,\ \forall t \in [0, T],\ \eta(M, t) = 0$.

Using the lower bound

$$0 = \eta(M, t) \geq \frac{a\,\rho}{2} \left[\underline{\dot{U}}_{KA} - \underline{\dot{U}}_{DA}\right]^2_{|t} \geq 0$$

one gets

$$\dot{\underline{U}}_{KA} = \dot{\underline{U}}_{DA}$$

Consequently, $(\underline{U}_{KA}, \underline{\Gamma}_{DA})$ verifies the constitutive relation (7.69).

Similarly, one also has

$$0 = \eta(M, t) \geq \frac{(1-a)}{2} \int_0^t \mathrm{Tr}[(\sigma_{KA} - \sigma_{DA})(\varepsilon(\dot{\underline{U}}_{KA}) - \dot{\varepsilon}_{DA})]\, d\tau \geq 0$$

Therefore, taking into account the strict stability in DRUCKER's sense, one gets

$$\sigma_{KA} = \sigma_{DA} \quad \text{and} \quad \varepsilon(\dot{\underline{U}}_{KA}) = \dot{\varepsilon}_{DA}$$

It follows that $(\varepsilon(\dot{\underline{U}}_{KA}), \sigma_{DA})$ verifies the constitutive relation (7.68).

In conclusion, $s_{Ad} = (\underline{U}_{KA}, \sigma_{DA}, \underline{\Gamma}_{DA})$ is the exact solution of the reference problem. The converse of this property is obvious.

These properties lead us to give the following definition:

Definition

One calls error in constitutive relation associated with the admissible triplet $s_{Ad} = (\underline{U}_{KA}, \sigma_{DA}, \underline{\Gamma}_{DA})$ *the real number*

$$\mathbf{e}_{RdC}(s_{Ad}) = \left[\operatorname*{Sup}_{t \in [0, T]} \int_\Omega \eta(M, t)\, d\Omega \right]^{1/2} \tag{7.76}$$

REMARKS AND COMMENTS

1. Another option is to define the error by

$$\tilde{\mathbf{e}}_{RdC}(s_{Ad}) = \left[\frac{1}{T} \int_0^T \int_\Omega \eta(M, t)\, d\Omega\, dt \right]^{1/2} \tag{7.77}$$

2. Let us consider the linear dynamic case; in this situation, the constitutive relation (7.68) becomes $\sigma = \mathbf{K}\varepsilon(\underline{U})$ and therefore one has

$$\sigma_{KA} = \mathbf{K}\varepsilon(\underline{U}_{KA}) \quad \text{and} \quad \sigma_{DA} = \mathbf{K}\varepsilon_{DA}$$

After integration over time, one gets

$$\eta(M,t)=\frac{a\,\rho}{2}\left[\underline{\dot{U}}_{KA}-\underline{\dot{U}}_{DA}\right]^2_{|t}+\frac{(1-a)}{2}\,\mathrm{Tr}\Big[(\sigma_{KA}-\sigma_{DA})\mathbf{K}^{-1}(\sigma_{KA}-\sigma_{DA})\Big]_{|t}$$

In this case, the integral

$$\int_\Omega \eta(M,t)\,d\Omega$$

can be interpreted as a weighted combination of the kinetic energy associated with the velocity gap $\underline{\dot{U}}_{KA}-\underline{\dot{U}}_{DA}$ and the strain energy associated with the stress gap $\sigma_{KA}-\sigma_{DA}$, these energies being calculated at time t

$$\int_\Omega \eta(M,t)\,d\Omega=\frac{a}{2}\int_\Omega\left[\underline{\dot{U}}_{KA}-\underline{\dot{U}}_{DA}\right]^2_{|t}\rho\,d\Omega$$
$$+\frac{(1-a)}{2}\int_\Omega \mathrm{Tr}\Big[(\sigma_{KA}-\sigma_{DA})\mathbf{K}^{-1}(\sigma_{KA}-\sigma_{DA})\Big]_{|t}\,d\Omega$$

3. In [LADEVEZE - PELLE, 1989a], an initial error measure for linear dynamics was developed.

This measure corresponds to the special case where one requires $s_{Ad}=(\underline{U}_{KA},\sigma_{DA},\underline{\Gamma}_{DA})$ to verify (7.69),

$$\underline{\Gamma}_{DA}=\rho\,\underline{\ddot{U}}_{KA}$$

In this case, letting $a=0$, $\eta(M,t)$ reduces to the single term

$$\frac{1}{2}\mathrm{Tr}\Big[(\sigma_{KA}-\sigma_{DA})\mathbf{K}^{-1}(\sigma_{KA}-\sigma_{DA})\Big]_{|t}$$

and the corresponding error in constitutive relation is a direct extension of the error defined in the linear static case.

4. One can easily associate a relative error with the absolute error (7.76) by letting

$$\varepsilon=\mathbf{e}_{RdC}/D$$

where the denominator D is defined, for example, by

$$D^2=\sup_{t\in[0,T]}\int_\Omega d(M,t)\,d\Omega \tag{7.78}$$

with

$$d(M,t) = a\int_0^t (\underline{\Gamma}_{KA}\cdot\dot{\underline{U}}_{KA} + \underline{\Gamma}_{DA}\cdot\dot{\underline{U}}_{DA})\,d\tau$$

$$+ (1-a)\int_0^t \mathrm{Tr}[\sigma_{KA}\varepsilon(\dot{\underline{U}}_{KA}) + \sigma_{DA}\dot{\varepsilon}_{DA}]\,d\tau \qquad (7.79)$$

This quantity is nonzero if any of the quantities $\dot{\underline{U}}_{KA}$, $\dot{\underline{U}}_{DA}$, σ_{KA} or σ_{DA} is nonzero.

Of course, if one uses (7.77), one takes the definition

$$D^2 = \frac{1}{T}\int_0^T \int_\Omega d(M,t)\,d\Omega\,dt \qquad (7.80)$$

In the case of linear behavior, one has

$$d(M,t) = \frac{a\rho}{2}\left[\dot{\underline{U}}_{KA}^2 + \dot{\underline{U}}_{DA}^2\right]$$

$$+ \frac{(1-a)}{2}\mathrm{Tr}\left[\sigma_{KA}\mathbf{K}^{-1}\sigma_{KA} + \sigma_{DA}\mathbf{K}^{-1}\sigma_{DA}\right] \qquad (7.81)$$

In this case, one can also use the following variant:

$$\underset{\sim}{d}(M,t) = \frac{a\rho}{4}\left[\dot{\underline{U}}_{KA} + \dot{\underline{U}}_{DA}\right]^2$$

$$+ \frac{(1-a)}{4}\mathrm{Tr}\left[(\sigma_{KA} + \sigma_{DA})\mathbf{K}^{-1}(\sigma_{KA} + \sigma_{DA})\right] \qquad (7.82)$$

which, when the admissible triplet $s_{Ad} = (\underline{U}_{KA}, \sigma_{DA}, \underline{\Gamma}_{DA})$ is close to the exact solution, differs very little from (7.81).

5. Formally, one can retrieve the relative errors in DRUCKER's sense defined in Chapter 6 for nonlinear behavior in the quasi-static case by taking

$$a = 0$$

7.3.4 Discretization of the reference problem

The implementation of the constitutive relation error measure requires the construction of admissible triplets

$$s_{Ad} = (\underline{U}_{KA}, \sigma_{DA}, \underline{\Gamma}_{DA})$$

As in the other types of problems, the details of this construction depend on the way the reference problem is discretized. To make things clear and in order to simplify the presentation, we will consider the most classical situation and assume that the constitutive relation is linear elastic $\sigma = \mathbf{K}\varepsilon(\underline{U})$.

However, the following developments can be extended to nonlinear behavior with no major difficulty.

Space discretization

The discretization of Problem [(7.65) - (7.69)] by a finite element displacement method is classical. To simplify the notation, let us assume that the given displacements $\underline{U}_d$ on $\partial_1\Omega$ are zero. If one seeks an approximate solution in the form $\underline{U}_h = \mathbf{N}(M)\,\mathbf{q}(t)$, one ends up with the approximate problem:

$$\begin{aligned} &\textit{Find } t \mapsto \mathbf{q}(t) \textit{ such that} \\ &\mathbb{M}\ddot{\mathbf{q}}(t) + \mathbb{K}\mathbf{q}(t) = \mathbf{F}(t) \\ &\mathbf{q}(0) = \mathbf{q}_0 \quad \textit{and} \quad \dot{\mathbf{q}}(0) = \mathbf{v}_0 \end{aligned} \tag{7.83}$$

- $\mathbb{M}$ and $\mathbb{K}$ are the mass and stiffness matrices defined by (7.8);
- $\mathbf{F}(t)$ is the generalized load defined at each instant t by

$$\mathbf{F}(t) = \int_\Omega \mathbf{N}^T \underline{f}_d(t)\, d\Omega + \int_{\partial_2\Omega} \mathbf{N}^T \underline{F}_d(t)\, d\Gamma$$

- $\underline{U}_0 = \mathbf{N}(M)\,\mathbf{q}_0$ and $\underline{V}_0 = \mathbf{N}(M)\,\mathbf{v}_0$ are the initial conditions.

Time discretization

Problem (7.83) is usually solved by a numerical integration method. The simplest integration schemes used in structural mechanics are the central difference method (for fast dynamics) and the NEWMARK schemes. These methods are generally used with "stabilized" finite elements, an example of which can be found in [ZENG - COMBESCURE, 1998] for fast dynamics and crash problems.

Nevertheless, more sophisticated schemes are sometimes used in special situations, particularly for very stiff problems (GEAR's scheme [GEAR, 1971], PARK's scheme [PARK, 1975], the HHT scheme [GERADIN - RIXEN, 1991]...) and for coupled problems ([PARK - FELIPPA, 1983]).

All these schemes assume that the interval of interest $[0, T]$ has been partitioned into subintervals $[t_n, t_{n+1}]$ such that

$$0 = t_0 < t_1 < \ldots < t_n < t_{n+1} < \ldots < t_p = T \tag{7.84}$$

Let

$$\Delta t_n = t_{n+1} - t_n \tag{7.85}$$

For the sake of simplicity, we will assume that the subdivision is regular

$$\Delta t_n = T/p = \Delta t$$

A particular integration scheme provides at each time t_n an approximation $\mathbf{q}_n$ of the displacement $\mathbf{q}(t_n)$, an approximation $\mathbf{v}_n$ of the velocity $\dot{\mathbf{q}}(t_n)$, and an approximation $\mathbf{a}_n$ of the acceleration $\ddot{\mathbf{q}}(t_n)$. For example, NEWMARK's scheme is defined by the relations

$$\begin{aligned} \mathbf{q}_{n+1} &= \mathbf{q}_n + \Delta t\, \mathbf{v}_n + \frac{1}{2}(\Delta t)^2 \left[(1-2\beta)\, \mathbf{a}_n + 2\beta\, \mathbf{a}_{n+1}\right] \\ \mathbf{v}_{n+1} &= \mathbf{v}_n + \Delta t \left[(1-\alpha)\, \mathbf{a}_n + \alpha\, \mathbf{a}_{n+1}\right] \end{aligned} \tag{7.86}$$

where α and β are parameters.

One adds to these relations the equilibrium at time t_{n+1},

$$\mathbb{M}\mathbf{a}_{n+1} + \mathbb{K}\mathbf{q}_{n+1} = \mathbf{F}(t_{n+1}) \tag{7.87}$$

and the initial conditions

$$\mathbf{q}_0, \mathbf{v}_0 \quad \text{and} \quad \mathbf{a}_0 = \mathbb{M}^{-1} [\mathbf{F}(0) - \mathbb{K}\, \mathbf{q}_0] \tag{7.88}$$

The explicit central difference scheme, which is used extensively in fast dynamics, can be viewed as a particular case by taking

$$\alpha = \frac{1}{2} \quad \text{and} \quad \beta = 0 \tag{7.89}$$

and by replacing the consistent mass matrix $\mathbb{M}$ by a diagonalized mass matrix $\mathbb{M}_{dia}$.

NEWMARK's scheme is unconditionally stable if [BELYTSCHKO - HUGHES, 1986]

$$2\beta \geq \alpha \geq \frac{1}{2} \tag{7.90}$$

Otherwise, for

$$\alpha \geq \frac{1}{2} \quad \text{and} \quad \alpha > 2\beta \tag{7.91}$$

the scheme is stable if and only if the time step is chosen such that (CFL condition)

$$\Delta t \leq \Delta t_{crit} = \frac{1}{\sqrt{\lambda_{max}\left[\frac{\alpha}{2} - \beta\right]}} \tag{7.92}$$

where λ_{max} is the largest eigenvalue of the problem $\mathbb{K}\mathbf{q} = \lambda\mathbb{M}\mathbf{q}$ (or $\mathbb{K}\mathbf{q} = \lambda\mathbb{M}_{dia}\mathbf{q}$ in the explicit case).

A lower bound of the critical time step Δt_{crit} is classically obtained by replacing λ_{max} by

$$\tilde{\lambda}_{max} = \operatorname*{Sup}_{E} \lambda_{E,\, max}$$

where $\lambda_{E,\, max}$ is the largest eigenvalue of the local problem on element E

$$\mathbb{K}_E\, \mathbf{q}_E = \lambda_E\, \mathbb{M}_E\, \mathbf{q}_E \quad (\text{or } \mathbb{K}_E\, \mathbf{q}_E = \lambda_E\, \mathbb{M}_{dia,\, E}\, \mathbf{q}_E)$$

the subscript $\bullet_E$ denoting the corresponding element matrices.

The order of convergence q of this method is $q = 1$ in general, and $q = 2$ in the case $\alpha = 1/2$.

More precisely, one has

$$\operatorname*{Sup}_{1 \leq n \leq N} \left|\mathbf{q}(t_n) - \mathbf{q}_n\right| \leq C_{\mathrm{q}}(\Delta t)^q, \qquad \operatorname*{Sup}_{1 \leq n \leq N} \left|\dot{\mathbf{q}}(t_n) - \mathbf{v}_n\right| \leq C_{\mathrm{v}}(\Delta t)^q \tag{7.93}$$

where C_q and C_v are constants which, in general, cannot be evaluated and which depend on the exact solution $t \mapsto \mathbf{q}(t)$ of (7.83).

7.3.5 Construction of admissible triplets

The numerical solution provides, at each nodal instant t_n of the time discretization, approximations $\mathbf{q}_n$ of the displacement, $\mathbf{v}_n$ of the velocity and $\mathbf{a}_n$ of the acceleration at the mesh nodes.

The objective is, starting from this numerical solution and the problem's data, to reconstruct a triplet $s_{Ad} = (\underline{U}_{KA}, \sigma_{DA}, \underline{\Gamma}_{DA})$ which is admissible in the sense of the definition of Section 7.3.3.

Construction of σ_{DA} and $\underline{\Gamma}_{DA}$

The construction is carried out in two main stages. First, one constructs at each nodal instant t_n a stress field σ_{DA}^n and a field of quantities of acceleration $\underline{\Gamma}_{DA}^n$ which verify the equilibrium equations at time t_n exactly. Then, one extends these fields by interpolation over the entire time interval $[0, T]$ in order to obtain σ_{DA} and $\underline{\Gamma}_{DA}$.

Construction at a nodal instant

Initially, let us consider the situation where the numerical scheme being used respects the finite element equilibrium in the weak sense at each nodal instant, which corresponds to the use of a consistent mass matrix

$$\mathbb{M}\mathbf{a}_n + \mathbb{K}\mathbf{q}_n = \mathbf{F}(t_n) \tag{7.94}$$

i.e.

$$\forall \underline{U}_h^* \text{ of the finite element type and } KA \text{ with } 0,$$
$$\int_\Omega \mathrm{Tr}\left[\sigma_h^n \varepsilon(\underline{U}_h^*)\right] d\Omega + \int_\Omega \underline{\Gamma}_h^n \cdot \underline{U}_h^* \, d\Omega = \int_\Omega \underline{f}_d(t_n) \cdot \underline{U}_h^* \, d\Omega + \int_{\partial_2\Omega} \underline{F}_d(t_n) \cdot \underline{U}_h^* \, d\Gamma \tag{7.95}$$

where

$$\sigma_h^n = \mathbf{K}\varepsilon(\mathbf{N}(M)\mathbf{q}_n) \quad \text{and} \quad \underline{\Gamma}_h^n = \rho\, \mathbf{N}(M)\mathbf{a}_n \tag{7.96}$$

In order to construct σ_{DA}^n and $\underline{\Gamma}_{DA}^n$, one introduces, as in the static case, a strong extension condition:

For any shape function φ_i of the finite element type and for any element E *of the mesh*

$$\int_E (\sigma_{DA}^n - \sigma_h^n)\, \underline{\text{grad}}\, \varphi_i \, dE + \int_E (\underline{\Gamma}_{DA}^n - \underline{\Gamma}_h^n)\, \varphi_i \, dE = 0 \tag{7.97}$$

Using the admissibility conditions, one obtains

$$\int_{\partial E} \eta_E \hat{\underline{F}}\, \varphi_i \, d\Gamma = - \int_E \underline{f}_d(t_n)\, \varphi_i \, dE + \int_E \sigma_h^n \, \underline{\text{grad}}\, \varphi_i \, dE + \int_E \underline{\Gamma}_h^n \, \varphi_i \, dE$$

i.e. going back to the notations of Section 2.4.3,

$$\sum_{r=1}^{R} \left[\int_{\Gamma_E^r} \eta_E \, \hat{\underline{F}}\, \varphi_i \, d\Gamma \right] = \underline{Q}_E(i)$$

where, in the definition of $\underline{Q}_E(i)$, the interior load is now replaced by the difference $\underline{f}_d(t_n) - \underline{\Gamma}_h^n$,

$$\underline{Q}_E(i) = - \int_E (\underline{f}_d(t_n) - \underline{\Gamma}_h^n)\, \varphi_i \, dE + \int_E \sigma_h^n \, \underline{\text{grad}}\, \varphi_i \, dE$$

As in Chapter 2, the compatibility condition (for an interior node)

$$\sum_{k=1}^{N} \underline{Q}_{E_k}(i) = 0$$

is a consequence of the finite element equilibrium (7.95). Thus, the construction of the densities is similar to that performed in the static case. The densities thus constructed verify the following property:

For any rigid body displacement field $\underline{U}_S$,

$$\int_E (\underline{f}_d - \underline{\Gamma}_h^n) \cdot \underline{U}_S \, dE + \int_{\partial E} \eta_E \, \hat{\underline{F}} \cdot \underline{U}_S \, dS = 0 \tag{7.98}$$

Thus, field σ_{DA} can be constructed locally, element by element, by seeking a simple solution of the problem

$$\underline{\text{div}}\, \mathbb{\sigma}^n_{DA} + \underline{f}_d = \underline{\Gamma}^n_h \quad \text{in } E$$

$$\mathbb{\sigma}^n_{DA}\, \underline{n}_E = \eta_E\, \hat{\underline{F}} \quad \text{on } \partial E$$

which is the same as choosing

$$\underline{\Gamma}^n_{DA} = \underline{\Gamma}^n_h$$

In practice, many other choices are possible. Indeed, once the densities have been constructed, the objective is to determine, element by element, a pair $(\mathbb{\sigma}^n_{DA}, \underline{\Gamma}^n_{DA})$ such that

$$\underline{\text{div}}\, \mathbb{\sigma}^n_{DA} + \underline{f}_d = \underline{\Gamma}^n_{DA} \quad \text{in } E$$

$$\mathbb{\sigma}^n_{DA}\, \underline{n}_E = \eta_E\, \hat{\underline{F}} \quad \text{on } \partial E$$

In order to choose $\underline{\Gamma}^n_{DA}$, taking (7.98) into account, one needs to respect only the global equilibrium condition:

For any field of rigid body displacement field $\underline{U}_S$,

$$\int_E \underline{\Gamma}^n_{DA} \cdot \underline{U}_S\, dE = \int_E \underline{\Gamma}^n_h \cdot \underline{U}_S\, dE$$

This is the case, for example, if in addition to the condition (7.97) one imposes the extension condition:

For any shape function φ_i *of the finite element type, and for any element* E *of the mesh*

$$\int_E \underline{\Gamma}^n_{DA}\, \varphi_i\, dE = \int_E \underline{\Gamma}^n_h\, \varphi_i\, dE \tag{7.99}$$

The previous construction does not apply directly when the time integration uses the explicit central difference scheme because this scheme uses a diagonal mass matrix. Indeed, in this case, Equation (7.94) is replaced by

$$\mathbb{M}_{dia}\mathbf{a}_n + \mathbb{K}\mathbf{q}_n = \mathbf{F}(t_n) \tag{7.100}$$

To carry out the construction, one must modify (7.96) by defining $\underline{\Gamma}^n_h$ by its restriction $\underline{\Gamma}^n_{h|E}$ on each element E,

$$\underline{\Gamma}^n_{h|E} = \rho\, \mathbf{N}_E(M)\tilde{\mathbf{a}}_{n,E} \tag{7.101}$$

where $\mathbf{N}_E$ is the matrix of the shape functions on E and $\tilde{\mathbf{a}}_{n,E}$ is obtained by solving on E the small linear system (in which the indices $\bullet_E$ designate element matrices)

$$\mathbf{M}_E\,\tilde{\mathbf{a}}_{n,E} = \mathbf{M}_{dia,E}\,\mathbf{a}_{n,E}$$

Indeed, with these choices, (7.100) is equivalent to:

$$\forall \underline{U}_h^* \textit{ of the finite element type and KA with } 0,$$
$$\int_\Omega \mathrm{Tr}\left[\sigma_h^n \varepsilon(\underline{U}_h^*)\right] d\Omega + \sum_E \int_E \underline{\Gamma}_{h|E}^n \cdot \underline{U}_{h|E}^* \, dE = \int_\Omega \underline{f}_d(t_n) \cdot \underline{U}_h^* \, d\Omega + \int_{\partial_2\Omega} \underline{F}_d(t_n) \cdot \underline{U}_h^* \, d\Gamma \tag{7.102}$$

Replacing (7.97) by

$$\int_E (\sigma_{DA}^n - \sigma_h^n)\, \underline{\mathrm{grad}}\, \varphi_i \, dE + \int_E (\underline{\Gamma}_{DA}^n - \underline{\Gamma}_{h|E}^n)\, \varphi_i \, dE = 0$$

the previous construction can be applied without any change.

REMARK. In (7.102), one has

$$\sum_E \int_E \underline{\Gamma}_{h|E}^n \cdot \underline{U}_{h|E}^* \, dE = \int_\Omega \underline{\Gamma}_h^n \cdot \underline{U}_h^* \, d\Omega$$

However, the quantity of acceleration $\underline{\Gamma}_h^n$ is no longer equal to $\rho\,\mathbf{N}(M)\mathbf{a}_n$. In particular, $\underline{\Gamma}_h^n$ is not continuous from one element to the next.

Extension over $[0, T]$

Once σ_{DA}^n and $\underline{\Gamma}_{DA}^n$ have been constructed at each nodal instant, one constructs σ_{DA} and $\underline{\Gamma}_{DA}$ at any time by letting, $\forall t \in [t_n, t_{n+1}]$,

$$\sigma_{DA}(M, t) = \frac{t - t_n}{\Delta t}\,(\sigma_{DA}^{n+1}(M) - \sigma_{DA}^n(M)) + \sigma_{DA}^n(M)$$
$$\underline{\Gamma}_{DA}(M, t) = \frac{t - t_n}{\Delta t}\,(\underline{\Gamma}_{DA}^{n+1}(M) - \underline{\Gamma}_{DA}^n(M)) + \underline{\Gamma}_{DA}^n(M) \tag{7.103}$$

Thus, assuming the loads to be linear in time over each interval $[t_n, t_{n+1}]$, one obtains admissible fields.

Construction of $\underline{U}_{KA}$

As before, the way to construct the field $\underline{U}_{KA}$ is far from unique; there exist many variants.

First, we will present a construction in a very general situation; then, we will present a construction which exploits the specific nature of the explicit scheme.

The general case

At each nodal instant, the quantities $\mathbf{q}_n$ and $\mathbf{a}_n$ verify an equilibrium condition in the weak sense. This leads us to carry out the construction of the field $\underline{U}_{KA}$ without making reference to these quantities. More precisely, the kinematically admissible field $\underline{U}_{KA}$ is defined from the velocity field $\underline{V}_{KA}$ by letting

$$\underline{V}_{KA}(M, t) = \mathbf{N}(M)\, \mathbf{v}_{KA}(t)$$

with

$$\forall t \in [t_n, t_{n+1}] : \quad \mathbf{v}_{KA}(t) = \frac{\mathbf{v}_{n+1} - \mathbf{v}_n}{\Delta t}\,(t - t_n) + \mathbf{v}_n$$

Then, the displacement is obtained by time integration, taking into account the initial condition

$$\underline{U}_{KA}(M, t) = \int_0^t \underline{V}_{KA}(M, \tau)\, d\tau + \underline{U}_0(M)$$

Thus, one obtains

$$\underline{U}_{KA}(M, t) = \mathbf{N}(M)\, \mathbf{q}_{KA}(t) \tag{7.104}$$

with

$$\forall t \in [t_n, t_{n+1}] : \quad \mathbf{q}_{KA}(t) = \frac{\mathbf{v}_{n+1} - \mathbf{v}_n}{2\,\Delta t}\,(t - t_n)^2 + \mathbf{v}_n\,(t - t_n) + \mathbf{q}_{KA}(t_n)$$

The constants $\mathbf{q}_{KA}(t_n)$ are defined by recurrence

$$\mathbf{q}_{KA}(t_{n+1}) = \mathbf{q}_{KA}(t_n) + \frac{\Delta t}{2}\,(\mathbf{v}_n + \mathbf{v}_{n+1}) \quad \text{and} \quad \mathbf{q}_{KA}(0) = \mathbf{q}_0$$

The specific case of the explicit scheme

In the explicit central difference scheme, the half-time steps

$$t_{n+1/2} = t_n + \Delta t/2 \quad (n = 0, 1, \ldots, p-1)$$

play a particular role. Indeed the explicit scheme can be defined by

$$\mathbf{v}_{n+1/2} = \Delta t\, \mathbf{a}_n + \mathbf{v}_{n-1/2}$$

$$\mathbf{q}_{n+1} = \mathbf{q}_n + \Delta t\, \mathbf{v}_{n+1/2}$$

$$\mathbf{a}_{n+1} = \mathbb{M}_{dia}^{-1}\left[\mathbf{F}(t_{n+1}) - \mathbb{K}\mathbf{q}_{n+1}\right]$$

and by the initial conditions

$$\mathbf{q}_0,$$

$$\mathbf{v}_{-1/2} = -\frac{\Delta t}{2}\mathbf{a}_0 + \mathbf{v}_0$$

$$\mathbf{a}_0 = \mathbb{M}_{dia}^{-1}\left[\mathbf{F}(0) - \mathbb{K}\mathbf{q}_0\right]$$

The velocity for the time step can be recovered by

$$\mathbf{v}_n = \frac{\mathbf{v}_{n+1/2} + \mathbf{v}_{n-1/2}}{2}$$

This enables one to construct a new velocity field by letting

$$\underline{\tilde{V}}_{KA}(M, t) = \mathbf{N}(M)\, \tilde{\mathbf{v}}_{KA}(t)$$

with

$$\text{for } t_n \le t \le t_{n+1/2}, \quad \tilde{\mathbf{v}}_{KA}(t) = 2\,\frac{\mathbf{v}_{n+1/2} - \mathbf{v}_n}{\Delta t}(t - t_n) + \mathbf{v}_n$$

$$\text{for } t_{n+1/2} \le t \le t_{n+1}, \quad \tilde{\mathbf{v}}_{KA}(t) = 2\,\frac{\mathbf{v}_{n+1} - \mathbf{v}_{n+1/2}}{\Delta t}(t - t_{n+1/2}) + \mathbf{v}_{n+1/2}$$

The velocity thus defined is continuous in time over $[0, T]$.

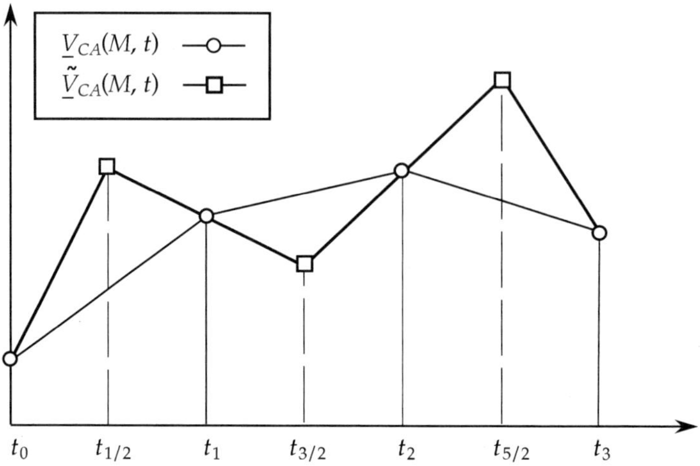

Figure 7.5. *Comparison of the two admissible velocity fields.*

The displacement field is obtained by time integration

$$\tilde{\underline{U}}_{KA}(M, t) = \mathbf{N}(M)\, \tilde{\mathbf{q}}_{KA}(t) \tag{7.105}$$

with

for $t_n \le t \le t_{n+1/2}$,

$$\tilde{\mathbf{q}}_{KA}(t) = \frac{\mathbf{v}_{n+1/2} - \mathbf{v}_n}{\Delta t}(t - t_n)^2 + \mathbf{v}_n\,(t - t_n) + \tilde{\mathbf{q}}_{KA}(t_n)$$

for $t_{n+1/2} \le t \le t_{n+1}$,

$$\tilde{\mathbf{q}}_{KA}(t) = \frac{\mathbf{v}_{n+1} - \mathbf{v}_{n+1/2}}{\Delta t}(t - t_{n+1/2})^2 + \mathbf{v}_{n+1/2}\,(t - t_{n+1/2}) + \tilde{\mathbf{q}}_{KA}(t_{n+1/2})$$

The constants $\tilde{\mathbf{q}}_{KA}(t_n)$ and $\ddot{\mathbf{q}}_{KA}(t_{n+1/2})$ are defined by recurrence

$$\tilde{\mathbf{q}}_{KA}(t_{n+1/2}) = \frac{\Delta t}{4}(\mathbf{v}_n + \mathbf{v}_{n+1/2}) + \tilde{\mathbf{q}}_{KA}(t_n)$$

$$\tilde{\mathbf{q}}_{KA}(t_{n+1}) = \frac{\Delta t}{4}(\mathbf{v}_{n+1/2} + \mathbf{v}_{n+1}) + \tilde{\mathbf{q}}_{KA}(t_{n+1/2})$$

with the initial condition

$$\tilde{\mathbf{q}}_{KA}(0) = \mathbf{q}_0$$

For very fast dynamic problems, this second construction yields a better error estimator.

Figure 7.5 shows, for a fixed point M and as a function of time, the difference between the fields $\underline{V}_{KA}$ and $\underline{\tilde{V}}_{KA}$ obtained by the two constructions.

REMARKS

1. Other techniques of construction of the admissible displacement field $\underline{U}_{KA}$ are possible. For example, one can construct the field $\underline{U}_{KA}$ over each interval as the third-degree H interpolant in time of the values $\underline{U}^n_{KA}$, $\underline{V}^n_{KA}$, $\underline{U}^{n+1}_{KA}$, $\underline{V}^{n+1}_{KA}$, or even the fifth-degree interpolant of the values $\underline{U}^n_{KA}$, $\underline{V}^n_{KA}$, $\underline{\Gamma}^n_{KA}$, $\underline{U}^{n+1}_{KA}$, $\underline{V}^{n+1}_{KA}$, $\underline{\Gamma}^{n+1}_{KA}$, with $\underline{\Gamma}^k_{KA} = \mathbf{N}(M)\,\mathbf{a}_k$. The study reported in [COMBE - PELLE, 1998] showed that these conditions do not always lead to satisfactory error estimators.

2. The previous construction techniques can be easily extended to the case of nonlinear behavior.

7.3.6 Error indicators in time and in space

The approach used to define error indicators in time and in space is very similar to that already presented in Chapter 6. We will describe the implementation of these indicators in the linear case only.

Time error indicator

Let us consider the following problem, discretized in space and continuous in time:

$$\begin{aligned}&\textit{Find } \underline{U}_h \textit{ of the finite element type, } \sigma_h \textit{ and } \underline{\Gamma}_h\textit{, such that}\\ &\forall t \in [0, T], \quad \underline{U}_h \textit{ is KA and verifies the initial conditions}\end{aligned} \tag{7.106}$$

$$\begin{aligned}&\forall t \in [0, T] \quad \textit{and} \quad \forall \underline{U}^*_h \in \boldsymbol{\mathcal{U}}_{h0},\\ &\int_\Omega \mathrm{Tr}\left[\sigma_h \varepsilon(\underline{U}^*_h)\right] d\Omega + \int_\Omega \underline{\Gamma}_h \cdot \underline{U}^*_h \, d\Omega\\ &\qquad = \int_\Omega \underline{f}_d \cdot \underline{U}^*_h \, d\Omega + \int_{\partial_2\Omega} \underline{F}_d \cdot \underline{U}^*_h \, dS\end{aligned} \tag{7.107}$$

$$\forall t \in [0, T], \quad \forall M \in \Omega,$$
$$\sigma_{h|t} = \mathbf{A}\left(\varepsilon(\dot{\underline{U}}_h)_{|\tau}; \tau \in [0, t]\right) \quad \textit{and} \quad \underline{\Gamma}_h = \rho\, \ddot{\underline{U}}_h \tag{7.108}$$

Definition

A triplet $s_{h,Ad} = (\underline{U}_{h,KA}, \sigma_{h,DA}, \underline{\Gamma}_{h,DA})$ *is said to be admissible with respect to Problem* [(7.106) - (7.108)] *if:*

- $\underline{U}_{h,KA}$ *is of the finite element type and verifies* (7.106);
- $\sigma_{h,DA}$ *and* $\underline{\Gamma}_{h,DA}$ *verify the equilibrium in the finite element sense* (7.107).

One associates with the field $\varepsilon_{h,KA} = \varepsilon\,(\underline{U}_{h,KA})$, through the constitutive relations, the stress field $\sigma_{h,KA}$ and the field of quantities of acceleration $\underline{\Gamma}_{h,KA}$,

$$\sigma_{h,KA} = \mathbf{A}\left(\varepsilon(\dot{\underline{U}}_{h,KA})_{|\tau}; \tau \in [0, t]\right) \quad \text{and} \quad \underline{\Gamma}_{h,KA} = \rho\, \ddot{\underline{U}}_{h,KA}$$

Similarly, through the inverse constitutive relation, one associates with the field $\sigma_{h,DA}$ a strain field $\varepsilon_{h,DA}$; by integrating $\underline{\Gamma}_{h,DA}$ in time and taking into account the initial conditions, one obtains successively a velocity field $\underline{V}_{h,DA}$ and a displacement field $\underline{U}_{h,DA}$.

The quality of $s_{h,Ad} = (\underline{U}_{h,KA}, \sigma_{h,DA}, \underline{\Gamma}_{h,DA})$ as an approximate solution of the space-discretized problem can be evaluated through the error in constitutive relation

$$\mathbf{e}_{h,RdC}(s_{h,Ad}) = \left[\operatorname*{Sup}_{t \in [0,T]} \int_\Omega \eta_h(M, t)\, d\Omega \right]^{1/2}$$

with

$$\eta_h(M, t) = a \int_0^t (\underline{\Gamma}_{h,KA} - \underline{\Gamma}_{h,DA}) \cdot (\dot{\underline{U}}_{h,KA} - \dot{\underline{U}}_{h,DA})\, d\tau$$
$$+ (1 - a) \int_0^t \mathrm{Tr}[(\sigma_{h,KA} - \sigma_{h,DA})(\varepsilon(\dot{\underline{U}}_{h,KA}) - \dot{\varepsilon}_{h,DA})]\, d\tau$$

For a given space discretization, this error measure enables one to appraise the errors due to the time discretization. Thus, let

$$\mathbf{i}_{time} = \mathbf{e}_{RdC}(s_{h,Ad})$$

A zero value of $\mathbf{i}_{time}$ means that $s_{h,Ad}$ is the exact solution of the space-discretized problem. This is therefore an indicator of the time error between the numerical model and the reference problem [(7.65) - (7.69)].

Elastic behavior case

The implementation of $\mathbf{i}_{temps}$ is simple. For the field $\underline{U}_{h,KA}$, it suffices to take the field $\underline{U}_{KA}$ defined in (7.104) (or, in the explicit case, the field $\tilde{\underline{U}}_{KA}$ defined in (7.105)). Concerning the fields $\sigma_{h,DA}$ and $\underline{\Gamma}_{h,DA}$, one chooses at each nodal instant t_n,

$$\sigma_{h,DA}(t_n) = \sigma_h^n \quad \text{and} \quad \underline{\Gamma}_{h,DA}(t_n) = \underline{\Gamma}_h^n \tag{7.109}$$

where σ_h^n and $\underline{\Gamma}_h^n$ are defined by (7.96). (In the explicit case, one uses for $\underline{\Gamma}_h^n$ the expression (7.101).)

The fields $\sigma_{h,DA}$ and $\underline{\Gamma}_{h,DA}$ are then constructed over each subinterval $[t_n, t_{n+1}]$ by linear time interpolation. At any instant, these fields verify the equilibrium (7.107) in the weak finite element sense.

In the case of linear elastic behavior, $\sigma = \mathbf{K}\,\varepsilon$, and if the fields $s_{Ad} = (\underline{U}_{KA}, \sigma_{DA}, \underline{\Gamma}_{DA})$ and $s_{h,Ad} = (\underline{U}_{h,KA}, \sigma_{h,DA}, \underline{\Gamma}_{h,DA})$ are defined by the methods described above one has the property:

Property

If $\underline{\Gamma}_{DA}$ *verifies the extension condition* (7.99), *one has*

$$\mathbf{i}_{time} \leq \mathbf{e}_{RdC}\,(s_{Ad})$$

Proof

In the case of linear elastic behavior, one has

$$\mathbf{e}^2_{RdC}\,(s_{Ad}) = \operatorname*{Sup}_{t\in[0,T]} \Theta^2 \quad \text{and} \quad \mathbf{i}^2_{time} = \operatorname*{Sup}_{t\in[0,T]} \Theta^2_h$$

with

$$\Theta^2 = a\,E_{kin}(\underline{\dot{U}}_{KA} - \underline{\dot{U}}_{DA}) + (1-a)\,E_{def}(\sigma_{KA} - \sigma_{DA})$$

$$\Theta^2_h = a\,E_{kin}(\underline{\dot{U}}_{h,KA} - \underline{\dot{U}}_{h,DA}) + (1-a)\,E_{def}(\sigma_{h,KA} - \sigma_{h,DA})$$

where

$$E_{kin}(\,\cdot\,) = \frac{1}{2}\int_\Omega (\,\cdot\,)^2\,\rho\,d\Omega \quad \text{and} \quad E_{def}(\,\cdot\,) = \frac{1}{2}\int_\Omega \mathrm{Tr}(\,\cdot\,\mathbf{K}^{-1}\,\cdot\,)\,d\Omega$$

Observing that $\underline{U}_{KA} = \underline{U}_{h,KA}$ by construction and, therefore, $\boldsymbol{\sigma}_{KA} = \boldsymbol{\sigma}_{h,KA}$ and $\dot{\underline{U}}_{KA} = \dot{\underline{U}}_{h,KA}$, one obtains

$$\Theta^2 = \Theta_h^2 + a\left[E_{kin}(\dot{\underline{U}}_{h,DA} - \dot{\underline{U}}_{DA}) + P_1\right] + (1-a)\left[E_{def}(\boldsymbol{\sigma}_{h,DA} - \boldsymbol{\sigma}_{DA}) + P_2\right]$$

with

$$P_1 = \int_\Omega (\dot{\underline{U}}_{KA} - \dot{\underline{U}}_{h,DA}) \cdot (\dot{\underline{U}}_{h,DA} - \dot{\underline{U}}_{DA})\, \rho\, d\Omega$$

$$P_2 = \int_\Omega \mathrm{Tr}\left[(\boldsymbol{\sigma}_{KA} - \boldsymbol{\sigma}_{h,DA})\mathbf{K}^{-1}(\boldsymbol{\sigma}_{h,DA} - \boldsymbol{\sigma}_{DA})\right] d\Omega$$

Calculation of P_1

By construction, $\dot{\underline{U}}_{KA}(M,t) - \dot{\underline{U}}_{h,DA}(M,t) = \mathbf{N}(M)\,\dot{\mathbf{z}}(t)$. Therefore, for a fixed t,

$$\dot{\underline{U}}_{KA}(M,t) - \dot{\underline{U}}_{h,DA}(M,t) = \underline{U}_h^*(M)$$

where $\underline{U}_h^*$ is of the finite element type.

Besides, (7.103) and (7.109) yield, that for a fixed $t \in [t_n, t_{n+1}]$,

$$\dot{\underline{U}}_{DA}(M,t) = \sum_{i=0}^{n+1} \gamma_i(t)\, \underline{\Gamma}_{DA}^i(M) + \delta(t)$$

$$\dot{\underline{U}}_{h,DA}(M,t) = \sum_{i=0}^{n+1} \gamma_i(t)\, \underline{\Gamma}_h^i(M) + \delta(t)$$

where the $\gamma_i(t)$ are known. Therefore

$$\dot{\underline{U}}_{h,DA}(M,t) - \dot{\underline{U}}_{DA}(M,t) = \sum_{i=0}^{n+1} \gamma_i(t)\left[\underline{\Gamma}_h^i(M) - \underline{\Gamma}_{DA}^i(M)\right]$$

From (7.99), one deduces that $P_1 = 0$.

Calculation of P_2

By construction, $\boldsymbol{\sigma}_{KA}(M,t) - \boldsymbol{\sigma}_{h,DA}(M,t) = \mathbf{K}\boldsymbol{\varepsilon}(\mathbf{N}(M)\,\mathbf{z}(t))$. Therefore, for a fixed t,

$$\boldsymbol{\sigma}_{KA}(M,t) - \boldsymbol{\sigma}_{h,DA}(M,t) = \mathbf{K}\boldsymbol{\varepsilon}(\underline{U}_h^*(M))$$

where $\underline{U}_h^*$ is of the finite element type.

Thus, one has

$$P_2 = \int_\Omega \mathrm{Tr}\Big[(\sigma_{h,DA} - \sigma_{DA})\varepsilon(\underline{U}_h^*) \Big] \, d\Omega$$

Since $\sigma_{h,DA}$ verifies the equilibrium in the weak finite element sense and since σ_{DA} verifies the equilibrium in the strong sense, thus also in the weak sense, one obtains

$$\begin{aligned} P_2 &= -\int_\Omega (\underline{\Gamma}_{h,DA} - \underline{\Gamma}_{DA}) \cdot \underline{U}_h^* \; d\Omega \\ &= -\sum_{i=0}^{n+1} \dot{\gamma}_i(t) \int_\Omega (\underline{\Gamma}_h^i - \underline{\Gamma}_{DA}^i) \cdot \underline{U}_h^* \; d\Omega \end{aligned}$$

Therefore, taking (7.99) into account, $P_2 = 0$.

This leads to the inequality $\Theta_h^2 \le \Theta^2$. Thus, one obtains

$$\mathbf{i}_{time} \le \mathbf{e}_{RdC}\,(s_{Ad})$$

REMARK. Let us recall that in the linear behavior case Problem [(7.106) - (7.108)] discretized in space and continuous in time is equivalent to the differential equation and initial conditions (7.83):

$$\begin{aligned} &\textit{Find } t \mapsto \mathbf{q}(t) \textit{ such that} \\ &\mathbb{M}\ddot{\mathbf{q}}(t) + \mathbb{K}\mathbf{q}(t) = \mathbf{F}(t) \\ &\mathbf{q}(0) = \mathbf{q}_0 \quad \text{and} \quad \dot{\mathbf{q}}(0) = \mathbf{v}_0 \end{aligned} \tag{7.110}$$

Therefore, the indicator $\mathbf{i}_{time}$ enables one to measure the error over $[0, T]$ due to the use of a time integration scheme to find an approximate solution of (7.110).

Space error indicator

To define a space error indicator, one must define a problem which is discretized in time, but continuous in space. The detailed definition of this problem depends on the integration scheme used. To make things clear, we will assume that the scheme being used is NEWMARK's scheme (7.86); however, the method can be extended to other integration schemes.

Let us use notations similar to those of Chapter 6,

$$\Delta = \{t_0 = 0, t_1, t_2, \ldots, t_n, \ldots, t_p = T\}$$

$$\underline{U}^\Delta = \{\underline{U}^0, \underline{U}^1, \ldots, \underline{U}^n, \ldots, \underline{U}^p\}$$

$$\underline{V}^\Delta = \{\underline{V}^0, \underline{V}^1, \ldots, \underline{V}^n, \ldots, \underline{V}^p\}$$

$$\underline{A}^\Delta = \{\underline{A}^0, \underline{A}^1, \ldots, \underline{A}^n, \ldots, \underline{A}^p\}$$

$$\mathbb{\varepsilon}^\Delta = \{\mathbb{\varepsilon}(\underline{U}^0), \mathbb{\varepsilon}(\underline{U}^1), \ldots, \mathbb{\varepsilon}(\underline{U}^n), \ldots, \mathbb{\varepsilon}(\underline{U}^p)\} = \mathbb{\varepsilon}^\Delta(\underline{U}^\Delta)$$

$$\mathbb{\sigma}^\Delta = \{\mathbb{\sigma}^0, \mathbb{\sigma}^1, \ldots, \mathbb{\sigma}^n, \ldots, \mathbb{\sigma}^p\}$$

$$\underline{\Gamma}^\Delta = \{\underline{\Gamma}^0, \underline{\Gamma}^1, \ldots, \underline{\Gamma}^n, \ldots, \underline{\Gamma}^p\}$$

where $\underline{U}^n$ designates a displacement field, $\underline{V}^n$ a "velocity" field, $\underline{A}^n$ an "acceleration" field, $\mathbb{\sigma}^n$ a stress field and $\underline{\Gamma}^n$ a field of quantities of acceleration, all defined on Ω at time t_n.

The time-discretized problem can be formulated as follows:

Find $\underline{U}^\Delta, \underline{V}^\Delta, \underline{A}^\Delta, \mathbb{\sigma}^\Delta$ *and* $\underline{\Gamma}^\Delta$ *such that*

$\forall t_n \in \Delta, \quad \underline{U}^n$ *is KA,*

$$\underline{U}^{n+1} = \underline{U}^n + \Delta t\, \underline{V}^n + \frac{1}{2}(\Delta t)^2 \left[(1-2\beta)\, \underline{A}^n + 2\beta\, \underline{A}^{n+1}\right] \tag{7.111}$$

$$\underline{V}^{n+1} = \underline{V}^n + \Delta t \left[(1-\alpha)\, \underline{A}^n + \alpha\, \underline{A}^{n+1}\right] \tag{7.112}$$

$\forall t_n \in \Delta$ *and* $\forall \underline{U}^* \in \boldsymbol{\mathcal{U}}_0$,

$$\int_\Omega \mathrm{Tr}\left[\mathbb{\sigma}^n \mathbb{\varepsilon}(\underline{U}^*)\right] d\Omega + \int_\Omega \underline{\Gamma}^n \cdot \underline{U}^* d\Omega = \int_\Omega \underline{f}_d(t_n) \cdot \underline{U}^* d\Omega + \int_{\partial_2 \Omega} \underline{F}_d(t_n) \cdot \underline{U}^* dS \tag{7.113}$$

$$\mathbb{\sigma}^\Delta = \mathbf{A}^\Delta(\mathbb{\varepsilon}^\Delta(\underline{U}^\Delta)) \tag{7.114}$$

$$\forall t_n \in \Delta, \quad \underline{\Gamma}^n = \rho\, \underline{A}^n \tag{7.115}$$

with the initial conditions

$$\underline{U}^0 = \underline{U}_0, \quad \underline{V}^0 = \underline{V}_0 \quad \text{and} \quad \underline{A}^0 = \frac{1}{\rho}\underline{\Gamma}_0 \tag{7.116}$$

where $\underline{\Gamma}_0$ is defined by (7.70).

The discretized constitutive relation (7.114) is defined as in (6.68). In the elastic behavior case, this relation becomes simply

$$\forall t_n \in \Delta, \quad \sigma^n = \mathbf{K}\varepsilon(\underline{U}^n)$$

Definition

The fields $s^{\Delta}_{Ad} = (\underline{U}^{\Delta}_{KA}, \underline{V}^{\Delta}_{KA}, \underline{A}^{\Delta}_{KA}, \sigma^{\Delta}_{DA}, \underline{\Gamma}^{\Delta}_{DA})$ *are said to be admissible with respect to Problem* [(7.111) – (7.115)] *if:*

- $\underline{U}^{\Delta}_{KA}, \underline{V}^{\Delta}_{KA}, \underline{A}^{\Delta}_{KA}$ *verify* (7.111) *and* (7.112) *as well as the initial conditions* (7.116);
- $\sigma^{\Delta}_{DA}, \underline{\Gamma}^{\Delta}_{DA}$ *verify the equilibrium* (7.113) *and* $\underline{\Gamma}^{0}_{DA} = \underline{\Gamma}_0$.

Then, it is very easy to construct an error measure based on the stability of the discretized constitutive relation.

The fields $\underline{U}^{\Delta}_{KA}, \underline{V}^{\Delta}_{KA}, \underline{A}^{\Delta}_{KA}$ are associated, through (7.114) and (7.115), with the fields

$$\sigma^{\Delta}_{KA} = \mathbf{A}^{\Delta}(\varepsilon^{\Delta}(\underline{U}^{\Delta}_{KA})) \quad \text{and} \quad \underline{\Gamma}^{\Delta}_{KA} = \rho\, \underline{A}^{\Delta}_{KA}$$

The fields σ^{Δ}_{DA} are associated, through the inverse constitutive relation, with the strain fields $\varepsilon^{\Delta}_{DA}$.

The fields $\underline{\Gamma}^{\Delta}_{DA}$ are associated, through the relation (7.115), with the fields $\underline{A}^{\Delta}_{DA}$, then, through the relations (7.112), with the fields $\underline{V}^{\Delta}_{DA}$ and $\underline{U}^{\Delta}_{DA}$.

Then, let

$$\mathbf{e}^{\Delta}_{RdC}(s^{\Delta}_{DA}) = \left[\underset{1 \le m \le p}{\text{Sup}} \int_{\Omega} \eta^m \, d\Omega \right]^{1/2} \tag{7.117}$$

with

$$\eta^m = \frac{a}{2}\rho\,(\underline{V}^m_{KA} - \underline{V}^m_{DA})^2$$
$$+ (1-a)\sum_{n=1}^{m} (t_n - t_{n-1})\, \text{Tr}[(\sigma^{\theta n}_{SA} - \sigma^{\theta n}_{KA})(\dot{\varepsilon}^{n}_{SA} - \dot{\varepsilon}^{n}_{KA})$$

Using the same approach as in Section 6.3.4, one can show that $\mathbf{e}^{\Delta}_{RdC}(s^{\Delta}_{DA})$ is zero if and only if $s^{\Delta}_{Ad} = (\underline{U}^{\Delta}_{KA}, \underline{V}^{\Delta}_{KA}, \underline{A}^{\Delta}_{KA}, \sigma^{\Delta}_{DA}, \underline{\Gamma}^{\Delta}_{DA})$ is the exact solution of

Problem [(7.111) - (7.115)].

For a given time discretization, the measure $\mathbf{e}^{\Delta}_{RdC}(s^{\Delta}_{DA})$ enables one to evaluate the errors due to the space discretization.

Therefore, let

$$\mathbf{i}_{space} = \mathbf{e}^{\Delta}_{RdC}(s^{\Delta}_{DA})$$

Thus, one obtains an error indicator which enables one to evaluate the errors due to the space discretization.

The elastic behavior case

In the elastic behavior case, the most natural approach is to let

$$\mathbf{e}^{\Delta}_{RdC}(s^{\Delta}_{DA}) = \left[\operatorname*{Sup}_{1 \leq m \leq p} \int_{\Omega} \eta^{m}_{elas} \, d\Omega \right]^{1/2} \qquad (7.117)$$

with

$$\eta^{m}_{elas} = \frac{a}{2} \rho \, (\underline{V}^{m}_{KA} - \underline{V}^{m}_{DA})^2 + \frac{(1-a)}{2} \operatorname{Tr}\left[(\sigma^{m}_{KA} - \sigma^{m}_{DA}) \mathbf{K}^{-1} (\sigma^{m}_{KA} - \sigma^{m}_{DA}) \right]$$

REMARK. Expression (7.118) can also be viewed as a special case of (7.117) if $\sigma^{0}_{CA} = \sigma^{0}_{DA}$ by taking $\theta = 0.5$.

The implementation is easy. For the fields σ^{n}_{DA} and $\underline{\Gamma}^{n}_{DA}$, it suffices to take the values at time t_n of the fields σ_{DA} and $\underline{\Gamma}_{DA}$ constructed in (7.103). For the fields $\underline{U}^{n}_{KA}$, $\underline{V}^{n}_{KA}$ and $\underline{A}^{n}_{KA}$, one simply takes

$$\underline{U}^{n}_{KA}(M) = \mathbf{N}(M)\, \mathbf{q}_n , \quad \underline{V}^{n}_{KA}(M) = \mathbf{N}(M)\, \mathbf{v}_n , \quad \underline{A}^{n}_{KA}(M) = \mathbf{N}(M)\, \mathbf{a}_n$$

Thus, one obtains fields $s^{\Delta}_{Ad} = (\underline{U}^{\Delta}_{KA}, \underline{V}^{\Delta}_{KA}, \underline{A}^{\Delta}_{KA}, \sigma^{\Delta}_{DA}, \underline{\Gamma}^{\Delta}_{DA})$ which are admissible for the space-continuous, time-discretized problem.

7.3.7 Mass diagonalization error indicator

The concept of error in constitutive relation also enables one to build an error measure to evaluate the errors due to the diagonalization of the mass matrix required in the explicit scheme.

For the sake of simplicity, we will examine the elastic behavior case. Let us consider as the reference problem the discretized problem with a consistent mass matrix:

Find $\left\{\mathbf{q}_n, \mathbf{v}_n, \mathbf{a}_n, \boldsymbol{\Sigma}_n, \boldsymbol{\Gamma}_n\right\}_{n=0,1,2,\ldots,p}$ *such that:*

- *For any integer* $n < p$,

$$\mathbf{q}_{n+1} = \mathbf{q}_n + \Delta t\,\mathbf{v}_n + \frac{(\Delta t)^2}{2}\,\mathbf{a}_n$$
$$\mathbf{v}_{n+1} = \mathbf{v}_n + \frac{\Delta t}{2}\,(\mathbf{a}_n + \mathbf{a}_{n+1}) \tag{7.118}$$

- *For any integer* $n \leq p$,

$$\boldsymbol{\Gamma}_n + \boldsymbol{\Sigma}_n = \mathbf{F}(t_n) \tag{7.119}$$

$$\boldsymbol{\Sigma}_n = \mathbb{K}\mathbf{q}_n$$
$$\boldsymbol{\Gamma}_n = \mathbb{M}\mathbf{a}_n \tag{7.120}$$

with the given initial values

$$\mathbf{q}_0, \mathbf{v}_0, \quad \mathbf{a}_0 = \mathbb{M}^{-1}\left[\mathbf{F}(0) - \mathbb{K}\mathbf{q}_0\right], \quad \boldsymbol{\Sigma}_0 = \mathbb{K}\mathbf{q}_0, \quad \boldsymbol{\Gamma}_0 = \mathbb{M}\mathbf{a}_0 \tag{7.121}$$

Definition

$\overset{\Delta}{s}_{h,Ad} = \left\{\mathbf{q}_{n,KA}, \mathbf{v}_{n,KA}, \mathbf{a}_{n,KA}, \boldsymbol{\Sigma}_{n,DA}, \boldsymbol{\Gamma}_{n,DA}\right\}_{n=0,1,2,\ldots,p}$ *is said to be admissible with respect to Problem* [(7.118) - (7.121)] *if:*

- $\left\{\mathbf{q}_{n,KA}, \mathbf{v}_{n,KA}, \mathbf{a}_{n,KA}\right\}_{n=0,1,2,\ldots,p}$ *verifies* (7.118) *with the initial data*

  $\mathbf{q}_{0,KA} = \mathbf{q}_0, \quad \mathbf{v}_{0,KA} = \mathbf{v}_0, \quad \mathbf{a}_{0,KA} = \mathbf{a}_0$;

- $\left\{\boldsymbol{\Sigma}_{n,DA}, \boldsymbol{\Gamma}_{n,DA}\right\}_{n=0,1,2,\ldots,p}$ *verifies* (7.119) *with the initial data*

  $\boldsymbol{\Sigma}_{0,DA} = \boldsymbol{\Sigma}_0, \quad \boldsymbol{\Gamma}_{0,DA} = \boldsymbol{\Gamma}_0$.

To construct a measure of the error in constitutive relation, one associates with $\overset{\Delta}{s}_{h,Ad}$:

- $\left\{\boldsymbol{\Sigma}_{n,KA}, \boldsymbol{\Gamma}_{n,KA}\right\}_{n=0,1,2,\ldots,p}$ such that

  $\boldsymbol{\Sigma}_{n,KA} = \mathbb{K}\mathbf{q}_{n,KA}$

  $\boldsymbol{\Gamma}_{n,KA} = \mathbb{M}\mathbf{a}_{n,KA}$

- $\{\mathbf{g}_{n,DA}, \mathbf{q}_{n,DA}, \mathbf{v}_{n,DA}, \mathbf{a}_{n,DA}\}_{n=0,1,2,\dots,p}$ such that

$$\mathbf{g}_{n,DA} = \mathbb{K}^{-1}\,\boldsymbol{\Sigma}_{n,DA}$$

$$\mathbf{a}_{n,DA} = \mathbb{M}^{-1}\,\boldsymbol{\Gamma}_{n,DA}$$

$$\mathbf{v}_{n+1,DA} = \mathbf{v}_{n,DA} + \frac{\Delta t}{2}\,(\mathbf{a}_{n,DA} + \mathbf{a}_{n+1,DA})$$

$$\mathbf{q}_{n+1,DA} = \mathbf{q}_{n,DA} + \Delta t\,\mathbf{v}_{n,DA} + \frac{(\Delta t)^2}{2}\,\mathbf{a}_{n,DA}$$

$$\mathbf{q}_{0,DA} = \mathbf{q}_0 \quad \text{and} \quad \mathbf{v}_{0,DA} = \mathbf{v}_0$$

Then, one lets

$$\mathbf{e}^{\Delta}_{h,RdC}(s^{\Delta}_{h,Ad}) = \left[\sup_{1\le n\le p}\ \eta_n\right]^{1/2}$$

with

$$\eta_n = \frac{a}{2}\,(\mathbf{v}_{n,KA} - \mathbf{v}_{n,DA})^T\mathbb{M}\,(\mathbf{v}_{n,KA} - \mathbf{v}_{n,DA}) + \frac{1-a}{2}\,(\mathbf{q}_{n,KA} - \mathbf{g}_{n,DA})^T\mathbb{K}\,(\mathbf{q}_{n,KA} - \mathbf{g}_{n,DA})$$

One can immediately see that $e^{\Delta}_{h,RdC}(s^{\Delta}_{h,Ad})$ is zero if and only if $s^{\Delta}_{h,Ad}$ is the exact solution of Problem [(7.118) - (7.121)].

Let us now consider the solution calculated with the explicit central difference scheme using a diagonalized mass matrix

$$\mathbf{q}_{n+1} = \mathbf{q}_n + \Delta t\,\mathbf{v}_n + \frac{(\Delta t)^2}{2}\,\mathbf{a}_n$$

$$\mathbf{v}_{n+1} = \mathbf{v}_n + \frac{\Delta t}{2}\,(\mathbf{a}_n + \mathbf{a}_{n+1})$$

$$\mathbb{M}_{dia}\,\mathbf{a}_n + \mathbb{K}\,\mathbf{q}_n = \mathbf{F}(t_n)$$

with the data $\mathbf{q}_0,\ \mathbf{v}_0,\ \mathbf{a}_0 = \mathbb{M}^{-1}\left[\mathbf{F}(0) - \mathbb{K}\,\mathbf{q}_0\right]$

Letting

$$\mathbf{q}_{n,KA} = \mathbf{q}_n\,, \quad \mathbf{v}_{n,KA} = \mathbf{v}_n\,, \quad \mathbf{a}_{n,KA} = \mathbf{a}_n$$

and

$$\boldsymbol{\Sigma}_{n,DA} = \mathbb{K}\,\mathbf{q}_n\,, \quad \boldsymbol{\Gamma}_{n,DA} = \mathbb{M}_{dia}\,\mathbf{a}_n$$

one obtains $\overset{\Delta}{s}_{h,Ad}$ admissible for Problem [(7.118) - (7.121)]. With these conditions, the calculation of

$$\dot{\mathbf{i}}_{dia} = \overset{\Delta}{\mathbf{e}}_{h,RdC}(\overset{\Delta}{s}_{h,Ad})$$

enables one to estimate the errors due to the use of $\mathbb{M}_{dia}$.

7.3.8 Simple examples

First example

The objective of this example is to study the time evolution of the error in constitutive relation in the framework of fast dynamics. In order to do that, we will consider a one-d.o.f. dynamic problem in elasticity.

The sample problem

Let us consider the following differential equation, in which the unknown is the function $t \mapsto q(t)$:

$$\forall t \in [0, T], \quad \ddot{q}(t) + \frac{\pi^2}{4} q(t) = f(t) \; ; \quad q(0) = 1 \; ; \quad \dot{q}(0) = 0$$

The prescribed $f(t)$ is an impulse-type force defined by its graph in Figure 7.6. The final time is $T = 6$.

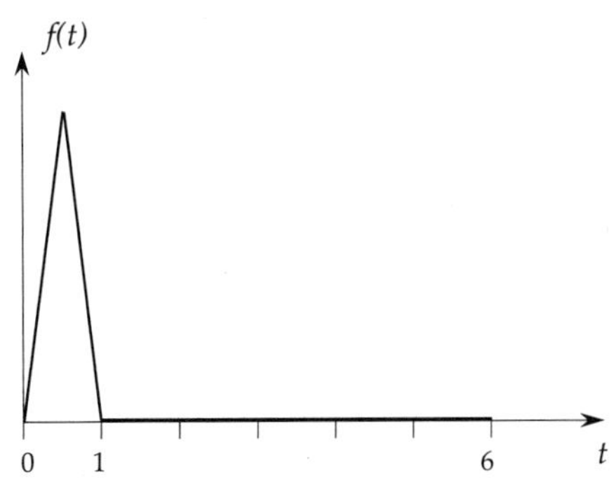

Figure 7.6. *Impulse-type force.*

Using the formalism of Section 7.3.1, this problem can be expressed as:

Find $q(t)$, $\sigma(t)$ *and* $\Gamma(t)$ *such that:*

- *Initial conditions*

$$q(0) = 1 \quad and \quad \dot{q}(0) = 0$$

- *Equilibrium equations*

$$\forall t \in [0, T], \quad \Gamma(t) + \sigma(t) = f(t)$$

- *Constitutive relations*

$$\forall\, t \in [0, T], \quad \Gamma(t) = \ddot{q}(t) \quad and \quad \sigma(t) = \frac{\pi^2}{4}\, q(t)$$

Quality of the constitutive relation error measure

The calculations were performed using the central difference scheme for a time step $\Delta t = 0.5$. One should note that for the problem being considered the stability condition of the scheme is $\Delta t \leq 1.27$.

Here, the error in constitutive relation is defined by

$$\mathbf{e}_{RdC} = \left[\operatorname*{Sup}_{t \in [0, T]} \eta(t) \right]^{1/2} \quad \text{or} \quad \tilde{\mathbf{e}}_{RdC} = \left[\frac{1}{T} \int_0^T \eta(t)\, dt \right]^{1/2}$$

with

$$\eta(t) = \frac{1}{4}\left[\dot{\tilde{U}}_{KA}(t) - \dot{U}_{DA}(t) \right]^2 + \frac{1}{\pi^2}\left[\sigma_{KA}(t) - \sigma_{DA}(t) \right]^2$$

In order to compare the error in constitutive relation with the error actually made, let us introduce the exact errors

$$\mathbf{e}_{ex}(t_n) = \left[\operatorname*{Sup}_{1 \leq i \leq n} \left[\frac{1}{4}(\dot{q}_{ex}(t_i) - v_i)^2 + \frac{\pi^2}{16}(q_{ex}(t_i) - q_i)^2 \right] \right]^{1/2}$$

and

$$\tilde{\mathbf{e}}_{ex}(t_n) = \left[\frac{\Delta t}{T} \sum_{1 \leq i \leq n} \left[\frac{1}{4}(\dot{q}_{ex}(t_i) - v_i)^2 + \frac{\pi^2}{16}(q_{ex}(t_i) - q_i)^2 \right] \right]^{1/2}$$

These exact errors are defined at each nodal instant of the discretization. To facilitate the graphical representations, these can be extended between two nodal instants by linear interpolation.

Figure 7.7 shows the evolutions of $\mathbf{e}_{RdC}$ and $\mathbf{e}_{ex}$ as functions of time, and Figure 7.8 shows the evolutions of $\tilde{\mathbf{e}}_{RdC}$ and $\tilde{\mathbf{e}}_{ex}$. One can observe that the estimate of the error in constitutive relation is very good.

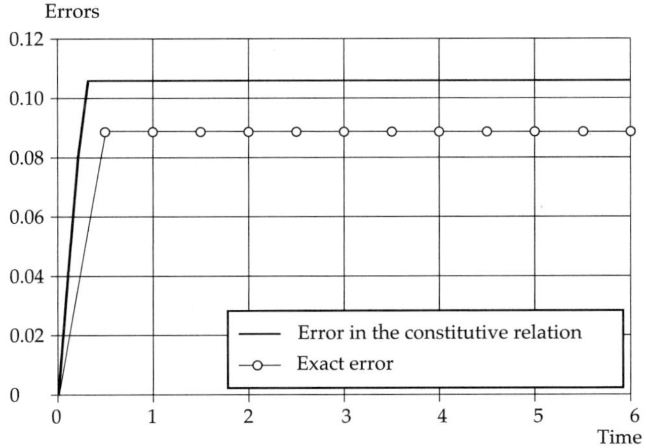

Figure 7.7. *Errors as functions of time (sup in time).*

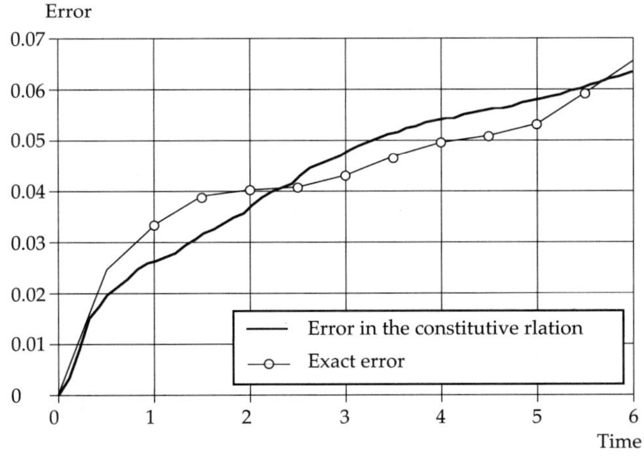

Figure 7.8. *Errors as functions of time (time integral).*

Besides, the study conducted in [COMBE - PELLE, 1998] showed that the error in constitutive relation converges, as does the exact error, in $O(\Delta t^2)$. Figure 7.9 shows that this result is confirmed numerically.

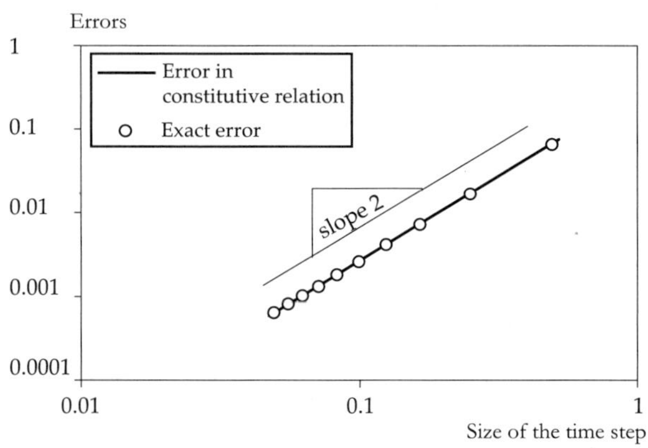

Figure 7.9. *Convergence rate of the error in constitutive relation.*

Comparison with other estimators

Now, still considering the one-d.o.f. problem, we will present a comparison of the constitutive relation error estimator with other indicators proposed in the literature. These indicators are based on the numerical evaluation of the consistency error $e_{c,\,i+1}$ over the time step $[t_i, t_{i+1}]$. Then, error E_{i+1} is obtained by the weighted summation of these consistency errors. For the problem considered here, the following indicators were tested:

- the ZIENKIEWICZ - XIE [ZIENKIEWICZ - XIE, 1991] indicator

$$e_{c,\,i+1} = \frac{(\Delta t)^2 (a_{i+1} - a_i)}{6}$$

and

$$E_{i+1} = \sum_{n=1}^{i+1} \left[\frac{\pi^2}{8} e_{c,\,n}^2 \right]^{1/2}$$

- the WIBERG - LI indicator [WIBERG - LI, 1993], which introduces a consistency error in the displacements and a consistency error in the velocities defined, respectively, by

$$e_{c,\,i+1} = -\frac{(\Delta t)^3(\xi_{i+1} + 3\,\xi_i)}{24}$$

$$\tilde{e}_{c,\,i+1} = -\frac{(\Delta t)^2(\xi_{i+1} - \xi_i)}{12}$$

where the ξ_i are defined by the recurrence

$$\xi_{i+1} = \frac{2\,(a_{i+1} - a_i)}{\Delta t} - \xi_i$$

$$\xi_0 = \dot{f}(0) - \frac{\pi^2}{4}v_0$$

Thus, the error is defined by

$$E_{i+1} = \sum_{n=1}^{i+1}\left[\frac{\pi^2}{8}e_{c,\,n}^2 + \frac{1}{2}\tilde{e}_{c,\,n}^2\right]^{1/2}$$

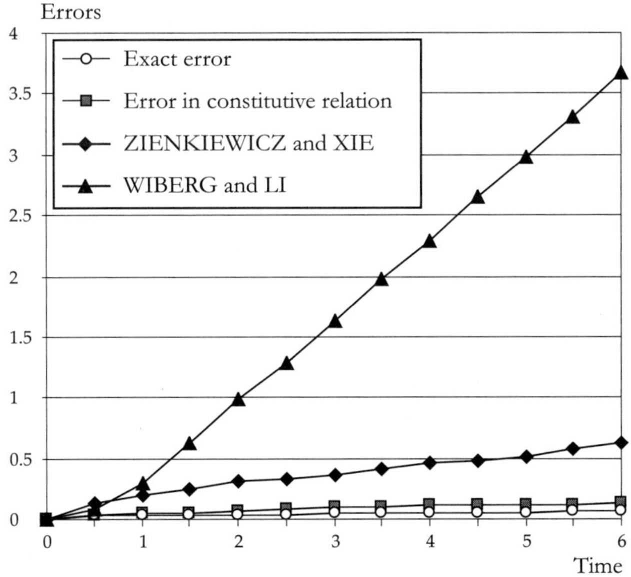

Figure 7.10. *Comparison of the indicators.*

Figure 7.10 shows the evolutions of the exact error, of the error in constitutive relation and of the two indicators above as functions of time. Clearly, the accumulation of errors over time is correct only for the constitutive relation error estimator.

Second example

This example concerns a compressed bar in linear dynamics (Figure 7.11).

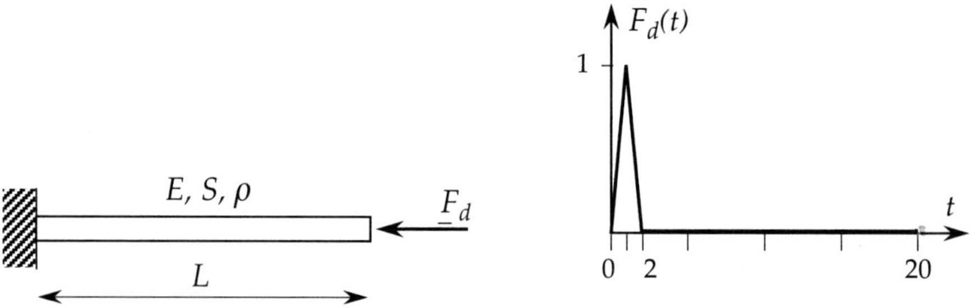

Figure 7.11. *Bar in compression.*

The different physical constants are set to the following values:

$$E = 1\,\mathrm{N\,m^{-2}}; \quad S = 1\,\mathrm{m^2}; \quad \rho = 1\,\mathrm{kg\,m^{-3}}; \quad L = 20\,\mathrm{m}$$

The duration $T = 20\,\mathrm{s}$ corresponds to the time needed for the wave to reach the built-in support. The computations were carried out for several regular meshes and several regular time discretizations.

Here, the stability limit is

$$\Delta t_{stab} = L / (\text{number of elements})$$

For the time step Δt considered, let $\beta = \Delta t / \Delta t_{stab}$

Figure 7.12 shows the evolution of the error in constitutive relation as a function of the number of elements and of the value of β, and Figure 7.13 shows the effectivity index obtained, again as a function of the number of elements and of β. One can see that for a time step chosen close to the stability limit ($0.8 \leq \beta \leq 1$) the constitutive relation error estimator is very good.

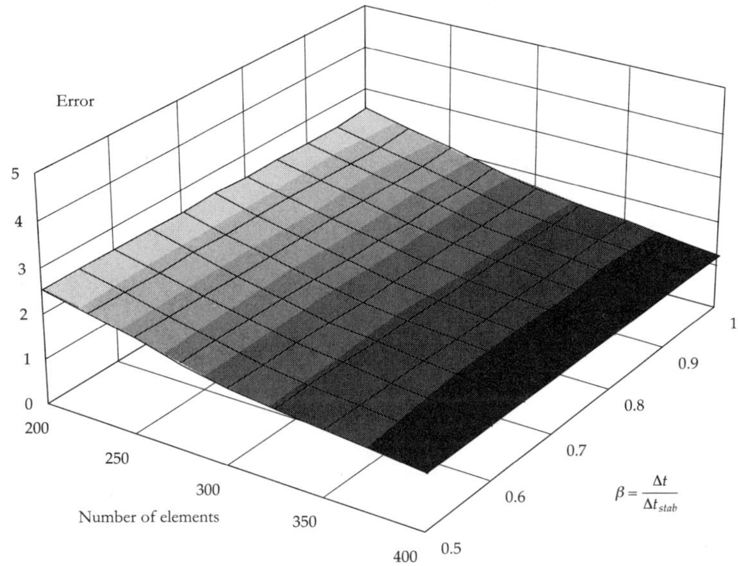

Figure 7.12. *Evolution of the error in constitutive relation.*

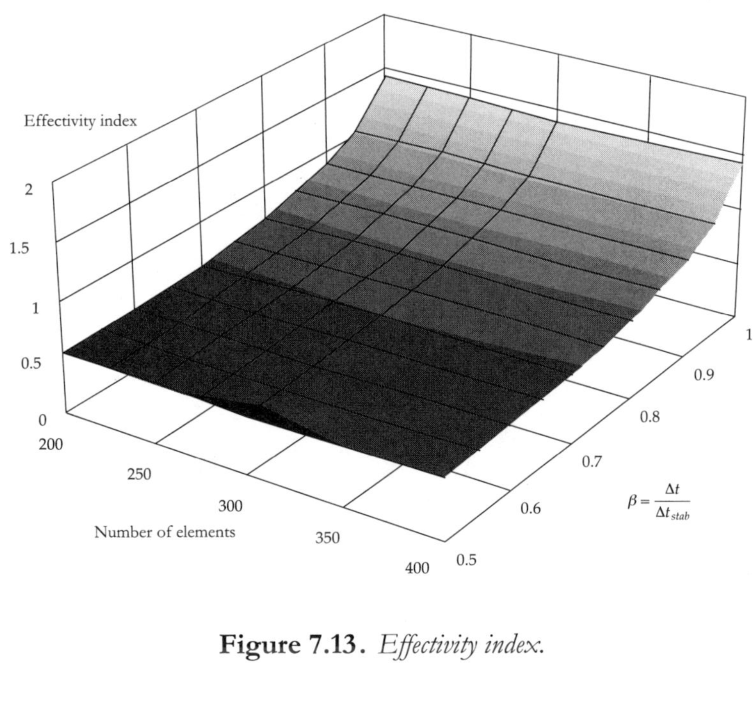

Figure 7.13. *Effectivity index.*

Chapter 8

Techniques for constructing admissible fields

8.1 Introduction

As we saw in the previous chapters, the implementation of the concept of error in constitutive relation requires the construction of admissible fields.

These construction techniques depend on the type of problem one is dealing with, and exploit the properties of the calculated finite element solutions. The main features of these construction techniques were presented in Chapter 2 in the case of linear elasticity problems. Let us recall that these admissible field construction techniques for linear problems also served as the bases of admissible field construction techniques for nonlinear problems (Chapter 6) and for dynamic problems (Chapter 7).

The objective of this chapter is to describe these techniques in more detail for the linear case.

The most widely used finite element methods usually provide an admissible displacement field. Therefore, most of this chapter focuses on the construction of admissible stress fields, i.e. of stress fields which verify the equilibrium equations exactly.

Let us recall that the construction method consists of two stages:

- construction at the interelement boundaries of inter-forces (or inter-fluxes in the thermal case) in equilibrium with each element's interior loads;
- construction element by element of a stress tensor (or a heat flux vector in the thermal case) in equilibrium with the inter-forces and the interior loads.

The main – and the most difficult – stage is the first. It was introduced for the first time for 2D elasticity and for plate theory in [LADEVEZE, 1975]. It was first developed for 2D linear thermal problems in [LADEVEZE, 1977]. It is on this problem, simpler than the elasticity problem but not trivial, that we describe the construction method in detail in the first part of this chapter, with the purpose of emphasizing the key aspects of the method. Another more geometric presentation can be found in [LADEVEZE - MAUNDER, 1996, 1997].

Elasticity problems are treated next. We finalize, for all node types, the construction of the force densities given in Chapter 2 for an interior nodal vertex. Then, we carry out the construction element by element of the admissible stress fields.

The analytical construction of admissible stress fields is developed for the most common 2D, axisymmetric and 3D finite elements. These constructions originated for 3-node elements in the works of LADEVEZE and GUEZEL [GUEZEL, 1982], for 6-node triangles, 4- and 8-node quadrilaterals and axisymmetric elements in the works of PELLE and ROUGEOT [PELLE - ROUGEOT, 1988, 1989], [ROUGEOT, 1989], and for 3D elements in the works of MARIN and PELLE [GASTINE - LADEVEZE - MARIN - PELLE, 1992].

Next, we present the new method for constructing densities based on the weak extension condition proposed in [LADEVEZE, 1994]. This method enables one to improve the quality of the constitutive relation error estimators significantly in difficult situations (particularly for elements with high aspect ratios) [LADEVEZE - ROUGEOT, 1997].

Then, we present two examples of situations where the construction of an admissible displacement field different from the finite element field is also necessary:

- incompressible or nearly incompressible elasticity calculations;
- analysis of elastic plates.

To conclude this chapter, we show in an example the principle of the proof of the inequalities (1.34) for the constitutive relation error measures.

More generally, the methods in which the equilibrium equations are verified a priori are called "TREFFTZ's methods". These methods are very powerful, although their implementation and use are not always easy. For finite element calculations based on such methods, the reader is referred to [FRAEIJS DE VEUBEKE, 1965], [FRAEIJS DE VEUBEKE - SANDER, 1968], [FRAEIJS DE VEUBEKE - HOGGE, 1970], [STEIN - AHMAD, 1977], [JIROUSEK, 1985], [JIROUSEK - WROBLESKI, 1996], [PEREIRA - ALMEIDA - MAUNDER, 1999], [KOMPIS - FRASTIA, 1997] and the review article [ZIENKIEWICZ, 1997]. There are specific problems for which these methods turn out to be even more effective [LADEVEZE, 1983], [HOCHARD - LADEVEZE - PROSLIER, 1993], [BOUBARCHENE - HOCHARD - POITOU, 1997], [MELENK - BABUSKA, 1997], [TEIXEIRAS DE FREITAS, 1997], [LADEVEZE - LOISEAU - DUREISSEIX, 2001], [LADEVEZE - ARNAUD - ROUCH - BLANZE, 2000].

8.2 The 2D thermal-type problem

8.2.1 Formulation of the problem

Let us consider a stationary heat conduction problem which can be expressed as follows:

Find a temperature field $u(M)$ *and a heat flux vector* $\underline{q}(M)$ *such that:*

- u *verifies the constraints*

$$u = u_d \quad on\ \partial_1\Omega \tag{8.1}$$

- $\underline{q}$ *verifies the thermal equilibrium equations*

$\forall u^*$ *zero on* $\partial_1\Omega$,

$$-\int_\Omega \underline{q} \cdot \underline{\text{grad}}\, u^*\, d\Omega + \int_\Omega f_d\, u^*\, d\Omega + \int_{\partial_2\Omega} F_d\, u^*\, d\Gamma = 0 \tag{8.2}$$

- $\underline{q}$ *and* $\underline{\text{grad}}\ u$ *verify the constitutive relation*

$$\underline{q} = \mathbf{K}\ \underline{\text{grad}}\ u \tag{8.3}$$

where u_d, f_d *and* F_d *are given and* $\mathbf{K}$ *is a tensor which depends on the material alone and is assumed to be positive definite.*

REMARK. Usually, it is the vector $-\underline{q}$ which is called the heat flux vector, and Fourier's law is written as

$$(-\underline{q}) = -\mathbf{K}\ \underline{\text{grad}}\ u$$

The choice we made here enables us to keep notations and properties which are consistent with those of the elasticity problem.

8.2.2 Finite element discretization

The "displacement" finite element method leads to the following approximate problem:

Find a field u_h *verifying* (8.1) *and such that:*

$\forall u_h^*$ *of the finite element type and zero on* $\partial_1\Omega$,

$$\int_\Omega \underline{\text{grad}}\ u_h \cdot \mathbf{K}(\underline{\text{grad}}\ u_h^*)\ d\Omega = \int_\Omega f_d\, u_h^*\ d\Omega + \int_{\partial_2\Omega} F_d\, u_h^*\ d\Gamma \tag{8.4}$$

Then, one calculates an approximation of the heat flux vector through the constitutive relation

$$\underline{q}_h = \mathbf{K}\ (\underline{\text{grad}}\ u_h)$$

Adjusting the definitions of Chapter 2 to this type of problem, a pair $(\hat{u}, \hat{\underline{q}})$ is admissible if $\hat{u}$ verifies (8.1) and $\hat{\underline{q}}$ verifies (8.2).

Thus, the corresponding error in constitutive relation is

$$\mathbf{e}_{RdC}(\hat{u}, \hat{\underline{q}}) = \left[\int_\Omega (\hat{\underline{q}} - \underline{\text{grad}}\ \hat{u})\mathbf{K}^{-1}(\hat{\underline{q}} - \underline{\text{grad}}\ \hat{u})\ d\Omega\right]^{1/2}$$

8.2.3 Construction of admissible fields

As in the case of elasticity problems, the field u_h is kinematically admissible. Thus, for simplicity's sake, one usually chooses $\hat{u} = u_h$. Then, the only difficulty in obtaining an admissible pair is in constructing a vector field $\hat{\underline{q}}_h$ which verifies the equations of thermal equilibrium (8.2) exactly. In order to facilitate the presentation, we will restate for this simple case the methods already presented in Chapter 2 for elasticity.

Review of the principles of the construction

In order to simplify the notations, we will consider only plane problems and assume that the discretization uses only 3-node triangles. Thus, the restrictions of the shape functions φ_i to a triangle E are the triangle's barycentric coordinates.

The construction is achieved by requiring $\hat{\underline{q}}_h$ to be an extension of the finite element solution (strong extension condition):

For any basis function φ_ι *of the finite element type and for any element* E *of the mesh*:

$$\int_E (\hat{\underline{q}}_h - \underline{q}_h) \cdot \underline{\text{grad}}\, \varphi_\iota \, dE = 0 \tag{8.5}$$

First, on the element sides, one constructs scalar densities $\hat{F}$ intended to represent the heat transfer across Γ as follows (η_E being defined as in Section 2.4.3),

$$\hat{\underline{q}}_h \cdot \underline{n}_E = \eta_E \hat{F} \quad \text{on } \partial E \tag{8.6}$$

Of course, on an edge $\Gamma \subset \partial_2\Omega$, one sets

$$\eta_E \hat{F} = F_d$$

Next, $\hat{\underline{q}}_h$ is constructed on each element as the solution of the following local problem:

$$\begin{aligned} &\text{div}\, \hat{\underline{q}}_h + f_d = 0 \quad \text{in } E \\ &\hat{\underline{q}}_h \cdot \underline{n}_E = \eta_E \hat{F} \quad \text{on } \partial E \end{aligned} \tag{8.7}$$

Construction of the densities $\hat{F}$

Preliminary calculations

The admissibility of $\hat{\underline{q}}_h$ leads to

$$\forall u^*, \quad \int_E \hat{\underline{q}}_h \cdot \underline{\text{grad}}\, u^* \, dE = \int_E f_d \, u^* \, dE + \int_{\partial E} \hat{\underline{q}}_h \cdot \underline{n}_E \, u^* \, d\Gamma$$

Letting $u^* = \varphi_i$, i.e. the shape function associated with a node i of E, one gets

$$\int_E \hat{\underline{q}}_h \cdot \underline{\text{grad}}\, \varphi_i \, dE = \int_E f_d \varphi_i \, dE + \int_{\partial E} \hat{\underline{q}}_h \cdot \underline{n}_E \varphi_i \, d\Gamma \tag{8.8}$$

Using (8.5), (8.6) and (8.8), one gets for each φ_i,

$$\int_{\partial E} \eta_E \hat{F} \varphi_i \, d\Gamma = Q_E(i) \tag{8.9}$$

with

$$Q_E(i) = \int_E \underline{q}_h \cdot \underline{\text{grad}}\, \varphi_i \, dE - \int_E f_d \varphi_i \, dE \tag{8.10}$$

which is a function of only the finite element solution and the data.

For a node i, the shape function φ_i is nonzero on the two edges Γ_1 and Γ_2 connected to i.

Therefore, one has

$$\int_{\Gamma_1} \eta_E \hat{F} \varphi_i \, d\Gamma + \int_{\Gamma_2} \eta_E \hat{F} \varphi_i \, d\Gamma = Q_E(i) \tag{8.11}$$

The relations (8.11), written for all the elements connected to a given node i, enable the determination of the projections of the densities $\hat{F}$ onto the shape functions of the element being considered.

More precisely, the method of calculation of these projections depends on the location of node i: node interior to Ω or node located at the boundary $\partial\Omega$; in the latter case, the method also depends on the types of boundary conditions applied along the edges connected to this node.

Case of an interior node

Let us now consider an interior node i. We will use the notations detailed in Figure 8.1. Letting

$$b_n(i) = \int_{\Gamma_n} \eta_{E_n} \hat{F} \varphi_i \, d\Gamma$$

and writing (8.11) for each of the elements E_n, one gets the linear system

$$\begin{cases} b_1(i) - b_2(i) = Q_{E_1}(i) \\ b_2(i) - b_3(i) = Q_{E_2}(i) \\ \vdots \qquad\qquad \vdots \\ b_N(i) - b_1(i) = Q_{E_N}(i) \end{cases} \tag{8.12}$$

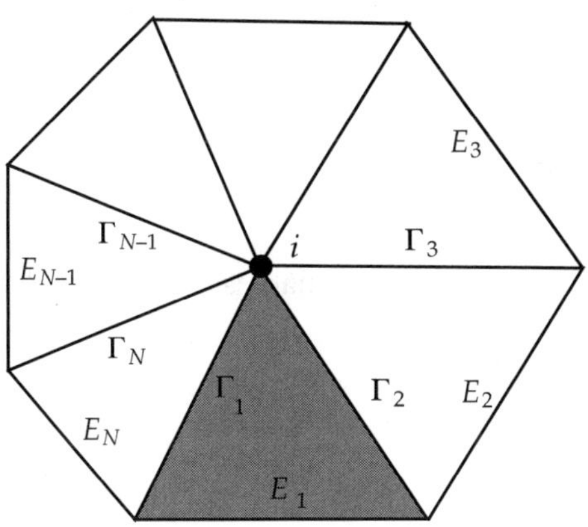

Figure 8.1. *Elements connected to an interior node.*

In this form, it is clear that this system has solutions if and only if the sum of the right-hand sides is zero

$$\sum_{n=1}^{N} Q_{E_n}(i) = 0 \tag{8.13}$$

From (8.10), one has

$$\sum_{n=1}^{N} Q_{E_n}(i) = \sum_{n=1}^{N} \left[\int_{E_n} q_h \cdot \underline{\text{grad}}\ \varphi_i \, dE - \int_{E_n} f_d \varphi_i \, dE \right]$$

Equation (8.13) follows immediately from the finite element equilibrium relation (8.4) with $u_h^* = \varphi_i$. Therefore, the system (8.12) always has an infinite number of solutions. Among these solutions, one selects that which minimizes

$$\sum_{n=1}^{N} \left[\frac{b_n(i) - b_n^m}{p_n} \right]^2 \tag{8.14}$$

where p_n are weighting coefficients and b_n^m is the average finite element heat flux from elements E_{n-1} and E_n,

$$b_n^m = \frac{1}{2} \int_{\Gamma_n} (\underline{q}_{h|E_{n-1}} + \underline{q}_{h|E_n}) \cdot \underline{n}_{E_n} \varphi_i \, d\Gamma$$

Let

$$\tilde{Q}_n = Q_{E_n}(i) - b_n^m + b_{n+1}^m \quad (\text{with: } n+1 = 1 \text{ if } n = N)$$

and

$$P_n = \sum_{k=n+1}^{N} \frac{1}{p_k^2}$$

Then, the minimization of (8.14) enables one to calculate $b_1(i)$ in the form

$$b_1(i) = b_1^m + \frac{1}{P_0} \sum_{k=1}^{N-1} \tilde{Q}_k P_k \tag{8.15}$$

and the other $b_n(i)$ by

$$b_n(i) = b_{n-1}(i) - Q_{E_{n-1}}(i) \tag{8.16}$$

which determines the projections of the densities onto the shape functions.

Case of a boundary node

For a boundary node, several cases must be considered depending on the types of boundary conditions along the edges on both sides of node *i*:

- prescribed temperature along both edges: interior node of $\partial_1\Omega$;
- prescribed heat flux along both edges: interior node of $\partial_2\Omega$;
- prescribed heat flux along one edge and prescribed temperature along the other: boundary node between $\partial_1\Omega$ and $\partial_2\Omega$.

First, let us consider the case of an interior node of $\partial_1\Omega$ (Figure 8.2).

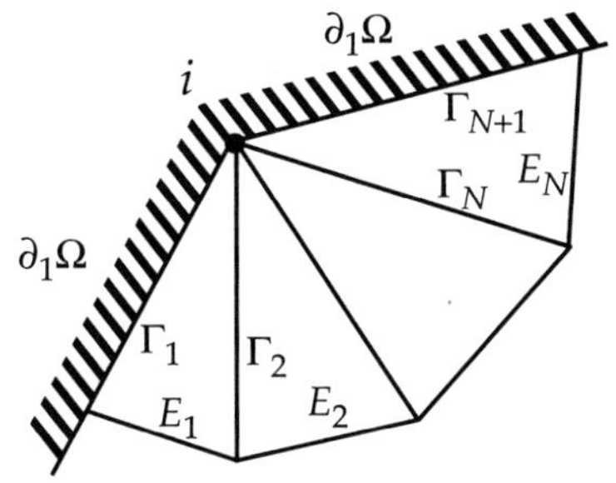

Figure 8.2. *Node inside* $\partial_1\Omega$.

Let us write

$$\text{for } n = 1, 2, \ldots, N: \quad b_n(i) = \int_{\Gamma_n} \eta_{E_n} \hat{F} \varphi_i \, d\Gamma$$

and

$$b_{N+1}(i) = -\int_{\Gamma_{N+1}} \eta_{E_N} \hat{F} \varphi_i \, d\Gamma$$

Expressing (8.11) for each element connected to i, one gets

$$\begin{cases} b_1(i) - b_2(i) = Q_{E_1}(i) \\ b_2(i) - b_3(i) = Q_{E_2}(i) \\ \vdots \qquad\qquad \vdots \\ b_N(i) - b_{N+1}(i) = Q_{E_N}(i) \end{cases} \tag{8.17}$$

which is a linear system of N equations with the N+1 unknowns $b_n(i)$. If one assigns $b_1(i)$ arbitrarily, all the other $b_n(i)$ are completely determined.

To eliminate this indeterminacy, one selects the solution which minimizes

$$\sum_{n=1}^{N+1} \left[\frac{b_n(i) - b_n^m}{p_n} \right]^2 \tag{8.18}$$

with, for $n = 2, 3, \ldots, N$,

$$b_n^m = \frac{1}{2} \int_{\Gamma_n} (\underline{q}_{h|E_{n-1}} + \underline{q}_{h|E_n}) \cdot \underline{n}_{E_n} \varphi_i \, d\Gamma \tag{8.19}$$

and

$$b_1^m = \int_{\Gamma_1} \underline{q}_{h|E_1} \cdot \underline{n}_{E_1} \varphi_i \, d\Gamma$$
$$b_{N+1}^m = -\int_{\Gamma_{N+1}} \underline{q}_{h|E_N} \cdot \underline{n}_{E_N} \varphi_i \, d\Gamma \tag{8.20}$$

The resolution of (8.18) enables one to calculate $b_1(i)$,

$$b_1(i) = b_1^m + \frac{1}{P_0} \sum_{k=1}^{N} \tilde{Q}_k P_k \tag{8.21}$$

with

$$\tilde{Q}_n = Q_{E_n}(i) - b_n^m + b_{n+1}^m \quad \text{for } n = 1, 2, \ldots, N$$
$$P_n = \sum_{k=n+1}^{N+1} \frac{1}{p_k^2} \quad \text{for } n = 0, 1, \ldots, N$$

which determines the projections of the densities.

Now, let us consider the case of an interior node of $\partial_2\Omega$ (Figure 8.3). One still gets (8.17); here, however, $b_1(i)$ and $b_{N+1}(i)$ are known data

$$b_1(i) = \int_{\Gamma_1} F_d \varphi_i \, d\Gamma; \quad b_{N+1}(i) = -\int_{\Gamma_{N+1}} F_d \varphi_i \, d\Gamma \tag{8.22}$$

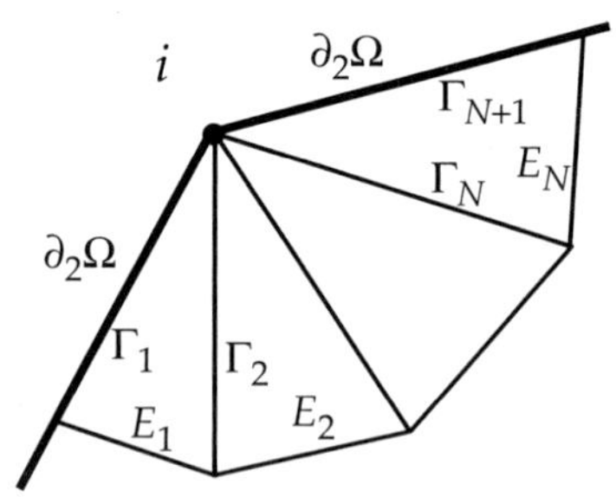

Figure 8.3. *Node inside* $\partial_2\Omega$.

Then, (8.17) is a system of N equations with $N-1$ unknowns. It has a unique solution if and only if

$$b_1(i) - b_{N+1}(i) = \sum_{n=1}^{N} Q_{E_n}(i) \tag{8.23}$$

Again, this condition follows from the finite element equilibrium verified by $\underline{q}_h$. In this case the resolution of the system is immediate.

Finally, let us examine the case where i is a boundary node between $\partial_1\Omega$ and $\partial_2\Omega$ (Figure 8.4). As in the previous cases, one arrives at System (8.17). However, in this case, a single one of the quantities $b_1(i)$ or $b_{N+1}(i)$ is known.

For example, in the case of Figure 8.4, the heat flux is prescribed along Γ_1. Therefore, $b_1(i)$ is given by

$$b_1(i) = \int_{\Gamma_1} F_d \varphi_i \, d\Gamma$$

and the other $b_n(i)$ are calculated directly from Equations (8.17).

Of course, if the heat flux is prescribed along Γ_{N+1}, it is $b_{N+1}(i)$ which is given by

$$b_{N+1}(i) = -\int_{\Gamma_{N+1}} F_d \, \varphi_i \, d\Gamma$$

Therefore, on any element edge, one knows the projections of $\hat{F}$ onto the shape functions associated with that edge.

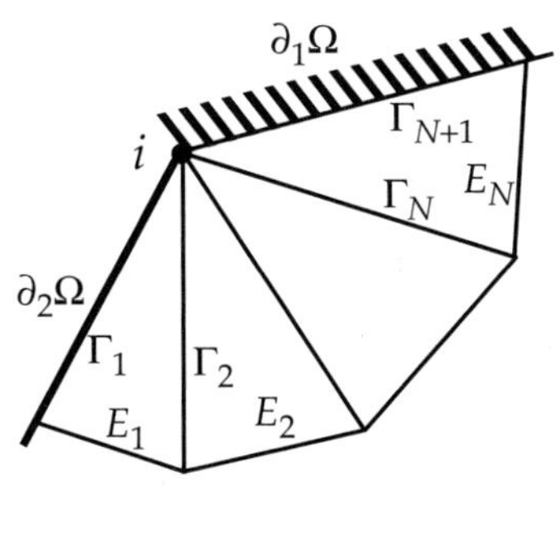

Figure 8.4. *Node common to* $\partial_1\Omega$ *and* $\partial_2\Omega$.

Calculation of the nodal values

Let Γ be an edge of an element E. One knows the projections

$$\int_\Gamma \hat{F}\,\varphi_i\,d\Gamma$$

for the two shape functions associated with the vertices of Γ.

Seeking $\hat{F}$ in the form

$$\hat{F} = \sum_i a_i \varphi_i$$

the nodal values a_i are obtained by solving a 2×2 linear system.

Equilibrium on each element

The admissibility of $\hat{\underline{q}}_h$ along with (8.5) yield, for any u_h^* of the finite element type,

$$\int_E \underline{q}_h \cdot \underline{\text{grad}}\, u^*\, dE = \int_E f_d\, u^*\, dE + \int_{\partial E} \eta_E\, \hat{F}\, u^*\, d\Gamma$$

In particular, for u_h^* constant, one gets

$$\int_E f_d\, dE + \int_{\partial E} \eta_E\, \hat{F}\, d\Gamma = 0 \tag{8.24}$$

which, for a thermal problem, expresses the equilibrium of element E. This property is a direct consequence of the extension condition (8.5).

REMARKS AND COMMENTS

1. In earlier works [LADEVEZE, 1977], the weighting coefficients p_n were chosen equal to 1. Since then, the many examples solved have shown that it is preferable to take $p_n = L_n = \text{mes}(\Gamma_n)$.

2. One must concede that it has not been possible to determine a cost function with a strong mechanical meaning, yet simple, in order to eliminate the indeterminacy of the projections of the densities onto the shape functions associated with the vertex nodes during the calculation. The cost function (8.14) used here works well for the thermal case. However, for elasticity problems, this cost function is clearly less effective in certain severe situations (particularly for elements with very high aspect ratios). This observation spurred the development of the improved construction of the densities given in Section 8.6.

Construction of the field $\hat{\underline{q}}_h$

Having determined the densities $\hat{F}$, one calculates the restriction $\hat{\underline{q}}_{h|E}$ of the field $\hat{\underline{q}}_h$ to each element by solving the local problem

$$\begin{aligned} &\operatorname{div} \hat{\underline{q}}_{h|E} + f_d = 0 \quad \text{in } E \\ &\hat{\underline{q}}_{h|E} \cdot \underline{n}_E = \eta_E \hat{F} \quad \text{on } \partial E \end{aligned} \tag{8.25}$$

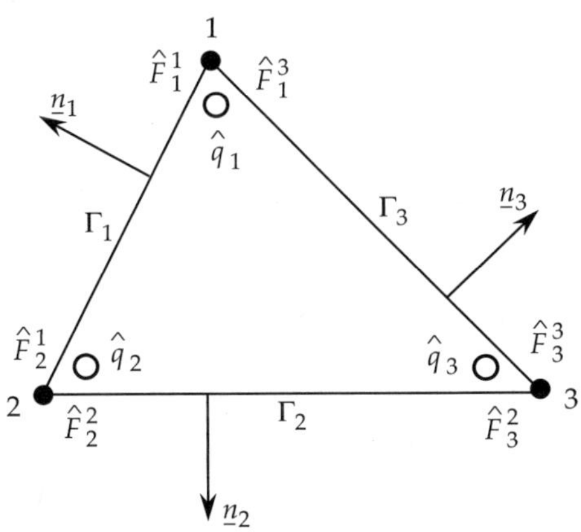

Figure 8.5. *Notations for a first-order triangle.*

For a first-order triangle, the densities are affine along each of the element's edges. Assuming constant heat-sources f_d on each element – which, in practice, is not very restrictive – one can construct a solution of (8.25) in the form

$$\hat{\underline{q}}_{h|E} = \sum_{i=1}^{3} \hat{\underline{q}}_i \lambda_i$$

where λ_i are the barycentric coordinates of the triangle E.

Let (Figure 8.5),

$$\eta_E \hat{F}_{|\Gamma_1} = \hat{F}_1^1 \lambda_1 + \hat{F}_2^1 \lambda_2$$

$$\eta_E \hat{F}_{|\Gamma_2} = \hat{F}_2^2 \lambda_2 + \hat{F}_3^2 \lambda_3$$

$$\eta_E \hat{F}_{|\Gamma_3} = \hat{F}_3^3 \lambda_3 + \hat{F}_1^3 \lambda_1$$

Then, using the second equation in (8.25), one determines the vectors $\hat{\underline{q}}_i$ through the covariant coordinates of two of the vectors $\underline{n}_i$,

$$\hat{\underline{q}}_1 \cdot \underline{n}_1 = \hat{F}_1^1 ; \quad \hat{\underline{q}}_1 \cdot \underline{n}_3 = \hat{F}_1^3$$

$$\hat{\underline{q}}_2 \cdot \underline{n}_2 = \hat{F}_2^2 ; \quad \hat{\underline{q}}_2 \cdot \underline{n}_1 = \hat{F}_2^1$$

$$\hat{\underline{q}}_3 \cdot \underline{n}_3 = \hat{F}_3^3 ; \quad \hat{\underline{q}}_3 \cdot \underline{n}_2 = \hat{F}_3^2$$

Then, it is easy to verify that the first equation in (8.25) is a consequence of the equilibrium condition (8.24).

8.3 2D elasticity problems

8.3.1 Construction of the densities in elasticity

The objective of this section is to detail, in the 2D case, the calculation method for the projections of the densities in elasticity which was developed in Chapter 2 for an interior node of Ω, and to complete this method by addressing the case of boundary nodes. We will reuse the notations already introduced in Section 2.4.3, except that for a 2D mesh the number of edges of an element which are connected to a vertex is always $R = 2$.

Case of an interior node

Vertex node

Let i be an interior vertex node of Ω. Using notations similar to those of the thermal case (Figure 8.1), and letting

$$\underline{b}_n(i) = \int_{\Gamma_n} \eta_{E_n} \hat{\underline{F}} \varphi_i \, d\Gamma$$

the system (2.27) becomes

$$\begin{cases} \underline{b}_1(i) - \underline{b}_2(i) = \underline{Q}_{E_1}(i) \\ \underline{b}_2(i) - \underline{b}_3(i) = \underline{Q}_{E_2}(i) \\ \vdots \qquad\qquad \vdots \\ \underline{b}_N(i) - \underline{b}_1(i) = \underline{Q}_{E_N}(i) \end{cases} \tag{8.26}$$

which is a system of the same type as (8.12) except that now the unknowns and the known data are vectors.

As in the thermal case, only $N-1$ equations are independent, and it follows from the finite element equilibrium that (8.26) has an infinite number of solutions.

Among all these solutions, one chooses that which minimizes

$$\sum_{n=1}^{N} \left[\frac{\underline{b}_n(i) - \underline{b}_n^m}{p_n} \right]^2 \tag{8.27}$$

with

$$\underline{b}_n^m = \frac{1}{2} \int_{\Gamma_n} (\sigma_{h|E_{n-1}} + \sigma_{h|E_n}) \underline{n}_{E_n} \varphi_i \, d\Gamma$$

The solution of this minimization problem is obtained through formulas similar to (8.15) and (8.16).

Nonvertex node

If node i is located on an interior edge Γ, it is common to only two elements E_1 and E_2.

In this case, (2.27) reduces to two equations

$$\int_{\Gamma} \eta_{E_1} \hat{\underline{F}} \varphi_i \, d\Gamma = \underline{Q}_{E_1}(i) \quad \text{and} \quad \int_{\Gamma} \eta_{E_2} \hat{\underline{F}} \varphi_i \, d\Gamma = \underline{Q}_{E_2}(i)$$

The first equation determines the projection being sought. Thanks to the equality $\eta_{E_1} + \eta_{E_2} = 0$ and to the finite element equilibrium $\underline{Q}_{E_1}(i) + \underline{Q}_{E_2}(i) = 0$, the second equation is also verified. In this case, the projection sought is unique.

Case of a boundary node

Vertex node

The details of this calculation depend on the boundary conditions prescribed along the edges connected to vertex i.

As for the thermal case, one can show that for an interior node i of $\partial_1\Omega$ the system (2.27) has an infinite number of solutions, and one chooses that which minimizes a quantity similar to (8.27). If i is inside $\partial_2\Omega$, or at the boundary between $\partial_1\Omega$ and $\partial_2\Omega$, the system has a unique solution.

REMARKS

1. In elasticity, other situations may be considered. For example, the normal displacement and the tangent force can be prescribed along an edge connected to i. The determination of the projections of the densities can be extended, with no difficulty, to any type of situation [ROUGEOT, 1989].

2. The construction of the densities works equally well for triangular elements, for quadrilateral elements, and for a mixture of triangles and quadrilaterals.

Nonvertex node

In this case, there is only one element E connected to node i. Depending on the type of boundary condition prescribed along the corresponding edge, either $\hat{\underline{F}}$ is known, or its projection is determined by the single equation derived from (2.27). Thus, the projections of the densities $\hat{\underline{F}}$ onto the shape functions are completely determined on each element edge.

Assuming that $\hat{\underline{F}}$ is a combination of the shape functions on Γ,

$$\hat{\underline{F}} = \sum_i \underline{A}_i \varphi_{i|\Gamma}$$

the density $\hat{\underline{F}}$ can be determined by solving a small linear system.

Modification of the densities

It is possible to modify locally, edge by edge, the densities constructed by the above method while preserving the equilibrium of each element. For example, on each edge Γ, one can add to $\underline{\hat{F}}$ any vector $\underline{\hat{H}}$ with zero resultant and moment

$$\int_\Gamma \underline{\hat{H}}\, d\Gamma = 0 \quad \text{and} \quad \underline{N} \cdot \int_\Gamma \underline{X} \wedge \underline{\hat{H}}\, d\Gamma = 0$$

where:

- $\underline{X}$ is a vector whose origin is a fixed point of the plane, e.g. the element's center of inertia G, and whose extremity is the current point M of Γ: $\underline{X} = \underline{GM}$;
- $\underline{N}$ is the unit vector normal to the plane such that the operator $\underline{N}\wedge$ is the rotation $+\pi/2$.

This technique was introduced in [COFFIGNAL, 1987] for 3-node triangles. In this case the density $\underline{\hat{F}}$ is affine along each edge Γ,

$$\underline{\hat{F}} = \underline{A} + (\lambda_1 - \lambda_2)\underline{B}$$

where $\underline{A}$ and $\underline{B}$ are constant vectors and λ_1 and λ_2 are the barycentric coordinates on Γ. Let us decompose $\underline{B}$ into

$$\underline{B} = b_n \underline{n} + b_t \underline{t}$$

where $\underline{n}$ and $\underline{t}$ are a normal vector and a tangent vector to Γ, respectively (Figure 8.6).

Then, the vector $(\lambda_1 - \lambda_2) b_t \underline{t}$ has zero resultant and moment. Therefore, $\underline{\hat{F}}$ can be replaced by $\underline{\hat{F}} - (\lambda_1 - \lambda_2) b_t \underline{t}$.

In some cases, this modification leads to a substantial improvement of the constitutive relation error measure.

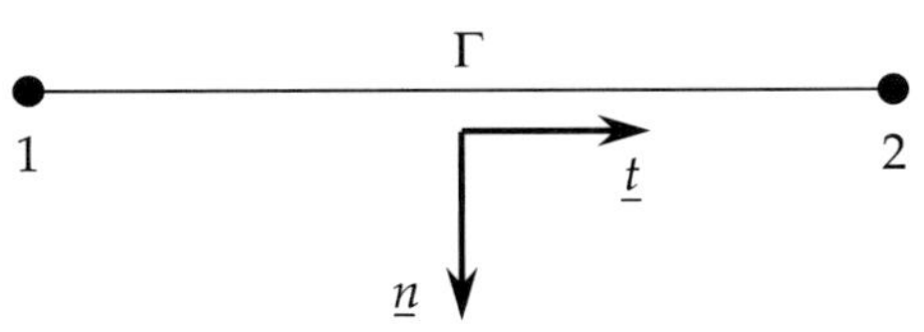

Figure 8.6. *Notations.*

REMARKS

1. If one uses such modifications, the strong extension conditions are no longer respected. We will return to this point in Section 8.6.

2. A more geometric presentation of the calculation of the densities of the inter-forces, which casts a new light on the method, can be found in [LADEVEZE - MAUNDER, 1996]. In that reference work, all possible modifications of the densities are systematically examined.

8.3.2 Construction of admissible stress fields

The objective of this section is to detail, for the most common finite elements (3- and 6-node triangles, 4- and 8-node quadrilaterals), the method of construction of a simple analytical solution $\hat{\sigma}_{h|E}$ of the local equations

$$\underline{\text{div}}\, \hat{\sigma}_h + \underline{f}_d = 0 \quad \text{in } E$$

$$\hat{\sigma}_h\, \underline{n}_E = \eta_E\, \underline{\hat{F}} \quad \text{on } \partial E$$

Preliminary remark

A priori, the first approach for obtaining a simple solution would be to seek a polynomial solution on each element E. In general, such a solution does not exist. Indeed, let us consider a vertex i of E (Figure 8.7).

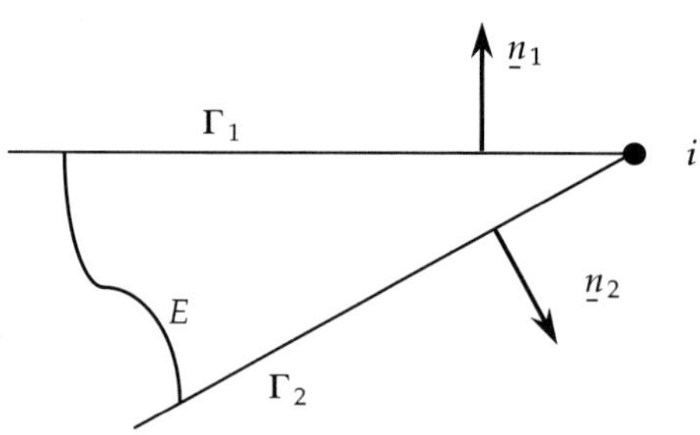

Figure 8.7. *Notations.*

For a regular stress field $\hat{\sigma}_{h|E}$, one must have

$$\left[\hat{\sigma}_{h|E}(i)\, \underline{n}_1\right] \cdot \underline{n}_2 = \left[\hat{\sigma}_{h|E}(i)\, \underline{n}_2\right] \cdot \underline{n}_1$$

Therefore, at i, $\hat{\underline{F}}_{|\Gamma_1}$ and $\hat{\underline{F}}_{|\Gamma_2}$ must verify the compatibility condition

$$\left[\eta_E \hat{\underline{F}}\right]_{|\Gamma_1} \cdot \underline{n}_2 = \left[\eta_E \hat{\underline{F}}\right]_{|\Gamma_2} \cdot \underline{n}_1$$

In general, the densities constructed do not verify these symmetry conditions. This leads us to partition the elements and to seek solutions which are polynomials within the subelements.

Construction for the 3-node triangle

Let us subdivide the triangle E into three subtriangles e_i as indicated in Figure 8.8, which also indicates the notations used in this section.

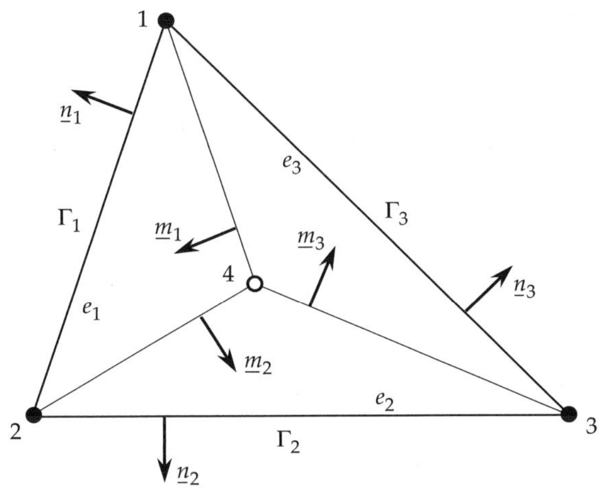

Figure 8.8. *Subdivision into subtriangles.*

Point 4 is usually the center of inertia of the triangle; however, any point strictly interior to the triangle can be used. Let

$$\lambda_i^i, \lambda_j^i \text{ and } \lambda_4^i \quad (i = 1, 2, 3 \text{ and } j = 2, 3, 1)$$

be the barycentric coordinates of the subtriangles.

It should be recalled that in this case the force densities constructed along the edges are affine

$$\left[\eta_E \hat{\underline{F}}\right]_{|\Gamma_i} = \hat{\underline{F}}_i^i \lambda_i^i + \hat{\underline{F}}_j^i \lambda_j^i$$

By considering only loads $\underline{f}_d$ which are constant on each element, one can seek $\hat{\sigma}_{h|E}$ as a field which is affine on each subtriangle

$$\hat{\sigma}_{h|e_i} = \hat{\sigma}_i = \sigma_i^i \lambda_i^i + \sigma_j^i \lambda_j^i + \sigma_4^i \lambda_4^i$$

where σ_i^i, σ_j^i and σ_4^i are symmetric 2×2 operators to be determined.

Calculation of the stresses at the vertices

At each vertex, one expresses the conditions of equilibrium with the force densities and the continuity of the stress vector along the interior edge.

For example, at Vertex 1, one gets

$$\sigma_1^1 \underline{n}_1 = \hat{\underline{F}}_1^1 ; \quad \sigma_1^3 \underline{n}_3 = \hat{\underline{F}}_1^3 \quad \text{and} \quad \sigma_1^1 \underline{m}_1 = \sigma_1^3 \underline{m}_1$$

which determines σ_1^1 and σ_1^3 completely. One proceeds in the same way to determine σ_2^2 and σ_2^1 at Vertex 2 and σ_3^3 and σ_3^2 at Vertex 3.

Calculation of the stresses at Point 4

The continuity conditions for the stress vector along the interior edges yield

$$(\sigma_4^1 - \sigma_4^3)\, \underline{m}_1 = 0 ; \quad (\sigma_4^2 - \sigma_4^1)\, \underline{m}_2 = 0 ; \quad (\sigma_4^3 - \sigma_4^2)\, \underline{m}_3 = 0$$

It is easy to verify that these three conditions are equivalent to the equality of the three operators

$$\sigma_4^1 = \sigma_4^2 = \sigma_4^3 = \sigma_4$$

To calculate σ_4, it suffices to write on each subtriangle the condition

$$\underline{\operatorname{div}}\, \hat{\sigma}_i + \underline{f}_d = 0$$

Thus, one has six scalar equations, only three of which are independent due to the global equilibrium of the element. These three equations enable one to determine σ_4.

Thus, taking into account the choices that were made, the field $\hat{\sigma}_{h|E}$ is uniquely determined.

Construction for the 6-node triangle

In this case, the constructed force densities are of degree 2 along each edge Γ_i; using a hierarchical basis, they can be written as

$$\left[\eta_E \hat{\underline{F}}\right]_{|\Gamma_i} = \hat{\underline{F}}_i^i \lambda_i^i + \hat{\underline{F}}_j^i \lambda_j^i + \lambda_i^i \lambda_j^i \hat{\underline{G}}_i$$

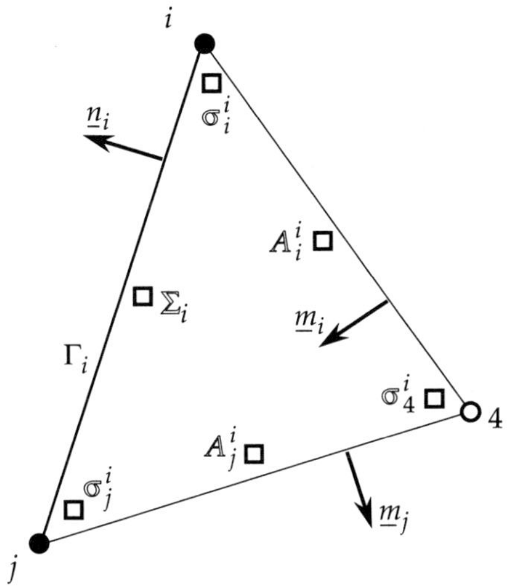

Figure 8.9. *Notations for subtriangle* e_i.

By considering only loads $\underline{f}_d$ which are linear on each element, one can seek $\hat{\sigma}_{h|E}$ in the form of a field of degree 2 per subtriangle (Figure 8.9),

$$\hat{\sigma}_{h|e_i} = \hat{\sigma}_i = \sigma_i^i \lambda_i^i + \sigma_j^i \lambda_j^i + \sigma_4^i \lambda_4^i + \Sigma_i \lambda_i^i \lambda_j^i + \lambda_4^i\left[\mathbb{A}_i^i \lambda_i^i + \mathbb{A}_j^i \lambda_j^i\right]$$

where σ_i^i, σ_j^i, σ_4^i, $\mathbb{A}_i^i$, $\mathbb{A}_j^i$ and Σ_i are symmetric 2×2 tensors to be determined.

Calculation of the stresses at the vertices

The calculation is performed as for the 3-node triangle.

Calculation of the stresses at the midpoints of the element's edges

Let

$$\underline{N}_i = \text{mes}(\Gamma_i)\, \underline{n}_i \quad \text{and} \quad \mathbf{N}_i = \underline{N}_i \underline{N}_i^T$$

Then, any symmetric 2×2 tensor $\mathbb{S}$ can be uniquely decomposed using the three tensors $\mathbf{N}_1$, $\mathbf{N}_2$ and $\mathbf{N}_3$,

$$\mathbb{S} = \sum_{p=1}^{3} S(p)\mathbf{N}_p$$

Therefore, let us seek $\mathbb{\Sigma}_i$ in the form

$$\mathbb{\Sigma}_i = \sum_{p=1}^{3} \Sigma_i(p)\mathbf{N}_p$$

The equality $\hat{\sigma}_h \, \underline{n}_E = \eta_E \hat{\underline{F}}$ written at the midpoint of edge Γ_i yields

$$\mathbb{\Sigma}_i \, \underline{N}_i = \text{mes}(\Gamma_i) \, \hat{\underline{G}}_i$$

which determines the two $\Sigma_i(p)$ for $p \neq i$. However, $\Sigma_i(i)$ remains undetermined.

Calculation of the stresses at the midpoints of the interior edges

Let

$$\underline{M}_i = \ell_i \, \underline{m}_i \, ; \quad \tilde{\underline{M}}_i = \underline{z} \wedge \underline{M}_i \quad \text{and} \quad \mathbf{M}_i = \tilde{\underline{M}}_i \tilde{\underline{M}}_i^T$$

where ℓ_i is the distance of vertex i to the center of inertia of the element.

Let us seek the operators $\mathbb{A}_i^i$ and $\mathbb{A}_j^i$ in the form

$$\mathbb{A}_i^i = \sum_{p=1}^{3} A_i^i(p) \, \mathbf{M}_p \quad \text{and} \quad \mathbb{A}_j^i = \sum_{p=1}^{3} A_j^i(p) \, \mathbf{M}_p$$

Then, by writing on each e_i the equations

$$\underline{\text{div}} \, \hat{\sigma}_h + \underline{f}_d = 0$$

and the continuity conditions along the interior edges, one gets

$$\mathbb{A}_i^i (\underline{N}_i - \underline{M}_j) + \mathbb{A}_j^i \underline{M}_i = \frac{2}{3} \text{mes}(E) \left[f_d(i) - f_d(4) \right] + \mathbb{\Sigma}_i \underline{M}_i$$

$$\mathbb{A}_j^i (\underline{N}_i + \underline{M}_i) - \mathbb{A}_i^i \underline{M}_j = \frac{2}{3} \text{mes}(E) \left[f_d(j) - f_d(4) \right] - \mathbb{\Sigma}_i \underline{M}_j$$

$$\left[\mathbb{A}_i^i - \mathbb{A}_i^k \right] \underline{M}_i = 0$$

where i, j, k is a circular permutation of 1, 2, 3 and $f_d(i), f_d(j), f_d(4)$ are the values of f_d at the vertices of e_i.

By expressing these equations in terms of the $A_i^i(p)$ and $A_j^i(p)$, one can show [PELLE - LADEVEZE - ROUGEOT, 1989] that the 18 $A_i^i(p)$ and $A_j^i(p)$ are the solutions of an 18×18 linear system of the form

$$\mathbf{SA} = \mathbf{b}$$

where the matrix $\mathbf{S}$ is independent of the element and, thus, may be inverted only once.

Calculation of the stresses at the center of inertia

The calculation is similar to that performed for the 3-node triangle.

Construction for the 4- and 8-node quadrilaterals

A similar approach can be used for 4- or 8-node quadrilaterals by subdividing each element into four subtriangles using the diagonals (Figure 8.10). Then, in each subtriangle, the field is sought of degree 1 (4-node quadrilateral) or 2 (8-node quadrilateral). The details of the calculations can be found in [ROUGEOT, 1989].

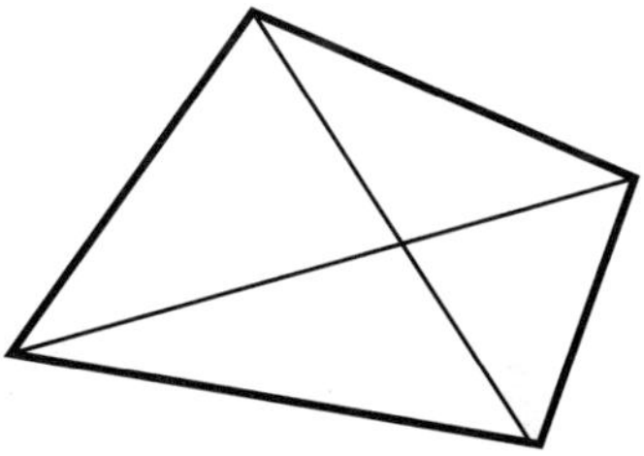

Figure 8.10. *Subdivision of a quadrilateral.*

REMARKS AND COMMENTS

1. For elements of degree 2, the analytical constructions presented above leave one free parameter per edge. To set these parameters, several strategies are possible.

For example, one can set these parameters by taking the equivalent quantity calculated from the finite element stress field. Another option consists in minimizing the error in constitutive relation locally on each element.

2. It is possible to extend the previous techniques to triangular or quadrangular elements of degree greater than 2. However, in this case, it seems preferable to use the approximate constructions proposed in Chapter 2.

8.4 Axisymmetric elasticity problems

Now, let us consider the situation where the geometry, the applied loads and the material are axially symmetric. The structure Ω is generated by a meridian S (Figure 8.11). A finite element calculation on Ω reduces to a finite element calculation on the plane geometry S. Thus, one uses plane finite elements E which, in reality, correspond to 3D axisymmetric elements $\tilde{E}$. Let $(\underline{e}_z, \underline{e}_r, \underline{e}_\theta)$ be the basis of polar coordinates and (z, r) the coordinates in the meridian plane $(\underline{e}_z, \underline{e}_r)$. A stress field σ can be written in the form

$$\sigma = \begin{bmatrix} \sigma_z & \sigma_{rz} & 0 \\ \sigma_{rz} & \sigma_r & 0 \\ 0 & 0 & \sigma_\theta \end{bmatrix}_{(\underline{e}_z, \underline{e}_r, \underline{e}_\theta)}$$

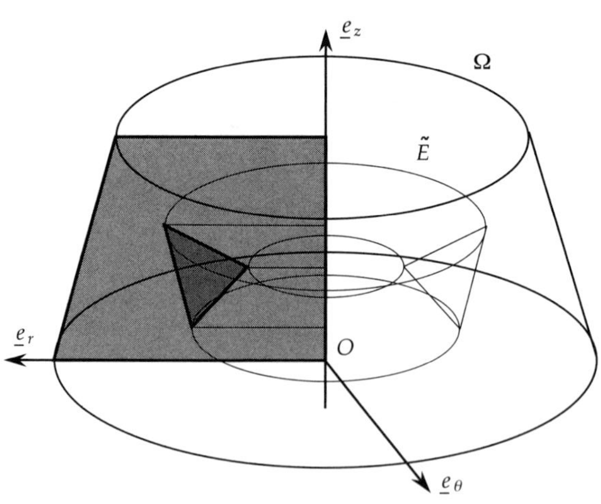

Figure 8.11. *Axisymmetric structure.*

Likewise, let $\underline{f}_d = f_z \, \underline{e}_z + f_r \, \underline{e}_r$

In particular, if $\eta_E \, \hat{\underline{F}}$ is the force density along the edges of an element, the equilibrium equations can be expressed as

$$\begin{aligned} &\operatorname{div}(\mathbb{C}\,\underline{e}_z) + r f_z = 0 \quad \text{in } E \\ &\mathbb{C}\,\underline{n}_E = \eta_E \, r \, \hat{\underline{F}} \qquad \text{on } \partial E \end{aligned} \tag{8.28}$$

$$\sigma_\theta = \operatorname{div}(\mathbb{C}\,\underline{e}_r) + r f_r \quad \text{in } E \tag{8.29}$$

where

$$\mathbb{C} = r \begin{bmatrix} \sigma_z & \sigma_{rz} \\ \sigma_{rz} & \sigma_r \end{bmatrix}_{(\underline{e}_z,\, \underline{e}_r)} \tag{8.30}$$

Equations (8.28) present definite analogies with the equations of 2D elasticity, although the form is simpler here.

Besides, having determined the operator $\mathbb{C}$, one obtains σ_θ through (8.29) regardless of the form of f_r. Finally, one can note that the densities $\eta_E \, \hat{\underline{F}}$ are multiplied by r.

8.4.1 Construction of the densities

The axial symmetry simplifies the equilibrium conditions on each element. Indeed, for (8.28) to have solutions it suffices that the following single scalar condition be verified:

$$\int_E r f_z \, dE + \int_{\partial E} \eta_E \, r \, \hat{\underline{F}} \cdot \underline{e}_z \, d\Gamma = 0 \tag{8.31}$$

Therefore, in the axisymmetric case, the extension conditions needed for the construction of a density field are reduced to

For any element E *of the mesh*
and for any basis function φ_i *of the finite element type*

$$\int_E \left[(\hat{\mathbb{C}}_h - \mathbb{C}_h) \, \underline{e}_z \right] \cdot \underline{\text{grad}} \; \varphi_i \, dE = 0 \tag{8.32}$$

where $\hat{\mathbb{C}}_h$ *and* $\mathbb{C}_h$ *are defined using* $\hat{\boldsymbol{\sigma}}_h$ *and* $\boldsymbol{\sigma}_h$ *as in* (8.30).

Then, two types of methods can be envisaged [PELLE - LADEVEZE - ROUGEOT, 1989].

Quasi-plane method

One prescribes an extension condition which is stronger than (8.32):

For any element E *of the mesh*
and for any basis function φ_i *of the finite element type*

$$\int_E \left[(\hat{\mathbb{C}}_h - \mathbb{C}_h) \, \underline{\text{grad}} \; \varphi_i + (\hat{\sigma}_\theta - \sigma_{h,\theta}) \, \varphi_i \, \underline{e}_r \right] dE = 0$$

Then, the calculation of the densities is performed in a way very similar to that of the densities in 2D. It suffices to observe that it is the projections of $r \, \underline{\hat{F}}$ which are determined, and that the $\underline{Q}_E(i)$ are given by

$$\underline{Q}_E(i) \cdot \underline{e}_z = \int_E r f_z \, \varphi_i \, dE - \int_E \left[\mathbb{C}_h \underline{e}_z \right] \cdot \underline{\text{grad}} \; \varphi_i \, dE$$

$$\underline{Q}_E(i) \cdot \underline{e}_r = \int_E r f_r \, \varphi_i \, dE - \int_E \left[\mathbb{C}_h \underline{e}_r \right] \cdot \underline{\text{grad}} \; \varphi_i \, dE - \int_E \sigma_{h,\theta} \, \varphi_i \, dE$$

Improved method

One uses the extension condition (8.32) to determine the $r \, \underline{\hat{F}} \cdot \underline{e}_z$. Having carried out this calculation, one determines the nodal values of $r \, \underline{\hat{F}} \cdot \underline{e}_r$ such that the regularity condition at the vertices of the elements is verified as closely as possible. For example, let us consider an interior node and use the notations of Figure 8.1. Let

$$r \, \underline{\hat{F}}_{|\Gamma_n}(i) = A_n \, \underline{e}_z + B_n \, \underline{e}_r$$

The A_n are known and the B_n are to be determined. This can be done by solving the minimization problem

$$\underset{\{B_n\}}{\text{Min}} \sum_{n=1}^{N} G_n^2$$

with

$$G_n = \eta_{E_n} \left[A_n \, \underline{e}_z + B_n \, \underline{e}_r \right] \cdot \underline{n}_{E_n} - \eta_{E_n} \left[A_{n+1} \, \underline{e}_z + B_{n+1} \, \underline{e}_r \right] \cdot \underline{n}_{E_n}$$

8.4.2 Construction of the stress field

We will consider only the case of the 3-node triangle. The method is the same for the other elements [PELLE - LADEVEZE - ROUGEOT, 1989].

In order to perform an analytic construction of stress fields $\hat{\mathbb{C}}_h$ which are affine within each subtriangle, one must set $f_z = 0$. However, one should remember that the component f_r is arbitrary.

Using the notations of Figure 8.8 again, let

$$r\left[\eta_E \underline{\hat{F}}\right]_{|\Gamma_i} = \underline{\hat{F}}_i^i \lambda_i^i + \underline{\hat{F}}_j^i \lambda_j^i$$

On each subtriangle, the field $\hat{\mathbb{C}}_h$ is sought in the form

$$\hat{\mathbb{C}}_{h|e_i} = \hat{\mathbb{C}}_i = \mathbb{C}_i^i \lambda_i^i + \mathbb{C}_j^i \lambda_j^i + \mathbb{C}_4^i \lambda_4^i$$

Calculation of the stresses at the vertices

At each vertex, one expresses the conditions of equilibrium with the force densities and the continuity of the stress vector along the interior edge. Thus, $\mathbb{C}_1^1$, $\mathbb{C}_1^3$, $\mathbb{C}_2^2$, $\mathbb{C}_2^1$, $\mathbb{C}_3^3$ and $\mathbb{C}_3^2$ are determined completely.

Calculation of the stresses at Point 4

The conditions of continuity of the stress vector across the interior edges show that

$$\mathbb{C}_4^1 = \mathbb{C}_4^2 = \mathbb{C}_4^3 = \mathbb{C}_4$$

Applying to each subtriangle the condition

$$\underline{\text{div}}\, \hat{\mathbb{C}}_i \cdot \underline{e}_z = 0$$

one gets three scalar equations which involve only the two unknowns $\mathbb{C}_{4,z}$ and $\mathbb{C}_{4,rz}$. These equations are not independent because of the equilibrium condition (8.31). Thus, one determines $\mathbb{C}_{4,z}$ and $\mathbb{C}_{4,rz}$, but $\mathbb{C}_{4,r}$ remains undetermined. Therefore, the field $\hat{\mathbb{C}}_{h|E}$ is known as a function of $\mathbb{C}_{4,r}$. In order to know $\hat{\sigma}_{h|E}$, $\hat{\sigma}_\theta$ remains to be calculated using

$$\hat{\sigma}_\theta = \text{div}\,(\hat{\mathbb{C}}_{h|E}\, \underline{e}_r) + r f_r$$

In order to set the unknown $\mathbb{C}_{4,r}$, one can, for example, minimize the error in constitutive relation locally on the element.

8.4.3 Specific treatment at nodes located on the axis

For elements which have one or two vertices on the axis of symmetry, it is necessary to verify that the field $\hat{\sigma}_{h|E}$ is sufficiently regular to be statically admissible.

More precisely, the following conditions must be verified:

$$\underline{\text{div}}\, \hat{\sigma}_{h|E} \text{ square integrable} \tag{8.33}$$

$$\hat{\sigma}_{h|E} \text{ square integrable and with finite energy} \tag{8.34}$$

For sufficiently regular interior loads $\underline{f}_d$, the property (8.33) is verified because of the equilibrium equation

$$\underline{\text{div}}\, \hat{\sigma}_{h|E} + \underline{f}_d = 0$$

Besides, condition (8.34) can be written for each subtriangle as

$$\int_{e_i} \left[\text{Tr}\left(\frac{1}{r} \hat{\mathbb{C}}_{h|E} \hat{\mathbb{C}}_{h|E} \right) + r\, \hat{\sigma}_\theta^2 \right] dz\, dr < +\infty \tag{8.35}$$

On each subtriangle, one has by construction

$$\text{Tr}\left(\frac{1}{r} \hat{\mathbb{C}}_{h|E} \hat{\mathbb{C}}_{h|E} \right) = a + b\, r + c\, z + d\, \frac{z}{r} + e\, \frac{z^2}{r}$$

where a, b, c, d, e are constants on subtriangle e_i.

If element E has only one vertex on the axis, a simple calculation shows that (8.35) is verified. However, if E has two vertices on the axis, (8.35) is verified if and only if d and e are zero on the subtriangle e_i which has an edge on the axis (Figure 8.12).

To verify these conditions, one may simply construct the field $\hat{\mathbb{C}}_i$ in such a way that

$$\hat{\mathbb{C}}_i = \mathbb{C}_4^i \lambda_4^i \tag{8.36}$$

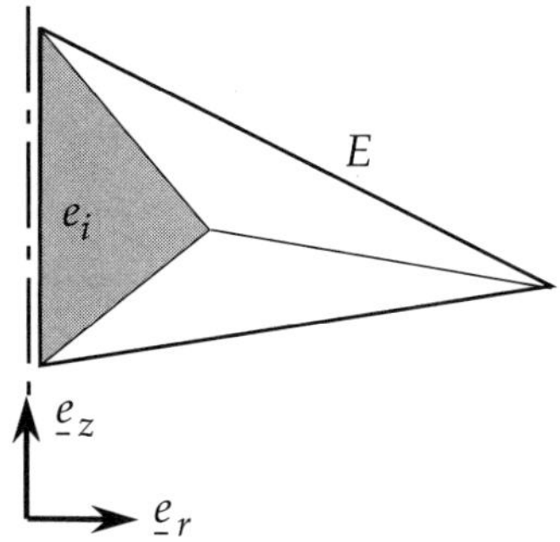

Figure 8.12. *Triangle with an edge on the axis.*

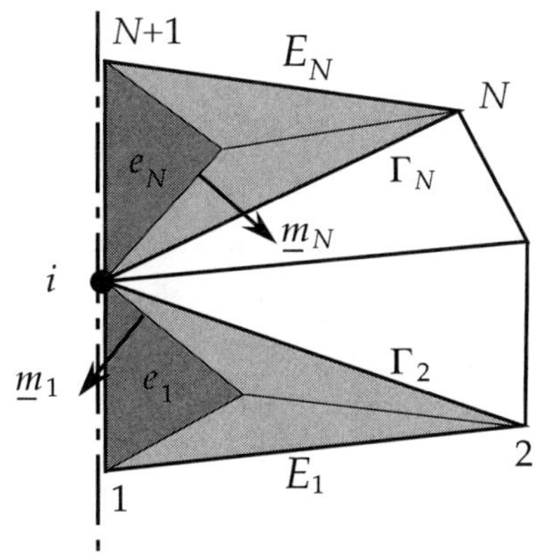

Figure 8.13. *Local modification of the densities.*

In order to verify (8.36), one considers densities $\hat{\underline{F}}$ which, initially, are a priori zero along the axis of symmetry, and one constructs these densities by one of the methods described in Section 8.4.1. Then, one modifies the densities locally so that they verify (8.36). Let us consider the case of a vertex *i* located on the axis of symmetry and surrounded by at least three elements, two of which have an edge on the axis (Figure 8.13). Since $\hat{\underline{F}}$ is zero along the axis, in order to achieve (8.36), one must modify the densities so that (with the notations defined in Figure 8.13),

$$\hat{\underline{F}}_{|\Gamma_2} \cdot \underline{m}_1 = 0 \quad \text{and} \quad \hat{\underline{F}}_{|\Gamma_N} \cdot \underline{m}_N = 0 \tag{8.37}$$

Thus, one ensures that fields $\hat{\mathbb{C}}_{h|e_1}$ and $\hat{\mathbb{C}}_{h|e_N}$ are zero at vertex i.

Let us consider the first relation in (8.37). The density $\hat{\underline{F}}_{|\Gamma_2}$ is of the form

$$r\,\hat{\underline{F}}_{|\Gamma_2} = a\ \underline{e}_z + b\ \underline{e}_r$$

where a and b are affine functions.

If $\underline{m}_1 \bullet \underline{e}_r \neq 0$, it suffices to replace the function b by

$$a\,\frac{\underline{m}_1 \bullet \underline{e}_z}{\underline{m}_1 \bullet \underline{e}_r}$$

If $\underline{m}_1 \bullet \underline{e}_r = 0$, one replaces b by 0 and the affine function $a = a_i\,\lambda_i + a_2\,\lambda_2$ by $\tilde{a} = \tilde{a}_2\,\lambda_2$, so that

$$\int_\Gamma \tilde{a}_2\,\lambda_2\,d\Gamma = \int_\Gamma (a_i\,\lambda_i + a_2\,\lambda_2)\,d\Gamma$$

in order for the equilibrium properties to be preserved.

Using the same procedure on Γ_N, one sets the fields $\hat{\mathbb{C}}_{h|e_1}$ and $\hat{\mathbb{C}}_{h|e_N}$ to zero at vertex i.

In the case of i being connected to only two elements (Figure 8.14), it suffices to modify the density along the common edge Γ as in the previous case for $\underline{m}_1 \bullet \underline{e}_r = 0$.

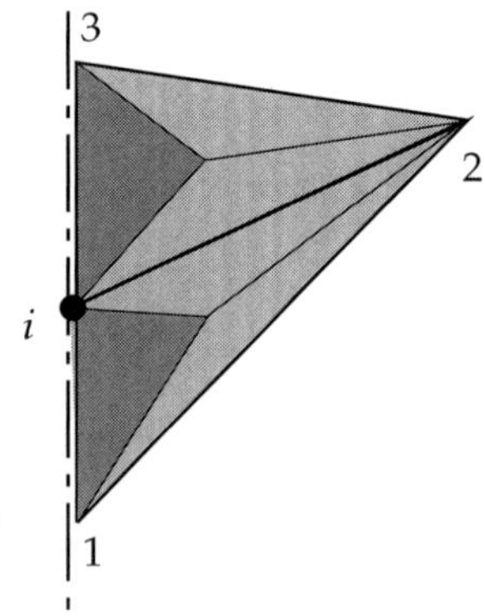

Figure 8.14.

The cases of vertices located on the axis and on the boundary $\partial\Omega$ can be dealt with in a very similar way, except for the special case represented in Figure 8.15 in which only a single element is connected to node *i*. Indeed, in this case, $\hat{\mathbb{C}}_{h|e_1}$ cannot be set to zero at vertex *i* unless the given $\underline{F}_d$ verifies

$$\underline{F}_d \cdot \underline{m} = 0$$

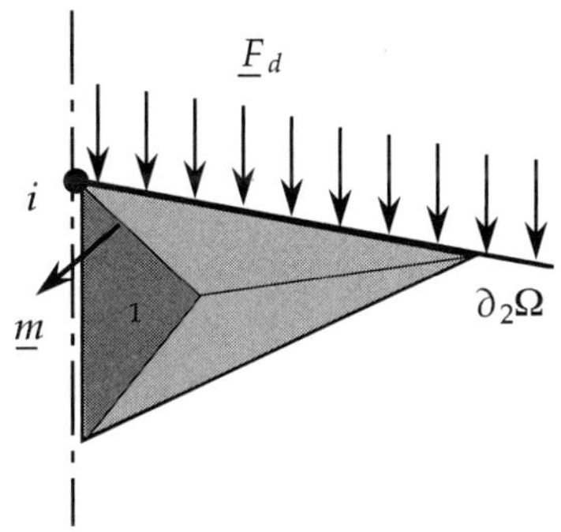

Figure 8.15.

In general, this property is not verified, except in the case of zero applied loads. Therefore, in constructing the mesh, one must exclude this type of situation by imposing that at least two elements be connected to node *i*.

8.5 3D elasticity problems

The construction of admissible stress field in 3D elasticity follows from the same principles as the construction in 2D elasticity.

8.5.1 Construction of the densities

We will point out the few specificities of the 3D case. Let us reuse the notations of Section 2.4.3. Here, for a mesh consisting of tetrahedra or bricks, one has $R = 3$.

Case of interior nodes

Case of a vertex node

For an interior vertex node i, the system (2.27) is of the form

$$\underline{b}_n^k(i) + \underline{b}_n^p(i) + \underline{b}_n^q(i) = \underline{Q}_{E_n}(i) \tag{8.38}$$

where Γ_k, Γ_p and Γ_q are the three sides of element E_n which are connected to node i.

The finite element equilibrium

$$\sum_{E_n} \underline{Q}_{E_n}(i) = 0$$

ensures the existence of an infinite number of solutions of the system (8.38). Among these solutions, one seeks that which minimizes

$$\sum_{\Gamma} \left[\frac{\underline{b}^{\Gamma}(i) - \underline{b}^{\Gamma}}{p^{\Gamma}} \right]^2$$

where the summation is applied to the sides Γ which have i as a vertex, and with

$$\underline{b}^{\Gamma}(i) = \int_{\Gamma} \eta_E \underline{\hat{F}}\, \varphi_i \, d\Gamma \quad \text{and} \quad \underline{b}^{\Gamma,m}(i) = \frac{1}{2} \int_{\Gamma} (\sigma_{h|E} + \sigma_{h|\tilde{E}}) \cdot \underline{n}_{\tilde{E}}\, \varphi_i \, d\Gamma$$

where E and $\tilde{E}$ are the two elements having Γ as a side and the p^{Γ} are weights (e.g. $p^{\Gamma} = \text{mes}(\Gamma)$).

Here, the minimization problem cannot easily be solved explicitly; therefore, one performs a numerical resolution.

Case of a node interior to an edge

This case is simpler and is formally the same as that of an interior vertex node in 2D, except that the vectors $\underline{\hat{F}}$ are of dimension 3 instead of 2. The system (2.27) has two unknowns per equation. The determination of the projections onto the shape functions attached to such a node is explicit.

Case of a node interior to a side

In this very simple case, as for a nonvertex node in 2D, (2.27) is reduced to two equivalent equations.

Case of boundary nodes

All node types can be dealt with in the same manner without difficulty.

Thus, as in the 2D case, one obtains on each element side the projections of the densities $\hat{\underline{F}}$ onto the shape functions, which enables one to determine $\hat{\underline{F}}$ on each side.

8.5.2 Construction of admissible stress fields

The objective is to construct an analytical solution $\hat{\sigma}_{h|E}$ of the local equations on each element E,

$$\underline{\text{div}}\, \hat{\sigma}_h + \underline{f}_d = 0 \quad \text{in } E$$

$$\hat{\sigma}_h \underline{n}_E = \eta_E \hat{\underline{F}} \qquad \text{on } \partial E$$

Construction for the 4-node tetrahedron

The analytical construction is an extension of that presented for the triangle. One subdivides the tetrahedron E into four subtetrahedra e_i using a strictly interior point, which is usually the center of inertia (Figure 8.16).

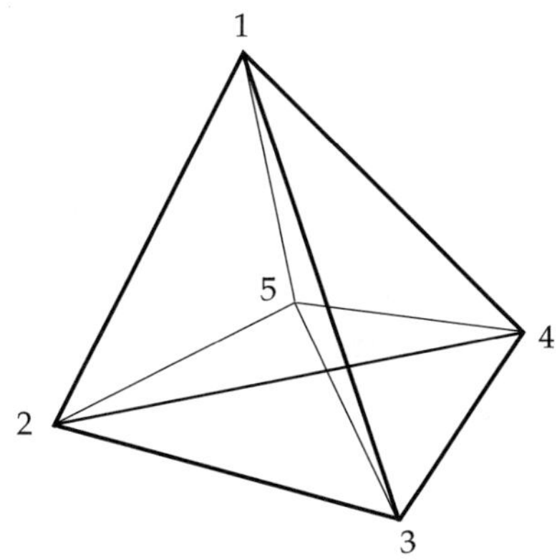

Figure 8.16. *Subdivision of a tetrahedron.*

One assumes that the interior loads are constant within each element. Then, one seeks an affine field on each subtetrahedron. Using the notations

$$\lambda_j^i,\ \lambda_k^i,\ \lambda_m^i \ \text{and}\ \lambda_5^i \qquad (j = i+1,\ k = i+2,\ m = i+3,\ \text{modulo } 4)$$

one has

$$\hat{\sigma}_{h|e_i} = \hat{\sigma}_i = \sigma_j^i\,\lambda_j^i + \sigma_k^i\,\lambda_k^i + \sigma_m^i\,\lambda_m^i + \sigma_5^i\,\lambda_5^i$$

where σ_j^i, σ_k^i, σ_m^i and σ_5^i are symmetric 3×3 operators to be determined.

Calculation of the stresses at the vertices

At each vertex i, one expresses the conditions of equilibrium with the force densities on the three outer sides and the continuity of the stress vector on the three interior sides. Thus, the operators σ_i^j, σ_i^k and σ_i^m are determined completely.

Calculation of the stresses at Point 5

The conditions of continuity of the stress vector across the interior sides yield

$$\sigma_5^1 = \sigma_5^2 = \sigma_5^3 = \sigma_5^4 = \sigma_5$$

In order to calculate the single operator σ_5, it suffices to express in each subtetrahedron the condition

$$\underline{\text{div}}\,\hat{\sigma}_i + \underline{f}_d = 0$$

Thus, one gets twelve scalar equations, only six of which are independent because of the global equilibrium of the element. These six equations enable one to determine σ_5. Thus, taking into account the choices that were made, the field $\hat{\sigma}_{h|E}$ is uniquely determined.

Other element types

The previous construction can be easily extended to 10-node tetrahedra. It can also be extended to brick elements by subdividing the bricks into tetrahedra, then applying the techniques developed for the tetrahedron.

8.6 Improved construction of the densities

As already mentioned, in constructing the densities (in particular for the interior vertices), it is not easy to find a cost function which is both simple and mechanically meaningful in order to eliminate the indeterminacies.

In some situations, particularly when the mesh consists of elements with high aspect ratios, the effectivity index, while remaining greater than 1, can become very large. Besides, in the case of 3-node triangular elements, the modification of the densities presented in Section 8.3.1 often improves the quality of the resulting admissible stress field significantly. In fact, this modification is equivalent to foregoing the extension condition associated with the vertex nodes. This remark led us to propose in [LADEVEZE, 1994] an improvement in the construction of the densities. The basic idea consists in removing, from the strong extension condition, the shape functions associated with the vertex nodes. For the sake of simplicity, we will describe the method in 2D in the case of triangular elements. However, it can be extended with no major difficulty to other types of 2D and 3D elements. Examples of applications can be found in [LADEVEZE - ROUGEOT, 1997].

8.6.1 Weak extension condition

The new extension condition is defined by:

For any element E *of the mesh*
and for any basis function φ_i *of the finite element type*
not associated with a vertex:

$$\int_E (\hat{\sigma}_h - \sigma_h)\, \underline{\text{grad}}\, \varphi_i \, dE = 0 \tag{8.39}$$

This condition is weaker than condition (2.19); it sets only the projections of the densities onto the shape functions associated with nonvertex nodes.

While keeping along each edge Γ densities of the form

$$\hat{\underline{F}}_{|\Gamma} = \sum \underline{A}_i \varphi_{i|\Gamma}$$

this new condition gives greater freedom in the construction of these densities, which will be used to minimize the error in constitutive relation.

8.6.2 Notations and preliminary results

Here, we will assume that the shape functions of degree p are described in hierarchical form [ZIENKIEWICZ - CRAIG, 1986].

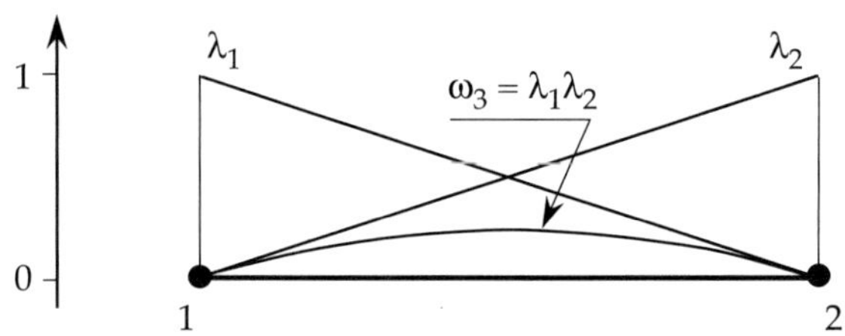

Figure 8.17. *Hierarchical shape functions of degree 2.*

For an edge Γ with endpoints 1 and 2, these functions are written as

$$\begin{aligned} &\lambda_1 \text{ and } \lambda_2 \text{ for the linear part} \\ &\{\omega_\alpha\}_{\alpha=3,4,\ldots,p+1} \text{ for the terms up to degree } p \end{aligned} \tag{8.40}$$

In the following discussion, we will systematically use the Greek indices $\alpha, \beta, \ldots$ to represent shape function numbers between 3 and $p+1$.

For example, for elements of degree 2, one has on an edge Γ (Figure 8.17),

$$\lambda_1, \lambda_2 \text{ and } \omega_3 = \lambda_1\lambda_2$$

We will also use the following properties:

Properties

1. *There exist two functions*

$$\psi_i = \lambda_i - \sum_{\alpha=3}^{p+1} a^i_\alpha \, \omega_\alpha \quad (i = 1, 2) \tag{8.41}$$

such that

$$\forall i = 1, 2 \quad \textit{and} \quad \forall \beta = 3, 4, \ldots, p+1, \quad \int_\Gamma \psi_i \, \omega_\beta \, d\Gamma = 0 \tag{8.42}$$

2. *Thus, the functions* ψ *such that*

$$\forall \beta = 3, 4, \ldots, p+1, \quad \int_\Gamma \psi\, \omega_\beta \, d\Gamma = 0 \tag{8.43}$$

can be written as

$$\psi = a_1 \, \psi_1 + a_2 \, \psi_2 \tag{8.44}$$

Proof

The proof is immediate. In the space of the polynomials of degree less than or equal to p, these properties express that functions ψ_1 and ψ_2 form a basis of a subspace of dimension 2 which is orthogonal to the subspace generated by the functions $\{\omega_\alpha\}_{\alpha = 3, 4, \ldots, p+1}$. Of course, orthogonality is taken in the sense of the scalar product

$$(f, g) \mapsto \int_\Gamma f g \, d\Gamma$$

To find a^i_α, it suffices to write the conditions (8.42) and solve the system obtained. For example, for functions of degree 2, one gets

$$\psi_1 = \lambda_1 - \frac{5}{2}\lambda_1\lambda_2 \quad \text{and} \quad \psi_2 = \lambda_2 - \frac{5}{2}\lambda_1\lambda_2 \tag{8.45}$$

8.6.3 Decomposition of the densities

With the notations introduced, $\hat{\underline{F}}_{|\Gamma}$ can be written as

$$\hat{\underline{F}}_{|\Gamma} = \hat{\underline{F}}_1 \lambda_1 + \hat{\underline{F}}_2 \lambda_2 + \sum_{\alpha=3}^{p+1} \hat{\underline{F}}_\alpha \omega_\alpha \tag{8.46}$$

Properties

1. $\hat{\underline{F}}_{|\Gamma}$ *has a unique decomposition of the form*

$$\hat{\underline{F}}_{|\Gamma} = \hat{\underline{H}}_{|\Gamma} + \hat{\underline{R}}_{|\Gamma} \tag{8.47}$$

with

$$\int_\Gamma \hat{\underline{H}}_{|\Gamma} \, \lambda_1 \, d\Gamma = \int_\Gamma \hat{\underline{H}}_{|\Gamma} \, \lambda_2 \, d\Gamma = 0 \tag{8.48}$$

$$\forall \beta = 3, 4, \dots, p+1, \quad \int_\Gamma \hat{\underline{R}}_{|\Gamma}\, \omega_\beta \, d\Gamma = 0 \tag{8.49}$$

2. $\hat{\underline{H}}_{|\Gamma}$ *has zero resultant and moment on* Γ.

3. $\hat{\underline{R}}_{|\Gamma}$ *is of the form*

$$\hat{\underline{R}}_{|\Gamma} = \underline{R}_1\, \psi_1 + \underline{R}_2\, \psi_2 \tag{8.50}$$

where $\underline{R}_1$ *and* $\underline{R}_2$ *are constant vectors which depend only on the coordinates* $\hat{\underline{F}}_1$, $\hat{\underline{F}}_2$ *and* $\hat{\underline{F}}_\alpha$ *of* $\hat{\underline{F}}$.

Proof

To prove the existence and uniqueness of $\hat{\underline{H}}_{|\Gamma}$, it suffices to observe that

$$\hat{\underline{H}}_{|\Gamma} = \hat{\underline{H}}_1 \lambda_1 + \hat{\underline{H}}_2 \lambda_2 + \sum_{\alpha=3}^{p+1} \hat{\underline{H}}_\alpha \omega_\alpha$$

is completely determined by the conditions

$$\int_\Gamma \hat{\underline{H}}_{|\Gamma}\, \lambda_1 \, d\Gamma = 0\,, \quad \int_\Gamma \hat{\underline{H}}_{|\Gamma}\, \lambda_2 \, d\Gamma = 0$$

and

$$\int_\Gamma \hat{\underline{H}}_{|\Gamma}\, \omega_\beta \, d\Gamma = \int_\Gamma \hat{\underline{F}}\, \omega_\beta \, d\Gamma$$

Then, $\hat{\underline{R}}_{|\Gamma}$ is determined by $\hat{\underline{R}}_{|\Gamma} = \hat{\underline{F}}_{|\Gamma} - \hat{\underline{H}}_{|\Gamma}$. This proves Property **1**. Property **2** follows immediately from (8.48). Finally, Property **3** follows from (8.49), (8.43) and (8.44).

For example, for shape functions of degree 2, $\hat{\underline{F}}_{|\Gamma}$ is of the form

$$\hat{\underline{F}}_{|\Gamma} = \hat{\underline{F}}_1 \lambda_1 + \hat{\underline{F}}_2 \lambda_2 + \hat{\underline{F}}_3 \lambda_1 \lambda_2$$

and one gets

$$\hat{\underline{H}}_{|\Gamma} = -\left[\frac{5}{2}(\hat{\underline{F}}_1 + \hat{\underline{F}}_2) + \hat{\underline{F}}_3\right]\left[\lambda_1 + \lambda_2\right] + \left[15(\hat{\underline{F}}_1 + \hat{\underline{F}}_2) + 6\hat{\underline{F}}_3\right] \lambda_1 \lambda_2$$

$$\hat{\underline{R}}_{|\Gamma} = \left[\frac{1}{2}(7\hat{\underline{F}}_1 + 5\hat{\underline{F}}_2) + \hat{\underline{F}}_3\right] \psi_1 + \left[\frac{1}{2}(5\hat{\underline{F}}_1 + 7\hat{\underline{F}}_2) + \hat{\underline{F}}_3\right] \psi_2$$

8.6.4 New construction of the densities

Principle of the method

Let us consider a family of densities $\hat{\underline{F}}$ such that on each edge Γ of the mesh one has

$$\hat{\underline{F}}_{|\Gamma} = \hat{\underline{F}}_1\lambda_1 + \hat{\underline{F}}_2\lambda_2 + \sum_{\alpha=3}^{p+1} \hat{\underline{F}}_\alpha\omega_\alpha$$

Definition

$\hat{\underline{F}}$ is said to be $*$-admissible if there exists at least one admissible stress field $\hat{\sigma}$ such that:

- *For any element E and for any edge Γ of ∂E,*

$$(\hat{\sigma}_{|E}\,\underline{n}_E)_{|\Gamma} = \eta_E\,\hat{\underline{F}}_{|\Gamma} \tag{8.51}$$

- *$\hat{\sigma}$ verifies the weak extension condition (8.39).*

Let us denote by $\mathbf{F}_{Ad}^{p}$ the set of these densities. One has the property:

Property

For a family of $*$-admissible densities $\hat{\underline{F}}$, for any edge Γ, the decomposition $\hat{\underline{H}}_{|\Gamma} + \hat{\underline{R}}_{|\Gamma}$ is such that:

- *$\hat{\underline{H}}_{|\Gamma}$ is known as a function of the data and of σ_h;*
- *$\hat{\underline{R}}_{|\Gamma} = \underline{R}_{1|\Gamma}\,\psi_1 + \underline{R}_{2|\Gamma}\,\psi_2$ where $\underline{R}_{1|\Gamma}$ and $\underline{R}_{2|\Gamma}$ are constant vectors which verify, for any element E and for any rigid body displacement $\underline{U}_S$ on E,*

$$\int_{\partial E} \eta_E\,\hat{\underline{R}}\cdot\underline{U}_S\,d\Gamma + \int_E \underline{f}_d\cdot\underline{U}_S\,dE = 0 \tag{8.52}$$

Proof

The weak extension condition determines the projections

$$\int_\Gamma \hat{\underline{F}}\,\omega_\alpha\,d\Gamma \tag{8.53}$$

in terms of the data and of σ_h. Therefore, the same is true for $\hat{\underline{H}}_{|\Gamma}$.

The knowledge of the projections (8.53) also enables one to calculate the

$\hat{\underline{F}}_\alpha$ as functions of $\hat{\underline{F}}_1$ and $\hat{\underline{F}}_2$, these two vectors remaining arbitrary. Then, using the property (8.50), one gets the form of $\hat{\underline{R}}_{|\Gamma}$. Finally, condition (8.52) expresses on each element E the equilibrium of the densities $\hat{\underline{F}}$ with the applied loads $\underline{f}_d$.

REMARKS

1. In particular, condition (8.51) contains the conditions

$$\text{For any edge } \Gamma \text{ included in } \partial_2\Omega, \qquad \eta_E \, \hat{\underline{F}}_{|\Gamma} = \underline{F}_{d\,|\Gamma}$$

2. The exponent p used in the notation $\mathfrak{F}^{\,p}_{Ad}$ is intended for one to keep in mind that the densities are chosen of degree p on each edge.

The set of the stress fields for which there exists $\hat{\underline{F}}$ $*$-admissible satisfying the definition (8.51) is denoted $\boldsymbol{S}^{p}_{Ad}$. Therefore, the best admissible $\hat{\sigma}_h$ possible is the solution of the problem

$$\operatorname*{Min}_{\hat{\sigma} \in \boldsymbol{S}^{p}_{Ad}} \mathbf{e}_{RdC}(\underline{U}_h, \hat{\sigma}) \tag{8.54}$$

which is equivalent to

$$\operatorname*{Min}_{\hat{\sigma} \in \boldsymbol{S}^{p}_{Ad}} \left[\frac{1}{2} \int_\Omega \operatorname{Tr}\left[\hat{\sigma} \mathbf{K}^{-1} \hat{\sigma} \right] d\Omega - \int_{\partial_1\Omega} \hat{\sigma}\, \underline{n} \cdot \underline{U}_d \, d\Gamma \right] \tag{8.55}$$

This problem is a global problem defined on domain Ω.

Nevertheless, it can be solved at reasonable cost by introducing two families of local problems defined on each element E, denoted $\boldsymbol{\mathcal{P}}^{H}_{E}$ and $\boldsymbol{\mathcal{P}}^{R}_{E}$, because each is related to one of the terms of the decomposition (8.47).

Problem $\boldsymbol{\mathcal{P}}^{H}_{E}$

Let us consider on the edges Γ of ∂E a family of densities $\hat{\underline{H}}_{|\Gamma}$ with zero resultant and moment, which, as we recall, is unique for $\hat{\underline{F}} \in \mathfrak{F}^{\,p}_{Ad}$.

Besides, let us associate with the interior load $\underline{f}_d$ the load $\underline{f}^{H}_{E}$ with zero resultant and moment on E defined by

$$\underline{f}^{H}_{E} = \underline{f}_{d\,|E} - \langle \underline{f}_d \rangle_E \tag{8.56}$$

where $\langle \underline{f}_d \rangle_E$, which has the same resultant and moment as $\underline{f}_d$, is given by

$$\langle \underline{f}_d \rangle_E = \frac{1}{\text{mes}(E)} \int_E \underline{f}_d \, dE + \frac{\underline{N} \wedge \underline{X}}{\int_E \underline{X}^2 \, dE} \int_E (\underline{N} \wedge \underline{X}) \cdot \underline{f}_d \, dE \tag{8.57}$$

Let us recall that $\underline{X} = \underline{GM}$ (G being the center of inertia and M any point of E) and that $\underline{N}$ is the unit vector normal to the plane such that $\underline{N}\wedge$ is the $+\pi/2$ rotation.

Therefore, the set of the stress fields $\hat{\sigma}_E$ which are admissible on E for the loads $\eta_E \underline{\hat{H}}_{|\partial E}$ and $\underline{f}_E^H$ is nonempty. We will denote this set $\boldsymbol{S}_{Ad|E}^H$.

Thus, one has $\hat{\sigma}_E \in \boldsymbol{S}_{Ad|E}^H$ if and only if:

For any displacement $\underline{U}^*$ regular on E,

$$\int_E \text{Tr}\left[\hat{\sigma}_E \varepsilon(\underline{U}^*)\right] dE = \int_E \underline{f}_E^H \cdot \underline{U}^* \, dE + \int_{\partial E} \eta_E \underline{\hat{H}} \cdot \underline{U}^* \, d\Gamma \tag{8.58}$$

The problem $\boldsymbol{p}_E^H$ is then:

Find $\hat{\sigma}_E^H$ solution of the minimization problem

$$\min_{\hat{\sigma}_E \in \boldsymbol{S}_{Ad|E}^H} \frac{1}{2} \int_E \text{Tr}\left[\hat{\sigma}_E \mathbf{K}^{-1} \hat{\sigma}_E\right] dE \tag{8.59}$$

The displacement field associated with $\hat{\sigma}_E^H$ is defined except for a rigid body displacement field. We will denote by $\underline{\hat{U}}_E^H$ the particular field which has the same resultant and the same moment as $\underline{U}_{h|E}$ on E, and by $\hat{\sigma}^H$ the stress field defined on each element by

$$\hat{\sigma}^H_{|E} = \hat{\sigma}_E^H$$

Problem $\boldsymbol{p}_E^R$

Let us consider the densities $\underline{\hat{R}}_{|\Gamma}$ on the edges Γ of ∂E On each edge, these densities are defined by two constant vectors $\underline{R}_{1|\Gamma}$ and $\underline{R}_{2|\Gamma}$. Let $\mathbf{R}_E$ be the column of these constant vectors for an element E.

Let

$$\underline{f}_E^R = -\frac{1}{\text{mes}(E)} \int_{\partial E} \eta_E \underline{\hat{R}} \, d\Gamma - \frac{\underline{N} \wedge \underline{X}}{\int_E \underline{X}^2 \, dE} \int_{\partial E} (\underline{N} \wedge \underline{X}) \cdot \eta_E \underline{\hat{R}} \, d\Gamma \tag{8.60}$$

By construction, one has:

For any rigid body displacement $\underline{U}_S$ *on* E,

$$\int_{\partial E} \eta_E \underline{\hat{R}} \cdot \underline{U}_S \, d\Gamma + \int_E \underline{f}_E^R \cdot \underline{U}_S \, dE = 0 \tag{8.61}$$

Thus, the set of the stress fields $\hat{\sigma}_E$ which are admissible on E for the loads $\eta_E \underline{\hat{R}}_{|\partial E}$ and $\underline{f}_E^R$ is nonempty.

We will denote this set $\boldsymbol{\mathcal{S}}_{Ad|E}^R$.

One has $\hat{\sigma}_E \in \boldsymbol{\mathcal{S}}_{Ad|E}^R$ if and only if:

For any displacement $\underline{U}^*$ *regular on* E

$$\int_E \text{Tr}\left[\hat{\sigma}_E \varepsilon(\underline{U}^*)\right] dE = \int_E \underline{f}_E^R \cdot \underline{U}^* \, dE + \int_{\partial E} \eta_E \underline{\hat{R}} \cdot \underline{U}^* \, d\Gamma \tag{8.62}$$

The problem $\boldsymbol{\mathcal{P}}_E^R$ is:

Find $\hat{\sigma}_E^R$ *the solution of the minimization problem*

$$\underset{\hat{\sigma}_E \in \boldsymbol{\mathcal{S}}_{Ad|E}^R}{\text{Min}} \; \frac{1}{2} \int_E \text{Tr}\left[\hat{\sigma}_E \mathbf{K}^{-1} \hat{\sigma}_E\right] dE \tag{8.63}$$

The expression (8.60) shows that $\underline{f}_E^R$ is a linear function of $\underline{\hat{R}}$ (and therefore of $\mathbf{R}_E$) on ∂E.

Thus, the solution $\hat{\sigma}_E^R$ of (8.63) is also a linear function of $\mathbf{R}_E$ which we will denote $\hat{\sigma}_E^R(\mathbf{R}_E)$.

Let $\mathbf{R}$ be the assembled vector of the $\mathbf{R}_E$; we will denote by $\hat{\sigma}^R(\mathbf{R})$ the stress field defined on each element E by

$$\hat{\sigma}^R(\mathbf{R})_{|E} = \hat{\sigma}_E^R(\mathbf{R}_E)$$

Determination of the best $\hat{\sigma}_h$

Let us consider $\hat{\underline{F}} \in \mathbf{F}_{Ad}^{p}$. One has:

Property

The field $\hat{\sigma}$ defined by $\hat{\sigma} = \hat{\sigma}^H + \hat{\sigma}^R(\mathbf{R})$ is admissible if and only if on each element E the $\mathbf{R}_E$ are chosen such that

$$\underline{f}_E^R = \langle \underline{f}_d \rangle_E \tag{8.64}$$

Proof

By construction, one has on ∂E,

$$\hat{\sigma}_E^H \underline{n}_E = \eta_E \hat{\underline{H}} \quad \text{and} \quad \hat{\sigma}_E^R \underline{n}_E = \eta_E \hat{\underline{R}}$$

and therefore

$$\hat{\sigma}_E \underline{n}_E = \eta_E \hat{\underline{F}}$$

Besides, one has on E,

$$\underline{\text{div}}\, \hat{\sigma}_E^H + \underline{f}_E^H = 0 \quad \text{and} \quad \underline{\text{div}}\, \hat{\sigma}_E^R + \underline{f}_E^R = 0$$

Therefore

$$\underline{\text{div}}\, \hat{\sigma}_E + \underline{f}_{d|E} = 0$$

if and only if

$$\underline{f}_{d|E} = \underline{f}_E^H + \underline{f}_E^R$$

i.e., taking (8.56) into account

$$\underline{f}_E^R = \langle \underline{f}_d \rangle_E$$

One should note that this last condition corresponds to three scalar conditions on $\mathbf{R}_E$ which we will express in the form

$$\mathbb{L}_E \mathbf{R}_E = \mathbf{b}_E \tag{8.65}$$

or, globally for the entire mesh

$$\mathbb{L}\mathbf{R} = \mathbf{b} \tag{8.66}$$

In order to obtain the solution of (8.55), taking into account the previous property, it suffices to minimize the complementary energy for the stress fields

$$\hat{\sigma}(\mathbf{R}) = \hat{\sigma}^H + \hat{\sigma}^R(\mathbf{R})$$

with $\mathbf{R}$ subjected to the constraints $\mathbb{L}\mathbf{R} = \mathbf{b}$.

Thus,

$\hat{\sigma}_h$ is the solution of the minimization problem

$$\underset{\mathbf{R}\text{ such that }\mathbb{L}\mathbf{R}=\mathbf{b}}{\text{Min}} \left[\frac{1}{2}\int_\Omega \text{Tr}\left[\hat{\sigma}(\mathbf{R})\mathbf{K}^{-1}\hat{\sigma}(\mathbf{R})\right] d\Omega - \int_{\partial_1\Omega} \hat{\sigma}(\mathbf{R})\, \underline{n} \cdot \underline{U}_d \, d\Gamma\right] \tag{8.67}$$

Therefore, remembering that for $*$-admissible densities the part $\hat{\underline{H}}$ is completely determined by the weak extension condition, the stress $\hat{\sigma}^H$ is known in terms of the data and of the finite element solution σ_h.

Thus, Problem (8.67) is equivalent to

$$\underset{\mathbf{R}\text{ such that }\mathbb{L}\mathbf{R}=\mathbf{b}}{\text{Min}} \quad g(\mathbf{R}) \tag{8.68}$$

with

$$g(\mathbf{R}) = \frac{1}{2}\int_\Omega \text{Tr}\left[\hat{\sigma}^R(\mathbf{R})\mathbf{K}^{-1}\hat{\sigma}^R(\mathbf{R})\right] d\Omega$$
$$+ \int_\Omega \text{Tr}\left[\hat{\sigma}^R(\mathbf{R})\mathbf{K}^{-1}\hat{\sigma}^H\right] d\Omega - \int_{\partial_1\Omega} \hat{\sigma}^R(\mathbf{R})\, \underline{n} \cdot \underline{U}_d \, d\Gamma$$

One has the following property which is important from a practical standpoint:

Property

Except for constant terms, one has

$$g(\mathbf{R}) = \frac{1}{2}\mathbf{R}^T\mathbb{A}\mathbf{R} + \left[\mathbf{B}^H - \mathbf{B}_d\right]^T\mathbf{R} \tag{8.69}$$

where:

- the symmetric matrix $\mathbb{A}$ is obtained by assembling a set of element matrices $\mathbb{A}_E$ which depend only on the element and on the material;

- the vector $\mathbf{B}^H$ is obtained by assembling a set of element vectors $\mathbf{B}_E^H$ determined by solving the problems $\boldsymbol{p}_E^H$;
- the vector $\mathbf{B}_d$ is obtained by assembling a set of element vectors $\mathbf{B}_{d,E}$ which depend only on the element and on the data $\underline{U}_d$.

Proof

Let us first observe that

$$g(\mathbf{R}) = \sum_E \frac{1}{2}\int_E \mathrm{Tr}\left[\hat{\sigma}_E^R(\mathbf{R}_E)\mathbf{K}^{-1}\hat{\sigma}_E^R(\mathbf{R}_E)\right] dE$$
$$+ \sum_E \int_E \mathrm{Tr}\left[\hat{\sigma}_E^R(\mathbf{R}_E)\mathbf{K}^{-1}\hat{\sigma}_E^H\right] dE - \sum_E \int_{\partial_1\Omega \cap \partial E} \hat{\sigma}_E^R(\mathbf{R}_E)\underline{n}_E \cdot \underline{U}_d \, d\Gamma$$

Since $\hat{\sigma}_E^R(\mathbf{R}_E)$ depends linearly on $\mathbf{R}_E$, there exists an element matrix $\mathbb{A}_E$ such that

$$\int_E \mathrm{Tr}\left[\hat{\sigma}_E^R(\mathbf{R}_E)\mathbf{K}^{-1}\hat{\sigma}_E^R(\mathbf{R}_E)\right] dE = \mathbf{R}_E^T \mathbb{A}_E \mathbf{R}_E \tag{8.70}$$

Using the field $\hat{\underline{U}}_E^H$ such that $\hat{\sigma}_E^H = \mathbf{K}\varepsilon\,(\hat{\underline{U}}_E^H)$, one gets

$$\int_E \mathrm{Tr}\left[\hat{\sigma}_E^R(\mathbf{R}_E)\mathbf{K}^{-1}\hat{\sigma}_E^H\right] dE = \int_E \mathrm{Tr}\left[\hat{\sigma}_E^R(\mathbf{R}_E)\varepsilon(\hat{\underline{U}}_E^H)\right] dE$$
$$= \int_{\partial E} \hat{\underline{R}} \cdot \hat{\underline{U}}_E^H \, d\Gamma + \int_E \underline{f}_E^R \cdot \hat{\underline{U}}_E^H \, dE$$

The first term is of the form

$$\int_{\partial E} \hat{\underline{R}} \cdot \hat{\underline{U}}_E^H \, d\Gamma = \left[\mathbf{B}_E^H\right]^T \mathbf{R}_E \tag{8.71}$$

Taking into account $\underline{f}_E^R = \langle \underline{f}_d \rangle_E$, the second term is equal to

$$\int_E \underline{f}_E^R \cdot \hat{\underline{U}}_E^H \, dE = \int_E \langle \underline{f}_d \rangle_E \cdot \hat{\underline{U}}_E^H \, dE = \int_E \langle \underline{f}_d \rangle_E \cdot \underline{U}_h \, dE$$

Since this is a constant, it can be disregarded. Finally, one has

$$\int_{\partial_1\Omega \cap \partial E} \hat{\sigma}_E^R(\mathbf{R}_E)\,\underline{n}_E \cdot \underline{U}_d \, d\Gamma = \left[\mathbf{B}_{d,E}\right]^T \mathbf{R}_E \tag{8.72}$$

By combining (8.70), (8.71) and (8.72) one gets

$$g(\mathbf{R}) = \sum_E \left[\frac{1}{2} \mathbf{R}_E^T \mathbb{A}_E \mathbf{R}_E + \left[\mathbf{B}_E^H - \mathbf{B}_{d,E} \right]^T \mathbf{R}_E \right]$$

which, once assembled, yields the desired result.

In summary, the minimization problem to be solved is

$$\underset{\mathbf{R} \text{ such that } \mathbb{L}\mathbf{R} = \mathbf{b}}{\text{Min}} \left[\frac{1}{2} \mathbf{R}^T \mathbb{A} \mathbf{R} + \left[\mathbf{B}^H - \mathbf{B}_d \right]^T \mathbf{R} \right] \tag{8.73}$$

Introducing a Lagrange multiplier $\mathbf{\Lambda}$, one gets

$$\begin{bmatrix} \mathbb{A} & \mathbb{L}^T \\ \mathbb{L} & \mathbb{0} \end{bmatrix} \begin{bmatrix} \mathbf{R} \\ \mathbf{\Lambda} \end{bmatrix} = \begin{bmatrix} \mathbf{B}_d - \mathbf{B}^H \\ \mathbf{b} \end{bmatrix} \tag{8.74}$$

Practical implementation

Thus, in order to obtain the field $\hat{\boldsymbol{\sigma}}_h$ which enables one to calculate the error, one must solve first the local problems $\boldsymbol{\mathcal{P}}_E^H$ and $\boldsymbol{\mathcal{P}}_E^R$, then the global problem (8.73).

The approximate resolution of the local problems can be achieved in two ways:

- either by using the analytical construction techniques described in this chapter;
- or by seeking an approximate solution using a finite element method with a single element of degree $p+k$ (Chapter 2). In practice, one gets a good approximate solution by taking $k=3$. One can also use a submesh of the element composed of elements of degree p or $p+k$. For high-order elements, these are the only truly practical methods. Therefore, these approaches are used systematically today.

Regarding the global problem, it suffices to use an iterative method of the conjugate gradient type. In order to initialize the algorithm, one uses the initial set of densities calculated by the classical method. Furthermore, one restricts the resolution of (8.73) to the zones in which the initialization is insufficient, i.e. the zones which include high gradients or elements with very high aspect ratios.

REMARKS AND COMMENTS

1. In the case of elements of degree 1, the weak extension condition vanishes completely. Nevertheless, the construction of the stress field $\hat{\sigma}_h$ takes into account some characteristics of the finite element discretization used: the mesh, and the degree of the densities $\hat{\underline{F}}$ along the edges.

2. This new construction of the densities enables one to improve the effectivity index of the error in constitutive relation significantly, particularly in the presence of elements with high aspect ratios. Examples can be found in [LADEVEZE - ROUGEOT, 1997] and in Chapter 9.

3. This new construction has been presented in the framework of 2D elasticity, but it can be extended without great difficulty to axisymmetric elasticity and 3D elasticity.

4. This type of technique can also be adapted to nonlinear calculations. For example, an initial implementation for DRUCKER's error in plasticity can be found in [GALLIMARD - LADEVEZE - PELLE, 2000].

8.7 Incompressible or nearly incompressible elasticity

8.7.1 The reference problem

Some elastic materials are incompressible or nearly incompressible. Here, for the sake of simplicity, we will assume that the material is isotropic and incompressible ($\nu = 0.5$), but all the results we are about to present can be extended without difficulty to the case of nearly incompressible materials [GASTINE - LADEVEZE - MARIN - PELLE, 1992]. The reference problem is

Find $\underline{U}(M)$ *and* $\sigma(M)$ *such that:*

- *Kinematic constraints*

$$\underline{U} \in \boldsymbol{\mathcal{U}} \quad \text{and} \quad \underline{U}_{|\partial_1 \Omega} = \underline{U}_d \tag{8.75}$$

- *Incompressibility*

$$\text{div}\, \underline{U} = 0 \tag{8.76}$$

- *Equilibrium equations*

$\sigma \in \boldsymbol{S}$ *and* $\forall \underline{U}^* \in \boldsymbol{U}_0,$

$$\int_\Omega \text{Tr}\left[\sigma \varepsilon(\underline{U}^*)\right] d\Omega = \int_\Omega \underline{f}_d \cdot \underline{U}^* \, d\Omega + \int_{\partial_2\Omega} \underline{F}_d \cdot \underline{U}^* dS \tag{8.77}$$

- *Constitutive relation*

$$\sigma_D = 2\mu\varepsilon(\underline{U}) \tag{8.78}$$

where σ_D is the deviatoric stress and μ the shear modulus

$$\mu = \frac{E}{2(1+\nu)}$$

The computational difficulty lies in the fact that the part of the constitutive relation which connects the trace of the stress with the trace of the strain is singular

$$\text{Tr}(\sigma) = \frac{E}{1-2\nu}\text{Tr}(\varepsilon)$$

8.7.2 Error in constitutive relation

In order to define a suitable error in constitutive relation for this problem, we add to the usual admissibility conditions the incompressibility condition. More precisely, one has:

Definition

A pair $(\hat{\underline{U}}, \hat{\sigma})$ *is said to be admissible if* $\hat{\underline{U}}$ *verifies* (8.75) *and* (8.76) *and if* $\hat{\sigma}$ *verifies* (8.77).

Then, the associated error in constitutive relation is defined by

$$e_{RdC} = \hat{\sigma}_D - \frac{E}{1+\nu}\varepsilon(\hat{\underline{U}})$$

and the constitutive relation error measure by

$$\mathbf{e}_{RdC} = \left[\int_\Omega \operatorname{Tr}\left[\frac{1+\nu}{E} \left(\hat{\underline{\underline{\sigma}}}_D - \frac{E}{1+\nu} \underline{\underline{\varepsilon}}(\hat{\underline{U}}) \right)^2 \right] d\Omega \right]^{1/2} \tag{8.79}$$

8.7.3 Discretization of the problem

The numerical treatment of the incompressibility condition requires the use of specific techniques. Here, we will use a mixed formulation of the HERRMANN type for nearly incompressible media [HERRMANN, 1965] which, in the strictly incompressible case, reduces simply to the minimization of the potential energy with the constraint $div\, \underline{U} = 0$ (enforced by a LAGRANGE multiplier). More precisely, the displacement field solution of Problem [(8.75) - (8.78)] is the solution of the minimization problem

$$\underset{\underline{U}\ KA\ \text{and}\ \operatorname{div}\underline{U}=0}{\operatorname{Min}} \quad \frac{1}{2}\int_\Omega 2\mu \operatorname{Tr}[\underline{\underline{\varepsilon}}(\underline{U})\underline{\underline{\varepsilon}}(\underline{U})]\, d\Omega - \int_\Omega \underline{f}_d \cdot \underline{U}\, d\Omega - \int_{\partial_2\Omega} \underline{F}_d \cdot \underline{U}\, dS$$

The associated mixed formulation is:

Find $\underline{U} \in \boldsymbol{\mathcal{U}}$ such that $\underline{U}_{|\partial_1\Omega} = \underline{U}_d$ and $H \in \boldsymbol{H}$ yield an extremum of the Lagrangian

$$L(\underline{U}; H) = \frac{1}{2}\int_\Omega 2\mu \operatorname{Tr}[\underline{\underline{\varepsilon}}(\underline{U})\underline{\underline{\varepsilon}}(\underline{U})]\, d\Omega + \int_\Omega H \operatorname{div} \underline{U}\, d\Omega$$
$$- \int_\Omega \underline{f}_d \cdot \underline{U}\, d\Omega - \int_{\partial_2\Omega} \underline{F}_d \cdot \underline{U}\, dS$$

The multiplier H can be interpreted as a pressure. The associated discretized problem is:

Find $\underline{U}_h \in \boldsymbol{\mathcal{U}}_h$ and $H_h \in \boldsymbol{H}_h$ such that

$\forall \underline{U}_h^* \in \boldsymbol{\mathcal{U}}_h,$

$$\int_\Omega 2\mu \operatorname{Tr}[\underline{\underline{\varepsilon}}(\underline{U}_h)\underline{\underline{\varepsilon}}(\underline{U}_h^*)]\, d\Omega + \int_\Omega H_h \operatorname{div} \underline{U}_h^*\, d\Omega$$
$$- \int_\Omega \underline{f}_d \cdot \underline{U}_h^*\, d\Omega - \int_{\partial_2\Omega} \underline{F}_d \cdot \underline{U}_h^*\, dS = 0$$

$$\forall H_h^* \in \boldsymbol{H}_h, \quad \int_\Omega H_h^* \operatorname{div} \underline{U}_h \, d\Omega = 0$$

where $\boldsymbol{U}_h$ and $\boldsymbol{H}_h$ are the discretization subspaces chosen.

In the following, we will restrict our discussion to the case of plane problems discretized into triangles, for which the displacement field is of degree 2 and HERRMANN's variable of degree 1, the two fields $\underline{U}_h$ and H_h being continuous on the structure. For the treatment of other elements, particularly 3D elements, one can refer to [GASTINE - LADEVEZE - MARIN - PELLE, 1992].

REMARK. In the case where the variable H is discretized without enforcing interelement continuity, the associated degrees of freedom can be eliminated locally, which leads to a modified stiffness matrix; this is the $\overline{\mathbf{B}}$ method.

8.7.4 Construction of the admissible fields

Thus, in order to evaluate the error in constitutive relation (8.79), one must construct, using the data and the finite element solution $(\underline{U}_h, H_h)$, an admissible pair $(\hat{\underline{U}}_h, \hat{\sigma}_h)$.

Construction of the stress field

The construction of the stress field $\hat{\sigma}_h$ is similar to that carried out in elasticity for 6-node triangles. Indeed, by taking as the finite element stresses

$$\sigma_h = H_h \mathbf{I} + 2\mu\varepsilon(\underline{U}_h)$$

one can easily show that σ_h verifies equilibrium conditions which are identical to those of finite element displacement methods.

Construction of the displacement field

In principle, the construction of a field $\hat{\underline{U}}_h$ such that

$$\hat{\underline{U}}_h \text{ is } KA \quad \text{and} \quad \operatorname{div} \hat{\underline{U}}_h = 0$$

presents similarities with the construction of the stress fields.

Displacement extension conditions

One imposes that field $\underline{\hat{U}}_h$ extends field $\underline{U}_h$ in the following sense:

- *at each vertex* i,

$$\underline{\hat{U}}_h(i) = \underline{U}_h(i)$$

- *along each edge* Γ *with tangent unit vector* $\underline{t}$,

$$\underline{\hat{U}}_h \cdot \underline{t} = \underline{U}_h \cdot \underline{t}$$

- *for any element* E *and for any shape function* ω_i *of the variable* H,

$$\int_E (\underline{\hat{U}}_h - \underline{U}_h) \cdot \underline{\text{grad}}\ \omega_i \, dE = 0$$

For the types of elements considered here, the shape functions ω_i are the three barycentric coordinates λ_i of triangle E.

Using these extension conditions, the construction of $\underline{\hat{U}}_h$ is carried out in two steps.

First, one determines along each edge of the triangulation a density $\hat{u}$ representing a trace of normal displacement, such that

$$\int_{\partial E} \eta_E \hat{u} \, dE = 0 \tag{8.80}$$

Then, one constructs field $\underline{\hat{U}}_h$ element by element by seeking a simple solution of the problem

$$\underline{\hat{U}}_{h|E}(i) = \underline{U}_h(i) \quad \text{at the vertices}$$

$$\left.\begin{array}{l} \underline{\hat{U}}_{h|E} \cdot \underline{t}_E = \underline{U}_h \cdot \underline{t}_E \\ \underline{\hat{U}}_{h|E} \cdot \underline{n}_E = \eta_E \hat{u} \end{array}\right\} \quad \text{on } \partial E$$

$$\text{div}\, \underline{\hat{U}}_{h|E} = 0 \quad \text{in } E$$

where $\underline{t}_E$ is obtained by rotation of $\underline{n}_E$ by $+\pi/2$.

Construction of the densities

Using the extension condition and incompressibility, one gets

$$\int_{\partial E} \eta_E \hat{u}\, \lambda_i \, d\Gamma = \int_E \underline{U}_h \cdot \underline{\text{grad}}\ \omega_i \, dE \tag{8.81}$$

Starting from (8.81), exactly as in the case of thermal-type problems, one determines for each edge the projections

$$\beta_i = \int_\Gamma \eta_E \hat{u} \, \lambda_i \, d\Gamma \quad \text{for } i = 1, 2 \tag{8.82}$$

Similarly, the addition member-by-member of the three relations (8.81) for the three vertices of an element yields (8.80).

Taking into account the conditions

$$\eta_E \hat{u}(i) = \hat{\underline{U}}_{h|E}(i) \cdot \underline{n}_E = \underline{U}_h(i) \cdot \underline{n}_E \quad \text{for 1 and 2}$$

one can determine $\eta_E \hat{u}_{|\Gamma}$ uniquely in the form of a polynomial of degree 3. A simpler technique consists in observing that in order for the equilibrium (8.80) to be verified it suffices to impose that along each edge Γ,

$$\int_\Gamma \eta_E \hat{u} \, d\Gamma = \beta_1 + \beta_2$$

Then, one can determine $\eta_E \hat{u}_{|\Gamma}$ in the form of a polynomial of degree 2 along each edge. This is the solution that we will choose below.

The conditions

$$\hat{\underline{U}}_{h|E}(i) \cdot \underline{t}_E = \underline{U}_h(i) \cdot \underline{t}_E \quad \text{and} \quad \hat{\underline{U}}_{h|E} \cdot \underline{t}_E = \underline{U}_h \cdot \underline{t}_E$$

complete the determination of the trace of $\hat{\underline{U}}_{h|E}$ at the boundary of the element.

Construction of the displacement within the element

The problem consists in finding a simple solution of the problem

$$\begin{aligned} &\operatorname{div} \hat{\underline{U}}_{h|E} = 0 && \text{in } E \\ &\left.\begin{aligned} &\hat{\underline{U}}_{h|E} \cdot \underline{n}_E = \eta_E \, \hat{u} \\ &\hat{\underline{U}}_{h|E} \cdot \underline{t}_E = \underline{U}_h \cdot \underline{t}_E \end{aligned}\right\} && \text{on } \partial E \end{aligned}$$

where $\hat{u}$ is of degree 2 along the edges.

In order to obtain a simple analytical solution, one subdivides the triangle (Figure 8.8 and Figure 8.9) and seeks within each subtriangle e_i a displacement field of degree 2

$$\underline{\hat{U}}_{h|e_i} = \underline{\hat{U}}_i^i \lambda_i^i + \underline{\hat{U}}_j^i \lambda_j^i + \underline{\hat{U}}_4^i \lambda_4^i + \underline{\hat{V}}_i \lambda_i^i \lambda_j^i + \lambda_4^i \left[\underline{\hat{W}}_i^i \lambda_i^i + \underline{\hat{W}}_j^i \lambda_j^i \right]$$

By imposing that field $\underline{\ddot{U}}_{h|e_i}$ be continuous across the boundaries between subelements, one determines a unique field $\underline{\hat{U}}_h$.

8.7.5 Examples

First example

The first example is a short cantilever beam calculated in plane strain.

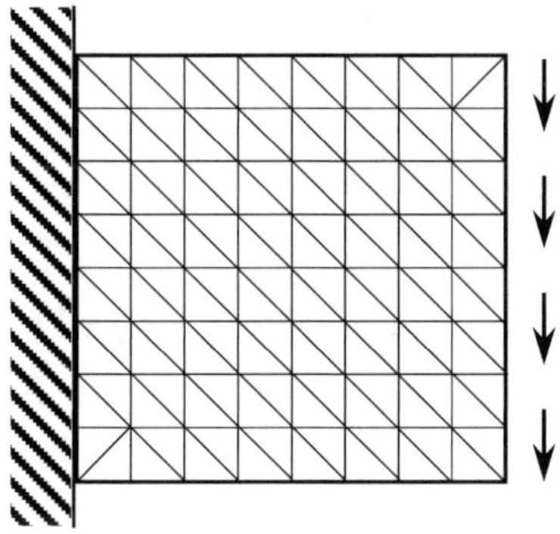

Figure 8.18. *Bending of a short beam.*

The mechanical problem is given in Figure 8.18. Figure 8.19 shows the evolutions of the error in constitutive relation as a function of Poisson's ratio ν for a calculation by a standard displacement formulation and for a calculation by HERRMANN's formulation. The calculations were performed on a single uniform mesh.

Clearly, with a standard displacement formulation the error deteriorates considerably as ν approaches 0.5, which is not the case with HERRMANN's formulation.

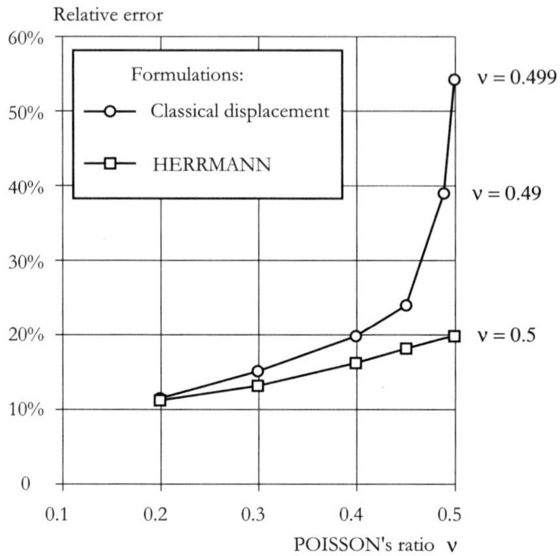

Figure 8.19. *Error in constitutive relation as a function of* **ν**.

Second example

The second example is, again, a short beam in bending (Figure 8.20), only this time with force boundary conditions prescribed such that the exact solution in stress is

$$\boldsymbol{\sigma}_{ex} = \sigma_0 \begin{bmatrix} \frac{2y}{L}\left(\frac{x}{L}-1\right) & \frac{y}{L}\left(1-\frac{y}{L}\right) \\ \frac{y}{L}\left(1-\frac{y}{L}\right) & 0 \end{bmatrix}_{(\underline{x},\, \underline{y})}$$

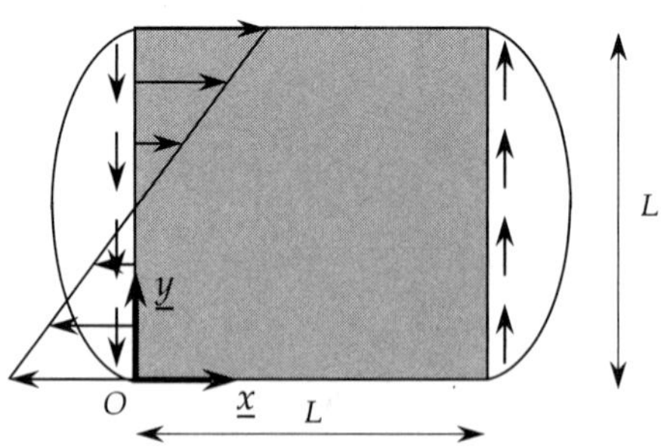

Figure 8.20. *The mechanical problem - known exact solution.*

Again, in this second example, we note (Figure 8.21) that with a displacement formulation the error deteriorates considerably as ν gets close to 0.5, which is not the case with HERRMANN's formulation.

In this example whose exact solution is known, one can calculate the exact errors in the stresses. Figure 8.22 shows the relative errors

$$\eta^{xx} = \frac{\sup\limits_{\Omega} \left| \sigma_{ex}^{xx} - \sigma_h^{xx} \right|}{\sup\limits_{\Omega} \left| \sigma_{ex}^{xx} \right|}; \quad \eta^{yy} = \frac{\sup\limits_{\Omega} \left| \sigma_{ex}^{yy} - \sigma_h^{yy} \right|}{\sup\limits_{\Omega} \left| \sigma_{ex}^{yy} \right|}; \quad \eta^{xy} = \frac{\sup\limits_{\Omega} \left| \sigma_{ex}^{xy} - \sigma_h^{xy} \right|}{\sup\limits_{\Omega} \left| \sigma_{ex}^{xy} \right|}$$

made using a standard displacement method, as well as the corresponding quantities

$$\eta_H^{xx} ; \eta_H^{yy} ; \eta_H^{xy}$$

for HERRMANN's formulation.

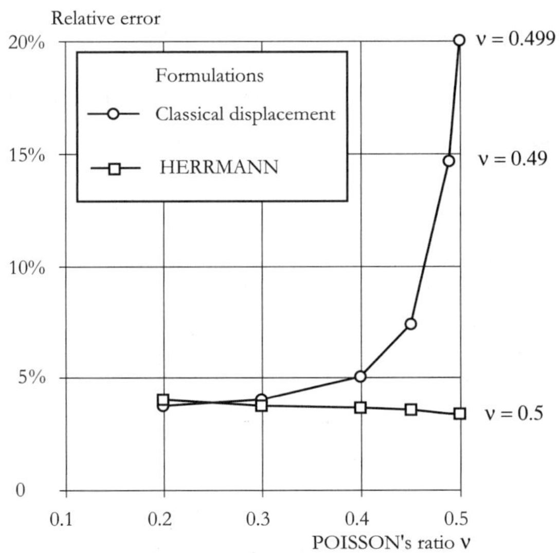

Figure 8.21. *Errors in the stresses as functions of* ν.

This final graph (Figure 8.22) shows that the evolutions of the constitutive relation error measures observed are, indeed, representative of the actual phenomena.

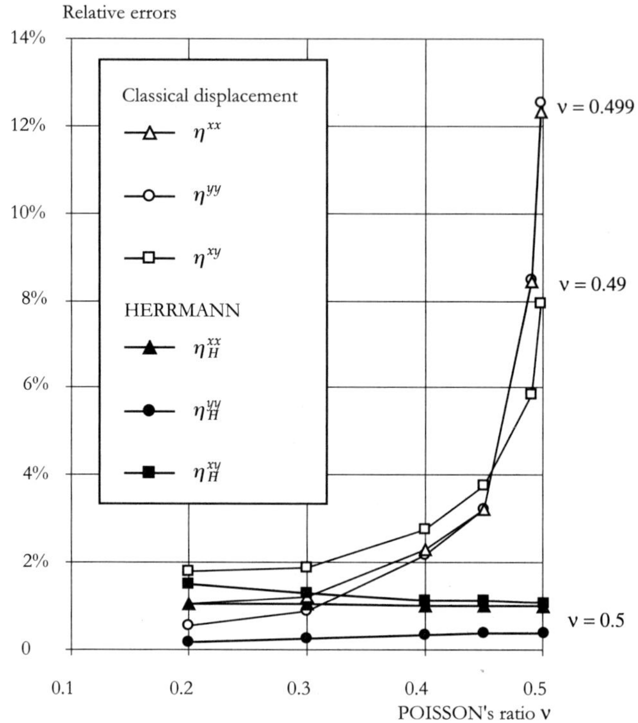

Figure 8.22. *Relative errors in the stresses.*

8.8 Elastic plates

The objective of this section is to show on a representative case that the concept of error in constitutive relation also enables one to evaluate the discretization errors inherent in the use of plate-type finite element models.

The first works in this area [LADEVEZE, 1975] presented the general concepts in the framework of the KIRCHHOFF - LOVE plate model and their implementation for rectangular elements. One can find in [BOISSE - PERRIN, 1995], [BOISSE - PERRIN - COFFIGNAL - HADJEB, 1999], the first application in the framework of the REISSNER - MINDLIN model for a circular plate.

[PELLE - BENOIT, 1995] and [BENOIT - COOREVITS - PELLE, 1999] present a systematic study of the implementation of the concept of error in constitutive relation for plates (with either REISSNER - MINDLIN's or KIRCHHOFF - LOVE's as the reference model) along with the field construction techniques for the most common elements and the procedures associated with mesh adaptation.

Other estimators based on residuals or stress smoothing have also been proposed [ZIENKIEWICZ - ZHU - GONG, 1989], [LEVI - NOUR OMID - STANLEY - SWENSON, 1989], [HOLZER - RANK - WERNER, 1990], [SELMAN - HINTON - ATAMAZ, 1990], [ONATE - CASTRO - KREINER, 1991], [BOUDI - BECKERS -ZHONG, 1992], [OKSTAD - MATHISEN, 1994], [STEPHEN - STEVEN, 1997], [RAMM - CIRAK, 1997].

Here, for the sake of simplicity, we will limit the discussion to the case of the calculation of an elastic, homogeneous and isotropic KIRCHHOFF - LOVE plate in bending using the DKT element [BATOZ, 1982].

8.8.1 The KIRCHHOFF - LOVE bending model

Let Ω be the plane domain occupied by the mean surface of the plate, $\partial_1\Omega$ the part of the boundary along which displacement-type conditions are prescribed, and $\partial_2\Omega$ the complementary part along which force-type conditions are prescribed. The half-thickness of the plate is denoted e.

One can refer to [VALID, 1995], [LIBAI - SIMMONDS, 1998] for more details on the theory of plates and shells, and to [BATOZ - DHATT, 1990] and [CRISFIELD, 1991] for the corresponding finite element methods.

To simplify the presentation, we will assume built-in conditions along $\partial_1\Omega$ and zero forces along $\partial_2\Omega$. We will designate the deflection by w, the bending moment by $\mathbb{M}$, and the bending strain operator by

$$\mathbb{X}(w) = -\,\mathbb{G}\text{rad}\,(\underline{\text{grad}}\; w)$$

The reference problem being considered can be expressed as:

Find $w(M)$ *and* $\mathbb{M}(M)$ *defined on* Ω *such that:*

- *Kinematic constraints*

$$w \textit{ regular}; \quad w = 0 \quad \textit{and} \quad \frac{\partial w}{\partial n} = \underline{\text{grad}}\; w \cdot \underline{n} = 0 \quad \textit{along}\ \partial_1\Omega \tag{8.83}$$

- *Equilibrium equations*

for any w^* regular which is zero along $\partial_1\Omega$:

$$\int_\Omega \mathrm{Tr}\big[\mathbb{M}\mathcal{X}(w^*)\big]\, d\Omega = \int_\Omega p_d\, w^*\, d\Omega \tag{8.84}$$

- *Constitutive relation*

$$\mathbb{M} = \mathbf{K}_f\, \mathcal{X}(w) \tag{8.85}$$

where p_d is a surface force density normal to the plane of the plate and $\mathbf{K}_f$ is the bending constitutive operator defined by

$$\mathbf{K}_f\, \mathcal{X} = \frac{2Ea^3}{3(1-\nu^2)}\big[\nu\, \mathrm{Tr}(\mathcal{X})\, \mathbf{I} + (1-\nu)\mathcal{X}\big] \tag{8.86}$$

with E the YOUNG modulus, ν the POISSON ratio, a the half-thickness of the plate and $\mathbf{I}$ the identity operator in 2D.

REMARK. In the kinematic constraints (8.83), strong $H^2(\Omega)$ -type regularity is required for the deflection. This explains, in particular, why it is difficult to construct conforming finite elements for the KIRCHHOFF - LOVE model.

8.8.2 Error in constitutive relation

According to the general approach, we will say that a pair $(\hat{w}, \hat{\mathbb{M}})$ is admissible if $\hat{w}$ verifies the kinematic constraints (8.83) and $\hat{\mathbb{M}}$ verifies the equilibrium equations (8.84). Then, the error in constitutive relation associated with an admissible pair $(\hat{w}, \hat{\mathbb{M}})$ is defined by

$$\mathbb{e}_{RdC}(\hat{w}, \hat{\mathbb{M}}) = \hat{\mathbb{M}} - \mathbf{K}_f \mathcal{X}(\hat{w}) \tag{8.87}$$

and the error measure by

$$\mathbf{e}_{RdC}(\hat{w}, \hat{\mathbb{M}}) = \left[\int_\Omega \mathrm{Tr}\Big[\big(\hat{\mathbb{M}} - \mathbf{K}_f \mathcal{X}(\hat{w})\big)\mathbf{K}_f^{-1}\big(\hat{\mathbb{M}} - \mathbf{K}_f \mathcal{X}(\hat{w})\big)\Big]\, d\Omega\right]^{1/2} \tag{8.88}$$

Naturally, as in Chapter 2, one can also define relative errors and local contributions.

8.8.3 Application to finite element calculation methods

Various finite element methods have been proposed in order to obtain approximate solutions of Problem [(8.83) - (8.85)]. A description of these methods can be found in [BATOZ - DHATT, 1990].

Let us consider the case where the approximations are obtained by a discretization using the "discrete KIRCHHOFF"-type triangular element DKT. This element is constructed starting from the REISSNER - MINDLIN formulation by neglecting the transverse shear energy and introducing the KIRCHHOFF conditions on the discrete level. Choosing a linear approximation of the deflection on the element, the finite element problem can be formulated as follows:

Find w_h, $\underline{\beta}_h$ and $\mathbb{M}_h$ such that:

- *The scalar field w_h and the vector field $\underline{\beta}_h$ are continuous on Ω and linear on each element* E ,

$$w_{h|E} = w_1\lambda_1 + w_2\lambda_2 + w_3\lambda_3$$
$$\underline{\beta}_{h|E} = \underline{\beta}_1\lambda_1 + \underline{\beta}_2\lambda_2 + \underline{\beta}_3\lambda_3 \tag{8.89}$$

- *For any such fields w_h^* and $\underline{\beta}_h^*$ which are zero on $\partial_1\Omega$,*

$$\int_\Omega \mathrm{Tr}\left[\mathbb{M}_h \varepsilon[\underline{\beta}_h^* + \underline{\tilde{\beta}}(w_h^*, \underline{\beta}_h^*)]\right] d\Omega = \int_\Omega p_d\, w_h^*\, d\Omega \tag{8.90}$$

- *$\mathbb{M}_h$ is of the form*

$$\mathbb{M}_h = \mathbf{K}_f\, \varepsilon[\underline{\beta}_h + \underline{\tilde{\beta}}(w_h, \underline{\beta}_h)] \tag{8.91}$$

with

$$\begin{aligned}\underline{\tilde{\beta}}(w_h, \underline{\beta}_h) = {} & \lambda_1\lambda_2\left[\frac{6(w_1 - w_2)}{\mathrm{mes}(\Gamma_1)}\,\underline{t}_1 - 3\mathbf{T}_1(\underline{\beta}_1 + \underline{\beta}_2)\right] \\ & + \lambda_2\lambda_3\left[\frac{6(w_2 - w_3)}{\mathrm{mes}(\Gamma_2)}\,\underline{t}_2 - 3\mathbf{T}_2(\underline{\beta}_2 + \underline{\beta}_3)\right] \\ & + \lambda_3\lambda_1\left[\frac{6(w_3 - w_1)}{\mathrm{mes}(\Gamma_3)}\,\underline{t}_3 - 3\mathbf{T}_3(\underline{\beta}_3 + \underline{\beta}_1)\right]\end{aligned}$$

where $\underline{t}_i$ is the unit tangent vector along the edge Γ_i (Figure 8.23), *and* $\mathbf{T}_i = \underline{t}_i\, \underline{t}_i^T$.

Thus, the resolution of the approximate problem yields a deflection which is not kinematically admissible and a bending moment field which is not statically admissible.

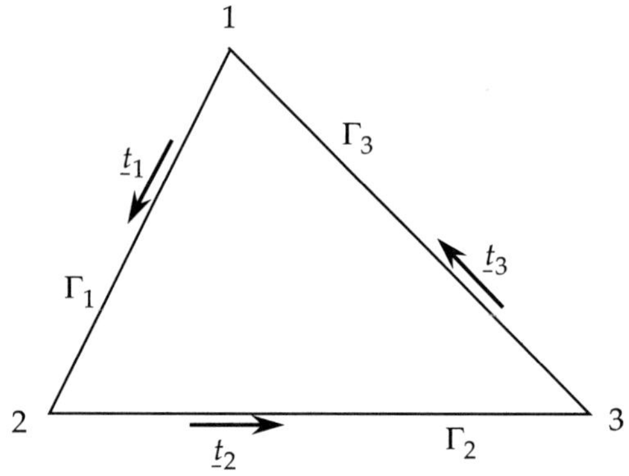

Figure 8.23. *Notations for the DKT element.*

Therefore, in this case, the implementation of the error in constitutive relation requires the reconstruction of a kinematically admissible field $\hat{w}_h$ which should especially have $H^2(\Omega)$ regularity, and of a field $\hat{\mathbb{M}}_h$ verifying the equilibrium equations exactly.

8.8.4 Construction of an admissible pair

Construction of a deflection $\hat{w}_h$

Starting from the nodal values of the approximate deflection w_h and rotation $\underline{\beta}_h$, one can easily reconstruct a field $\hat{w}_h$ using the technique developed in [CLOUGH - TOCHER, 1965] in order to construct a conforming element. This technique consists in subdividing the element into three subtriangles and seeking $\hat{w}_h$ as a polynomial of degree 3 on each subtriangle. Since the nodal values are known, one may simply use the shape functions of the element directly. The details of the calculation of these shape functions can be found in [BERNADOU - HASSAN, 1981].

Construction of a field $\hat{\mathbb{M}}_h$

The construction is carried out in a very classical manner in two steps. First, one determines a pair of densities $(\hat{P}, \underline{\hat{H}})$ along the interfaces, where $\hat{P}$ is a bending load normal to the plane of the plate and $\underline{\hat{H}}$ is an in-plane bending moment.

Of course, $(\hat{P}, \underline{\hat{H}})$ must be in equilibrium on each element E with the applied load p_d. Then, one seeks a simple solution of the equilibrium equations

$$\text{div}\,(\underline{\text{div}}\,\mathbb{M}_{|E}) + p_d = 0 \quad \text{in } E$$

$$\underline{\text{div}}\,\mathbb{M}_{|E} \cdot \underline{n}_E = \eta_E \hat{P} \quad \text{and} \quad \mathbb{M}_{|E}\,\underline{n}_E = \eta_E \underline{\hat{H}} \quad \text{along } \partial E$$

Construction of the densities $(\hat{P}, \underline{\hat{H}})$

The method is similar to that used in the thermal and elastic cases. One determines affine $\hat{P}$ and $\underline{\hat{H}}$ along each edge by introducing the following conditions, similar to strong extension conditions:

For any element, any w_h, and any $\underline{\beta}_h$ linear on E,

$$\int_{\partial E} \eta_E \hat{P}\, w_h \, dE + \int_{\partial E} \eta_E \underline{\hat{H}} \cdot \left[\underline{\beta}_h + \underline{\tilde{\beta}}(w_h, \underline{\beta}_h)\right] dE$$
$$= \int_E \text{Tr}\left[M_h \varepsilon(\underline{\beta}_h + \underline{\tilde{\beta}}(w_h, \underline{\beta}_h))\right] dE - \int_E p_d\, w_h \, dE \qquad (8.92)$$

One can show [PELLE - BENOIT, 1995] that the linear densities $(\hat{P}, \underline{\hat{H}})$ thus constructed are in equilibrium with the applied loads on each element.

Having constructed these densities, reusing the notations of Figure 8.7, one can modify them in such a way that

$$\left[\eta_E \underline{\hat{H}}\right]_{|\Gamma_1} \cdot \underline{n}_2 = \left[\eta_E \underline{\hat{H}}\right]_{|\Gamma_2} \cdot \underline{n}_1 \qquad (8.93)$$

Thus, one can construct a regular bending moment field without subdividing the element. However, the modified densities $(\hat{P}, \underline{\hat{H}})$ are still linear for the $\hat{P}$ part, but quadratic for the $\underline{\hat{H}}$ part.

Construction of $\hat{\mathbb{M}}_h$

If one uses unmodified densities, the analytical construction of $\hat{\mathbb{M}}_h$ can be accomplished by subdividing the elements as in the elastic case. The field $\hat{\mathbb{M}}_h$ is constructed with degree 2 on each subtriangle. In the case of modified densities, the elements are not subdivided and $\hat{\mathbb{M}}_h$ is constructed with degree 3 on each element.

8.8.5 Examples

We will give two examples of applications taken from [BENOIT - COOREVITS - PELLE, 1999]. For both of these examples, the analytical solutions are known; thus, it is possible to compare the estimated error with the actual error.

The first example concerns a circular plate of radius R, built-in along its border, and subjected to a uniform load p_d (Figure 8.24).

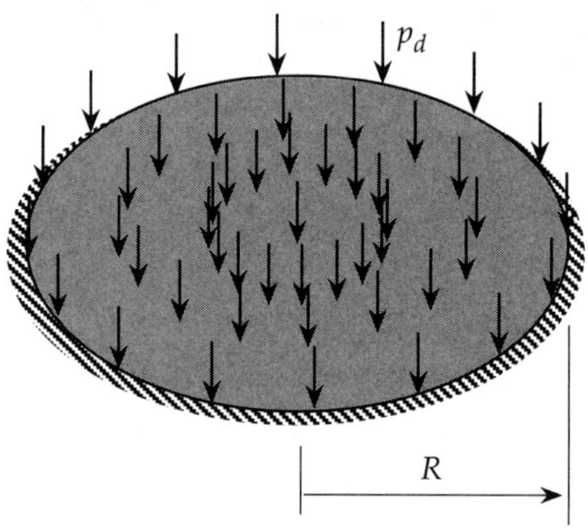

Figure 8.24. *Built-in circular plate in bending.*

According to the KIRCHHOFF - LOVE model, the exact deflection of a point located at a distance r from the center is given by

$$w(r) = \frac{3(1-\nu^2)}{2\,E\,a^3}\,\frac{p_d\,(R^2-r^2)^2}{64}$$

The second example concerns a simply-supported square plate with side L subjected to a uniform surface load p_d (Figure 8.25).

In this case, the exact deflection can be obtained by a double FOURIER series expansion of the form

$$w(x, y) = \frac{3(1-\nu^2)\,L}{2Ee^3\pi^4} \sum_{k=1}^{\infty} \sum_{n=1}^{\infty} \frac{16}{k\,n\,\pi^2} \sin\left(\frac{k\,\pi\,x}{L}\right) \sin\left(\frac{n\,\pi\,y}{L}\right)$$

where (x, y) are the Cartesian coordinates in the plane of the plate for a reference system whose origin is a corner of the plate and whose axes are parallel to the sides.

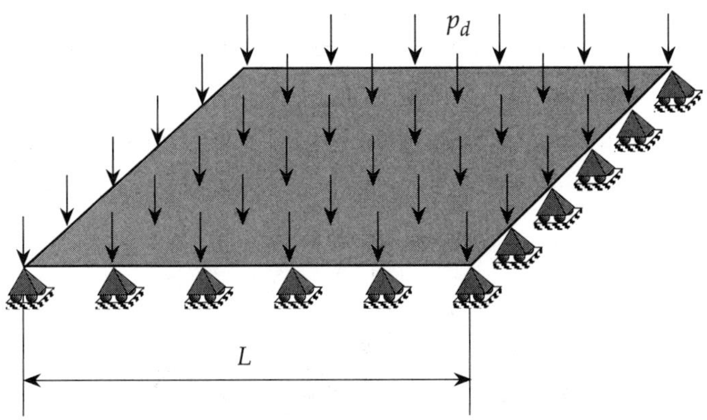

Figure 8.25. *Simply-supported square plate in bending.*

In each of these two examples, the evaluations of the exact error and of the error in constitutive relation with or without subdivision of the elements were carried out for regular meshes obtained from an initial, relatively coarse mesh by dividing the mesh size by 2 between two consecutive calculations.

The finite element calculations were carried out with the DKT element of the CASTEM 2000 code.

In these examples (Figure 8.26 and Figure 8.27), we do observe again that the discrete solution obtained with the DKT element converges toward the KIRCHHOFF - LOVE solution. The estimate of the exact error obtained by the constitutive relation error method is good in all cases. One should note that the construction without element subdivision, which gives more regular fields, improves the quality of the estimates significantly.

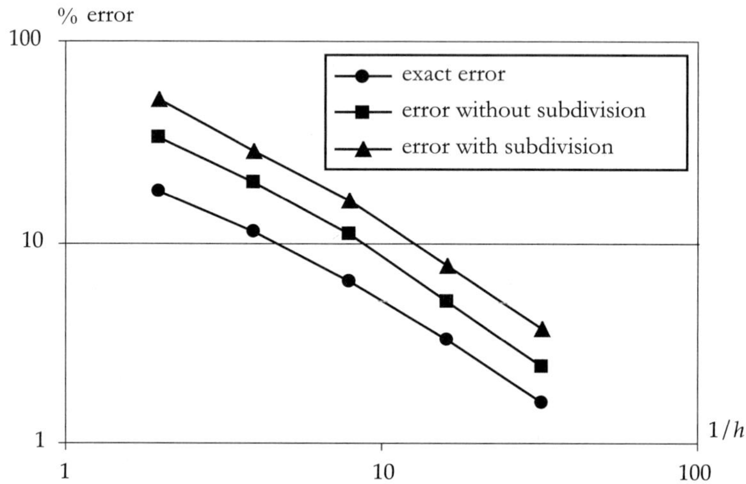

Figure 8.26. *Circular plate - Errors as functions of the element size.*

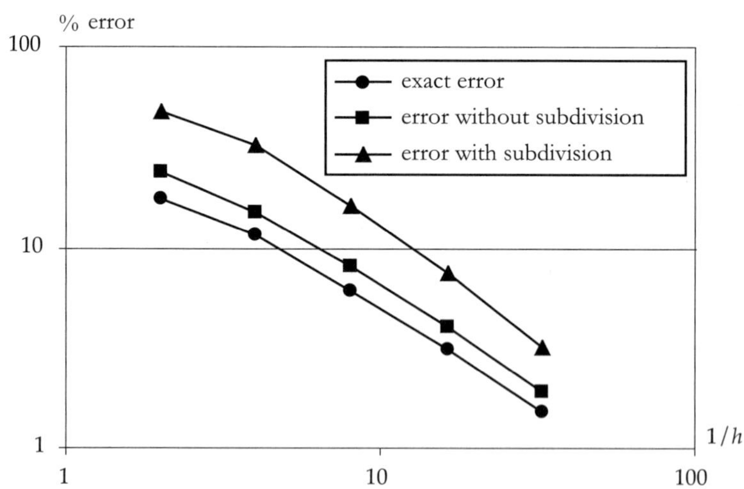

Figure 8.27. *Square plate - Errors as functions of the element size.*

8.9 Asymptotic behavior

As already pointed out in Section 2.4.7, the constitutive relation error measure is connected to the error in the solution through the inequalities

$$\| \underline{U}_{ex} - \underline{U}_h \|_{u,\Omega} \leq \mathbf{e}_{RdC} \leq C \| \underline{U}_{ex} - \underline{U}_h \|_{u,\Omega} \tag{8.94}$$

Let us recall that the inequality $\| \underline{U}_{ex} - \underline{U}_h \|_{u,\Omega} \leq \mathbf{e}_{RdC}$ follows from the PRAGER - SYNGE theorem.

The objective of this section is to prove the inequality

$$\mathbf{e}_{RdC} \leq C \| \underline{U}_{ex} - \underline{U}_h \|_{u,\Omega} \tag{8.95}$$

In the framework of thermal problems, this inequality was proven in [LADEVEZE - LEGUILLON, 1983]. To make things clear, we will discuss the particular context of 2D elasticity problems for 3-node triangles, but the method we are about to present can be extended with no major difficulty to other situations.

Let σ_h be the finite element solution and $\hat{\sigma}_h$ the equilibrated stress field constructed according to the method of Section 8.3 using the strong extension condition. We will also reuse the notations introduced in that section. The proof of (8.95) requires the proof of some preliminary properties.

Preliminary Property 1

On each element E *of the mesh, field* $\Sigma_E = (\hat{\sigma}_h - \sigma_h)_{|E}$ *is the unique solution of the problem*

$$\begin{aligned} &\underline{\mathrm{div}}\, \Sigma_E + \underline{f}_d = 0 \quad \textit{in } E \\ &\Sigma_E \underline{n}_E = \underline{\hat{G}} \quad \textit{on } \partial E \\ &\Sigma_E \quad \textit{affine on each subtriangle } e_i \\ &\Sigma_E \underline{m}_i \quad \textit{continuous across the interfaces between the } e_i \end{aligned} \tag{8.96}$$

with $\underline{\hat{G}} = \eta_E \underline{\hat{F}} - \sigma_h \underline{n}_E$

Proof

To prove this property, it suffices to observe that the problem which defines Σ_E is similar to the problem solved in Section 8.3.2 in order to find field $\hat{\sigma}_h$ in the 3-node triangle, except that the force densities are different.

Preliminary Property 2

For $\Gamma \subset \partial E$ *and* i *being a vertex of* E, *the six projections*

$$\int_\Gamma \hat{\underline{G}}\, \lambda_i \, d\Gamma$$

are linear functions of $\underline{f}_{d|\bar{E}}$ *(where* $\bar{E}$ *is any element connected to node* i*) and of the jumps* $[\![\sigma_h \underline{n}]\!]_{\bar{\Gamma}}$ *(for* $\bar{\Gamma}$ *connected to node* i *and not included in* $\partial_1\Omega$*).*

The jumps $[\![\sigma_h \underline{n}]\!]_{\bar{\Gamma}}$ are defined by:

- if $\bar{\Gamma} \not\subset \partial\Omega$,

$$[\![\sigma_h \underline{n}]\!]_{\bar{\Gamma}} = \sigma_{h|\bar{E}_1} \underline{n}_{\bar{E}_1} + \sigma_{h|\bar{E}_2} \underline{n}_{\bar{E}_2}$$

where $\bar{E}_1$ and $\bar{E}_2$ are two elements adjacent along $\bar{\Gamma}$;

- if $\bar{\Gamma} \subset \partial_2\Omega$,

$$[\![\sigma_h \underline{n}]\!]_{\bar{\Gamma}} = \underline{F}_d - \sigma_{h|\bar{E}} \underline{n}_{\bar{E}}$$

Proof

First, let us consider a node i inside Ω. Let

$$\underline{c}_n(i) = \int_{\Gamma_n} \hat{\underline{G}}\, \lambda_i \, d\Gamma$$

With the notations of Section 8.3.1, one has

$$\underline{c}_n(i) = \underline{b}_n(i) - \int_{\Gamma_n} \sigma_{h|E_n} \underline{n}_{E_n} \lambda_i \, d\Gamma$$

Using (8.26) and (8.27), one gets

$$\underline{c}_1(i) = \frac{1}{2} [\![\sigma_h \underline{n}]\!]_{\Gamma_1} + \frac{1}{P_0} \sum_{k=1}^{N-1} \tilde{\underline{Q}}_k P_k$$

and

$$\underline{c}_n(i) = \underline{c}_{n-1}(i) + \llbracket \sigma_h \underline{n} \rrbracket_{\Gamma_n} + \int_{E_{n-1}} \underline{f}_d \, \lambda_i \, dE$$

with

$$\tilde{\underline{Q}}_k = \underline{Q}_{E_k}(i) - \frac{1}{2}\int_{\Gamma_k} (\sigma_{h|E_{k-1}} + \sigma_{h|E_k}) \, \underline{n}_{E_k} \, d\Gamma$$
$$+ \frac{1}{2}\int_{\Gamma_{k+1}} (\sigma_{h|E_k} + \sigma_{h|E_{k+1}}) \, \underline{n}_{E_{k+1}} \, d\Gamma$$

i.e.

$$\tilde{\underline{Q}}_k = \frac{1}{2}\llbracket \sigma_h \underline{n} \rrbracket_{\Gamma_k} - \frac{1}{2}\llbracket \sigma_h \underline{n} \rrbracket_{\Gamma_{k+1}} - \int_{E_k} \underline{f}_d \, \lambda_i \, dE$$

Thus, in this case, Preliminary Property 2 follows immediately. The same approach enables one to prove this property for the nodes located on the boundary $\partial\Omega$.

Preliminary Property 3

There exists a constant C_1 *depending only on the element shapes and on the material such that*

$$\| \hat{\sigma}_h - \sigma_h \|^2_{\sigma, E} \leq C_1 \left[\sum_{\bar{E} \subset \tilde{\mathbf{E}}} \text{mes}(\bar{E}) \, \| \underline{f}_d \|^2_{\bar{E}} + \sum_{\bar{\Gamma} \subset \tilde{\mathbf{E}} \text{ and } \bar{\Gamma} \not\subset \partial_1\Omega} \ell^2_{\bar{\Gamma}} \, \| \sigma_h \underline{n} \|^2_{\bar{\Gamma}} \right]$$

where $\tilde{\mathbf{E}}$ *designates the set of the triangles having at least one common vertex with triangle* E.

Proof

The field Σ_E is determined on each subtriangle as

$$\Sigma_{E|e_i} = \Sigma^i_i \lambda^i_i + \Sigma^i_j \lambda^i_j + \Sigma^i_4 \lambda^i_4$$

Letting

$$\hat{\underline{G}}_{|\Gamma_i} = \underline{G}^i_i \, \lambda^i_i + \underline{G}^i_j \, \lambda^i_j$$

the explicit calculation of $\Sigma_{E|e_i}$ shows that Σ_E is a linear function of $\underline{f}_d$ and of the $\underline{G}_i^i$ and $\underline{G}_j^i$.

Finally, if one observes that the $\underline{G}_i^i$ and $\underline{G}_j^i$ are expressed linearly as functions of the projections

$$\int_{\Gamma_i} \hat{\underline{G}}_{|\Gamma_i} \lambda_i^i \, d\Gamma \quad \text{and} \quad \int_{\Gamma_i} \hat{\underline{G}}_{|\Gamma_i} \lambda_j^i \, d\Gamma$$

one obtains that Σ_E is a linear function of $\underline{f}_{d|\bar{E}}$ ($\bar{E}$ being any element connected to one of the nodes) and of the jumps $[\![\sigma_h \underline{n}]\!]_{\bar{\Gamma}}$ (for $\bar{\Gamma}$ connected to one of the nodes and not included in $\partial_1\Omega$).

Therefore,

$$\left\| \Sigma_E \right\|_{\sigma, E}^2 = \left\| \hat{\sigma}_h - \sigma_h \right\|_{\sigma, E}^2$$

is a quadratic form of the $\underline{f}_{d|\bar{E}}$ and of the jumps $[\![\sigma_h \underline{n}]\!]_{\bar{\Gamma}}$ since there exists a constant C_1 such that

$$\left\| \hat{\sigma}_h - \sigma_h \right\|_{\sigma, E}^2 \leq C_1 \left[\sum_{\bar{E} \subset \tilde{E}} \text{mes}(\bar{E}) \left\| \underline{f}_d \right\|_{\bar{E}}^2 + \sum_{\bar{\Gamma} \subset \tilde{E} \text{ and } \bar{\Gamma} \not\subset \partial_1\Omega} \ell_{\bar{\Gamma}}^2 \left\| [\![\sigma_h \underline{n}]\!] \right\|_{\bar{\Gamma}}^2 \right]$$

The detailed dimensional analysis of the various terms of Σ_E shows that this constant depends on the element shapes, but is independent of the element sizes. Therefore, for a family of regular meshes in the sense of [CIARLET, 1978], the constant C_1 is independent of h.

Preliminary Property 4

There exists a constant C_2 which depends only on the element shapes and on the material such that

$$\text{mes}(\bar{E}) \left\| \underline{f}_d \right\|_{\bar{E}}^2 \leq C_2 \left\| \sigma_{ex} - \sigma_h \right\|_{\sigma, \bar{E}}^2$$

Proof

Let us designate by 1, 2 and 3 the vertices of $\bar{E}$ and consider the displacement field defined on $\bar{E}$ by

$$\underline{U} = \lambda_1 \lambda_2 \lambda_3 \, \underline{f}_d \, \text{mes}(\bar{E})$$

One has

$$\int_{\Omega} \mathrm{Tr}\big[\sigma_{ex}\varepsilon(\underline{U})\big]\, d\Omega = \int_{\Omega} \underline{f}_d \cdot \underline{U}\, d\Omega + \int_{\partial_2\Omega} \underline{F}_d \cdot \underline{U}\, d\Gamma$$

and therefore,

$$\int_{\bar{E}} \mathrm{Tr}\big[\sigma_{ex}\ \varepsilon(\underline{U})\big]\, dE = \int_{\bar{E}} \underline{f}_d \bullet \underline{U}\, dE$$

A direct calculation also shows that

$$\int_{\bar{E}} \mathrm{Tr}\big[\sigma_h\varepsilon(\underline{U})\big]\, dE = 0$$

Thus, one gets

$$\int_{\bar{E}} \mathrm{Tr}\big[(\sigma_{ex} - \sigma_h)\varepsilon(\underline{U})\big]\, dE = \int_{\bar{E}} \underline{f}_d^2\, \lambda_1\lambda_2\lambda_3\, \mathrm{mes}(\bar{E})\, dE$$

then

$$\frac{\mathrm{mes}(\bar{E})}{60} \| \underline{f}_d \|_{\bar{E}}^2 \leq \| \sigma_{ex} - \sigma_h \|_{\sigma,\bar{E}} \| \underline{U} \|_{u,\bar{E}} \tag{8.97}$$

The calculation of $\| \underline{U} \|_{u,\bar{E}}$ shows that there exists a constant $\tilde{C}$ which depends only on the material and on the shape of triangle $\bar{E}$ such that

$$\| \underline{U} \|_{u,\bar{E}}^2 \leq \tilde{C} \| \underline{f}_d \|_{\bar{E}}^2\, \mathrm{mes}(\bar{E}) \tag{8.98}$$

Combining (8.97) and (8.98), one gets

$$\mathrm{mes}(\bar{E}) \| \underline{f}_d \|_{\bar{E}}^2 \leq C_2 \| \sigma_{ex} - \sigma_h \|_{\sigma,\bar{E}}^2$$

with $C_2 = 360\tilde{C}$.

Preliminary Property 5

For $\bar{\Gamma} \not\subset \partial\Omega$, there exists a constant C_3 which depends only on the shape of the elements and on the material such that

$$\ell_{\bar{\Gamma}}^2 \big[\!\big[\sigma_h \underline{n} \big]\!\big]_{\bar{\Gamma}}^2 \leq C_3 \sum_{k=1}^{2} \| \sigma_{ex} - \sigma_h \|_{\bar{E}_k}^2$$

where $\bar{E}_1$ and $\bar{E}_2$ are two elements adjacent along $\bar{\Gamma}$.

Proof

Let us designate by 1 and 2 the vertices of $\bar{\Gamma}$ and consider the displacement field defined on $\bar{E}_1 \cup \bar{E}_2$ by

$$\underline{U} = \lambda_1 \lambda_2 \left[\!\left[\sigma_h \underline{n} \right]\!\right]_{\bar{\Gamma}} \ell_{\bar{\Gamma}}$$

A calculation similar to the previous one shows that

$$\int_{\bar{E}_1 \cup \bar{E}_2} \mathrm{Tr}\left[(\sigma_{ex} - \sigma_h) \varepsilon(\underline{U}) \right] dE$$
$$= \int_{\bar{E}_1 \cup \bar{E}_2} \underline{f}_d \cdot \underline{U} \, dE - \int_{\Gamma} \left[\!\left[\sigma_h \underline{n} \right]\!\right]_{\bar{\Gamma}} \cdot \underline{U} \, d\Gamma$$

Therefore,

$$\frac{\ell_{\bar{\Gamma}}^2 \left[\!\left[\sigma_h \underline{n} \right]\!\right]_{\bar{\Gamma}}^2}{6} \leq \sum_{k=1}^{2} \left[\| \underline{f}_d \|_{\bar{E}_k} \| \underline{U} \|_{\bar{E}_k} + \| \sigma_{ex} - \sigma_h \|_{\sigma, \bar{E}_k} \| \underline{U} \|_{u, \bar{E}_k} \right] \tag{8.99}$$

One also has

$$\| \underline{U} \|_{\bar{E}_k}^2 = \int_{\bar{E}_k} \left| \lambda_1 \lambda_2 \left[\!\left[\sigma_h \underline{n} \right]\!\right]_{\bar{\Gamma}} \ell_{\bar{\Gamma}} \right|^2 dE = \frac{\mathrm{mes}(\bar{E}_k)}{90} \ell_{\bar{\Gamma}}^2 \left\| \sigma_h \underline{n} \right\|_{\bar{\Gamma}}^2 \tag{8.100}$$

and, as above,

$$\| \underline{U} \|_{u, \bar{E}_k}^2 \leq \tilde{C}_k \left\| \sigma_h \underline{n} \right\|_{\bar{\Gamma}}^2 \mathrm{mes}(\bar{E}_k) \tag{8.101}$$

where $\tilde{C}_k$ is a constant which depends only on the material and on the shape of triangle $\bar{E}_k$.

The combination of (8.99), (8.100), (8.101) and Preliminary Property 4 yields Preliminary Property 5.

Preliminary Property 6

For $\bar{\Gamma} \subset \partial_2 \Omega$, there exists a constant C_4 depending only on the element shapes and on the material such that

$$\ell_{\bar{\Gamma}}^2 \left[\!\left[\sigma_h \underline{n} \right]\!\right]_{\bar{\Gamma}}^2 \leq C_4 \left\| \sigma_{ex} - \sigma_h \right\|_{\bar{E}}^2$$

where $\bar{E}$ is the element bordered by $\bar{\Gamma}$.

The proof, which is very similar to the previous one, will not be given here.

Proof of **(8.95)**

Using the preliminary properties, one gets

$$\| \hat{\boldsymbol{\sigma}}_h - \boldsymbol{\sigma}_h \|^2_{\sigma, E} \leq \tilde{\tilde{C}} \| \boldsymbol{\sigma}_{ex} - \boldsymbol{\sigma}_h \|^2_{\sigma, \tilde{E}} = \tilde{\tilde{C}} \| \underline{U}_{ex} - \underline{U}_h \|^2_{u, \tilde{E}}$$

and the desired result is deduced by observing that for a regular family of meshes the term

$$\sum_E \| \underline{U}_{ex} - \underline{U}_h \|^2_{u, \tilde{E}}$$

contains each term $\| \underline{U}_{ex} - \underline{U}_h \|^2_{u, \tilde{E}}$ only a finite number of times, and that this number is bounded as the element's size tends to zero.

REMARKS AND COMMENTS

1. For a given mesh, the constants C_p, $\tilde{C}$, $\tilde{\tilde{C}}$ introduced above can be considered to be independent of the element being studied.

2. For a regular family of meshes [CIARLET, 1978], these constants are also independent of the mesh. So is the constant C of the inequality (8.95). It follows that the error in constitutive relation $\mathbf{e}_{RdC}$ has the same convergence rate as the error in the solution $\| \underline{U}_{ex} - \underline{U}_h \|_{u,\Omega}$.

3. The existence of the upper bound (8.95) has been proven for the case where the field $\hat{\boldsymbol{\sigma}}_h$ is determined analytically by the method of Section 8.3.2 using the densities based on the strong extension condition. If, using the same densities, the field is constructed using a finite element displacement method in each element (Section 2.4.3), Property (8.95) is still valid.

Indeed, let us denote $\tilde{\boldsymbol{\sigma}}_h$ the stress field obtained by an approximate numerical construction. In each element E, one has

$$\tilde{\boldsymbol{\sigma}}_{h|E} = \mathbf{K}(\underline{V}_{h,E})$$

where $\underline{V}_{h,E}$ is a displacement field of the finite element type defined on E.

Then, one has

$$\begin{aligned}\left\| \hat{\sigma}_h - \sigma_h \right\|^2_{\sigma, E} &= \left\| \hat{\sigma}_h - \tilde{\sigma}_{h|E} + \tilde{\sigma}_{h|E} - \sigma_h \right\|^2_{\sigma, E} \\ &= \left\| \hat{\sigma}_h - \tilde{\sigma}_{h|E} \right\|^2_{\sigma, E} + \left\| \tilde{\sigma}_{h|E} - \sigma_h \right\|^2_{\sigma, E} \\ &\quad + 2\int_E \mathrm{Tr}\left[(\hat{\sigma}_h - \tilde{\sigma}_{h|E})\varepsilon(\underline{V}_{h,E} - \underline{U}_h) \right] dE\end{aligned}$$

Noting that

$$\int_E \mathrm{Tr}\left[(\hat{\sigma}_h - \tilde{\sigma}_{h|E})\varepsilon(\underline{V}_{h,E} - \underline{U}_h) \right] dE = 0$$

one gets

$$\left\| \tilde{\sigma}_{h|E} - \sigma_h \right\|^2_{\sigma, E} \leq \left\| \hat{\sigma}_h - \sigma_h \right\|^2_{\sigma, E}$$

and therefore,

$$\left\| \tilde{\sigma}_h - \sigma_h \right\|_{\sigma, \Omega} \leq \left\| \hat{\sigma}_h - \sigma_h \right\|_{\sigma, \Omega} \leq C \left\| \underline{U}_{ex} - \underline{U}_h \right\|_{u,\Omega}$$

However, in this case, one can no longer prove the inequality

$$\left\| \underline{U}_{ex} - \underline{U}_h \right\|_{u,\Omega} \leq \left\| \tilde{\sigma}_h - \sigma_h \right\|_{\sigma, \Omega}$$

because the field $\tilde{\sigma}_h$ is not strictly admissible. Nevertheless, numerical tests show that in many common situations this inequality also remains verified.

4. If one imposes only the weak extension condition and determines the densities by solving the optimization problem (8.54) exactly, one always obtains by this technique a reduction of the error in constitutive relation; thus, Inequality (8.95) is still verified.

5. A proof based on the inequalities derived in [RAVIART - THOMAS, 1975, 1983] can be found in [DESTUYNDER - METIVET, 1996], [DESTUYNDER - COLLOT - SALAÛN, 1997].

Chapter 9

Estimation of local errors

9.1 Introduction

Most of the error estimators presented in the previous chapters do not really provide access to the local errors, for example, in the stresses. Indeed, all these methods give only a global energy estimation of the discretization error made. One must recognize that in the great majority of cases such global information is totally insufficient for dimensioning purposes in mechanical design: in many common situations, the dimensioning criteria involve local values (stresses, displacements, stress intensity factors…); other quantities, such as stress flow averages, may also be of interest. Therefore, it is also necessary to assess the quality of these local quantities when they are obtained by a finite element analysis.

This development of true estimators of the local errors is, from a research perspective, one of today's key issues, even though only few works have yet been devoted to this subject.

The first works are recent. The concept of pollution error was introduced by BABUSKA and STROUBOULIS [BABUSKA - STROUBOULIS - UPADHYAY - GANGARAJ, 1995]. This means that in the zone of interest the local error is the sum of a local contribution due to the local discretization, and an error due to the discretization in the rest of the domain, called a pollution error.

This work, along with [RANNACHER - SUTTMEIER, 1998], [PARASCHIVOIU - PERAIRE - PATERA, 1997], [PERAIRE - PATERA, 1998], [ODEN - PRUDHOMME, 1999], [CIRAK - RAMM, 1998], [LADEVEZE, 1998], [STROUBOULIS - BABUSKA - DATTA - COPPS - GANGARAJ, 2000], is based on the finite element calculation of a GREEN operator, which is a concept well known to mechanical engineers and used in many fields. We will return to the differences between these approaches later.

The objective of this chapter is to present, for linear problems, a general theory of the estimation of local errors which is an extension of the approach leading to the constitutive relation error estimators.

This theory, introduced by LADEVEZE [LADEVEZE, 1998] and developed in [LADEVEZE - ROUGEOT - BLANCHARD - MOREAU, 1999], uses the new technique of construction of admissible fields presented in Section 7.6.4, whose main advantage is that it yields local error estimators which have been shown experimentally to be upper bounds of the actual local errors in the stresses [LADEVEZE - ROUGEOT, 1997]. Therefore, if one is interested only in estimates of local stresses, it is unnecessary to activate the general theory: the contributions of the elements to the estimate of the global error constitute upper bounds of the local errors.

In the general case, we also present an iterative version as an extension of the iterative technique proposed by [BABUSKA - STROUBOULIS - GANGARAJ, 1999], [STROUBOULIS - BABUSKA - GANGARAJ, 2000], which does not use heuristics.

In the first part of the chapter, we present the remarkable property that the new error estimator is a local upper bound. Then, the general theory and its iterative version are developed and illustrated.

9.2 Reconsideration of the reference problem

In this chapter, we will consider again the linear elastic problem [(1.6) - (1.9)]:

Find a displacement field $\underline{U}(M)$ *and a stress field* $\sigma(M)$ *defined on* Ω *such that:*

- *Constraint equations*

$$\underline{U} \in \boldsymbol{\mathcal{U}} \quad \text{and} \quad \underline{U}_{|\partial_1\Omega} = \underline{U}_d \tag{9.1}$$

- *Equilibrium equations*

$$\sigma \in \boldsymbol{S} \quad \text{and} \quad \forall \underline{U}^* \in \boldsymbol{\mathcal{U}}_0$$

$$\int_\Omega \mathrm{Tr}\left[\sigma \varepsilon(\underline{U}^*)\right] d\Omega = \int_\Omega \underline{f}_d \cdot \underline{U}^* \, d\Omega + \int_{\partial_2\Omega} \underline{F}_d \cdot \underline{U}^* \, dS \tag{9.2}$$

- *Constitutive relation*

$$\sigma = \mathbf{K}\varepsilon(\underline{U}) \tag{9.3}$$

whose solution is the displacement-stress pair $(\underline{U}_{ex}, \sigma_{ex})$.

This problem is discretized by a classical displacement method defined by a mesh $\mathcal{T}_h$ and a discretization space $\boldsymbol{\mathcal{U}}_h$.

9.3 A heuristic property of the local stress errors

9.3.1 Local effectivity indexes

Let $(\underline{U}_h, \hat{\sigma}_h)$ be the admissible pair which enables one to evaluate the global error in constitutive relation

$$\mathbf{e}_{RdC} = \left\| \hat{\sigma}_h - \mathbf{K}\varepsilon(\underline{U}_h) \right\|_{\sigma,\Omega} \tag{9.4}$$

Let us remember the important upper bound property

$$\mathbf{e}_{ex} = \left\| \sigma_{ex} - \sigma_h \right\|_{\sigma,E} = \left\| \underline{U}_{ex} - \underline{U}_h \right\|_{u,\Omega} \leq \mathbf{e}_{RdC} \tag{9.5}$$

which shows that the global effectivity index is always greater than 1.

To estimate the level of the error made on the stress for a particular element, one can use the local contributions

$$\mathbf{e}_E = \left\| \hat{\sigma}_h - \mathbf{K}\varepsilon(\underline{U}_h) \right\|_{\sigma,E} \tag{9.6}$$

A priori, on the local level of an element E, $\mathbf{e}_E$ is not an upper bound of the local error and, thus, there is no guarantee that the local effectivity index

$$\gamma_E = \frac{\mathbf{e}_E}{\left\| \sigma_{ex} - \sigma_h \right\|_{\sigma,E}} \tag{9.7}$$

is greater than 1.

Nevertheless, it has been observed in numerous examples that when using the improved version of the error in constitutive relation (Section 7.6) the local effectivity indexes are indeed always greater than 1. Thus, one can observe that the upper bound

$$\left\| \sigma_{ex} - \sigma_h \right\|_{\sigma,E} \leq \bar{C} \left\| \hat{\sigma}_h - \mathbf{K}\varepsilon(\underline{U}_h) \right\|_{\sigma,E} \tag{9.8}$$

is verified numerically with $\bar{C} = 1$. Of course, in some specific situations, this result may not hold in zones where the local error is small or the stresses are low. Such a result has not yet been proved mathematically.

9.3.2 Examples

These examples are taken from the paper [LADEVEZE - ROUGEOT - BLANCHARD - MOREAU, 1999].

First example

Let us consider the mechanical problem described in Figure 9.1.

The mesh being used is given in Figure 9.2.

In order to obtain an accurate estimate of the actual error $\mathbf{e}_{ex}$, a second finite element calculation was performed with a very fine mesh obtained by subdividing each element of the initial mesh into 64 regular elements.

For this example, the global effectivity index is $\gamma = 2.99$. Figure 9.3 shows the histogram of the local effectivity indexes. We observe that the local effectivity indexes are all between 1.76 and 8.1.

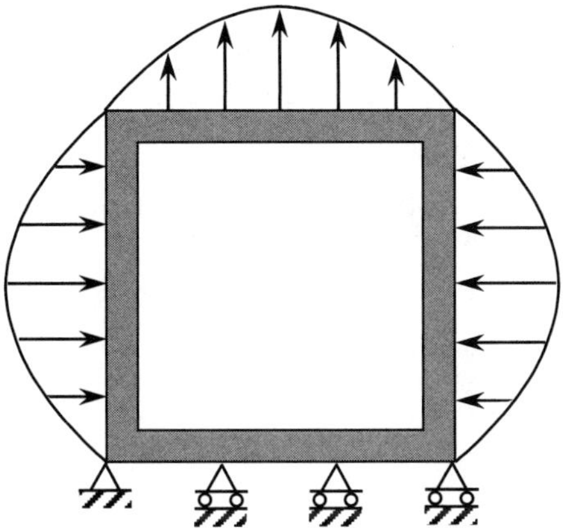

Figure 9.1. *Description of the first test case.*

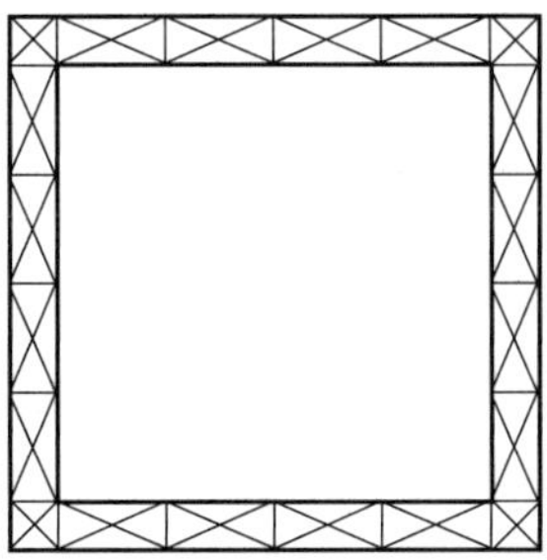

Figure 9.2. *Mesh.*

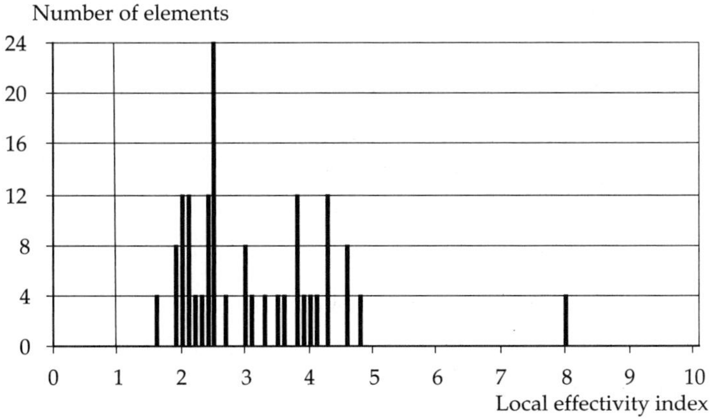

Figure 9.3. *Histogram of the error in constitutive relation.*

In comparison, the same example was processed with the ZZ2 estimator, which yielded the excellent global effectivity index $\gamma = 0.968$. However, Figure 9.4 and Figure 9.5 show that the local effectivity indexes are much more scattered since they vary between 0.3 and 35.3. Furthermore, for many elements, these indexes are clearly less than 1.

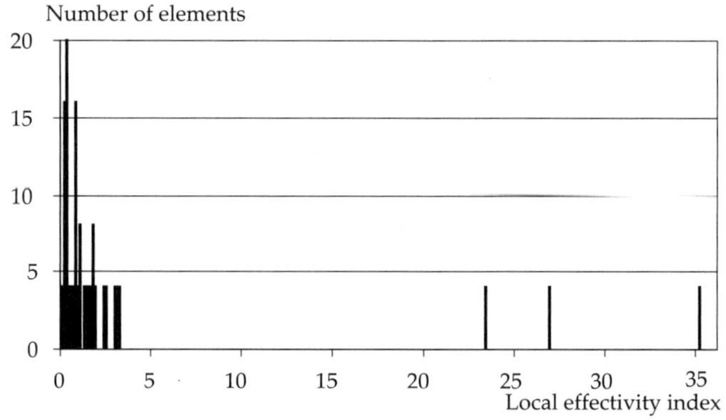

Figure 9.4. *Histogram for the ZZ2 indicator.*

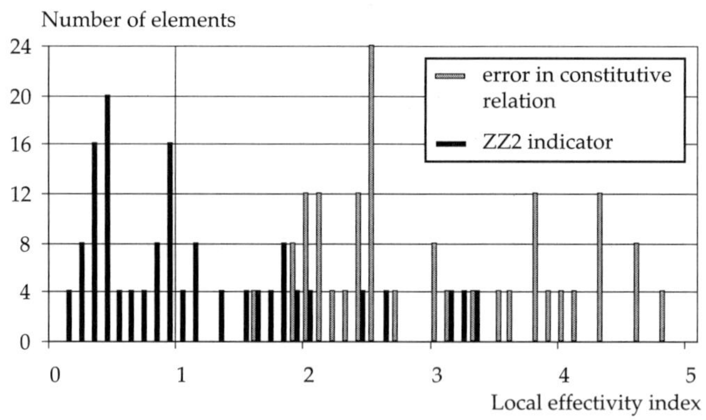

Figure 9.5. *Comparison for values close to 1.*

The results obtained for a second loading case (Figure 9.6) are of the same nature. For the error in constitutive relation, the local effectivity index γ_E is between 1.48 and 12.1.

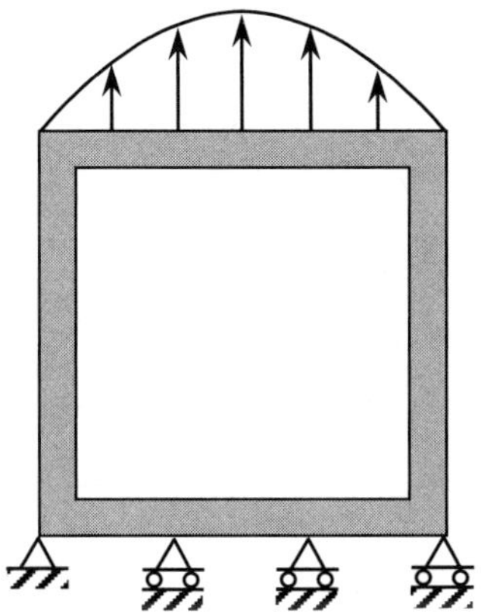

Figure 9.6. *Second loading case.*

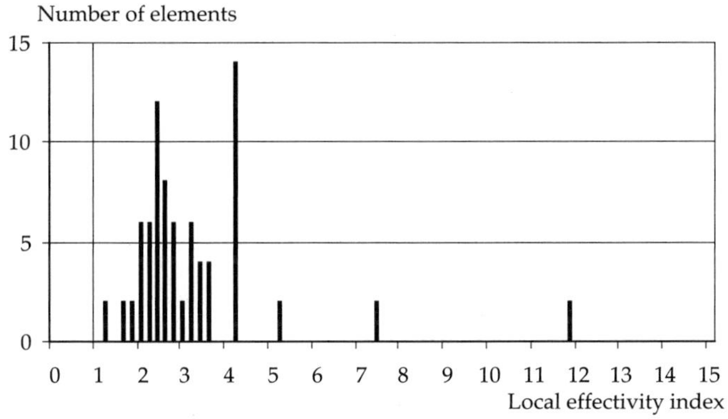

Figure 9.7. *Histogram of the error in constitutive relation.*

In Figure 9.7 and 9.8, one can observe an even greater tendency for the local indicators obtained with the ZZ2 indicator to be scattered ($0.08 \leq \gamma_E \leq 121.2$). Here again, for many elements (Figure 9.9), the ZZ2 indicator yields local effectivity indexes clearly less than 1.

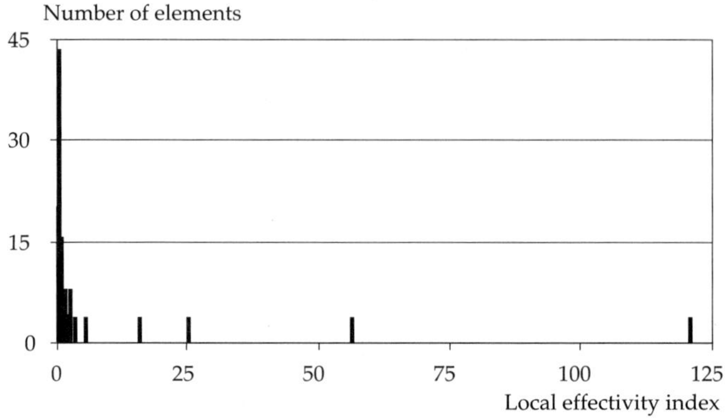

Figure 9.8. *Histogram for the ZZ2 indicator.*

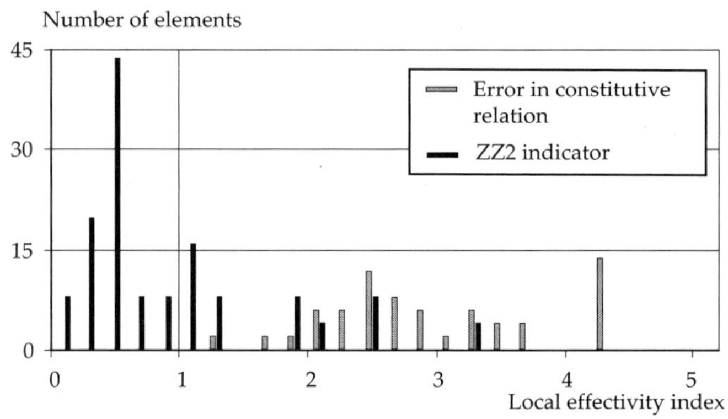

Figure 9.9. *Comparison for values close to 1.*

Second example

The second example is described in Figure 9.10.

Again, one can observe that for the error in constitutive relation the effectivity indexes are greater than 1 ($1.26 \le \gamma_E \le 3.33$); even more importantly, they are less scattered than for the ZZ2 indicator ($0.09 \le \gamma_E \le 7.62$).

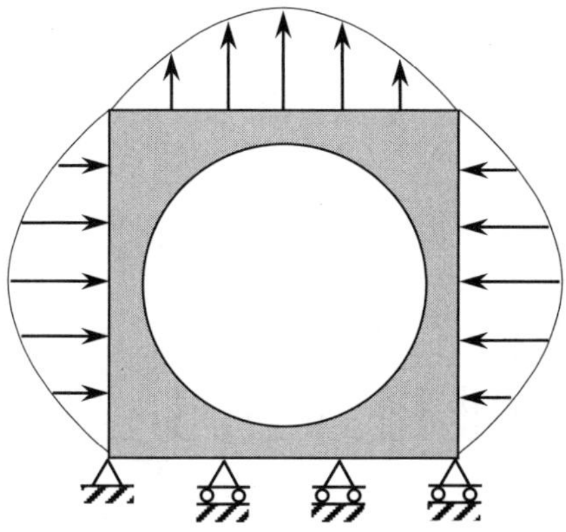

Figure 9.10. *Description of the second test case.*

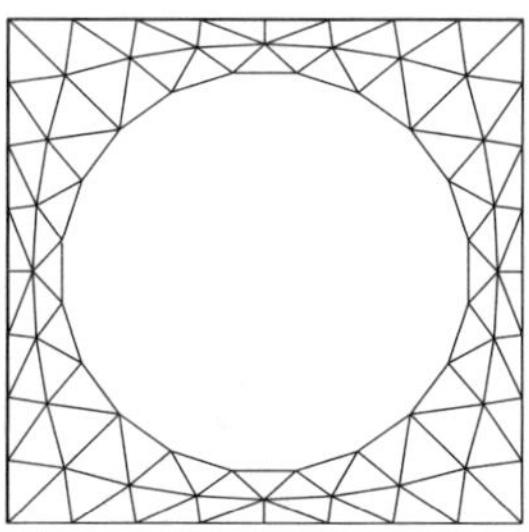

Figure 9.11. *Mesh.*

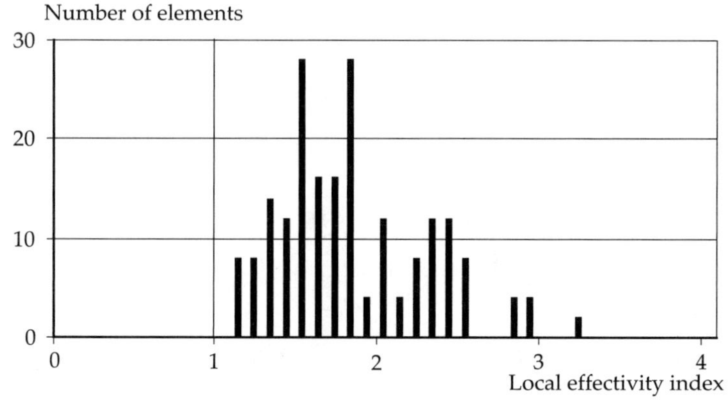

Figure 9.12. *Histogram of the error in constitutive relation.*

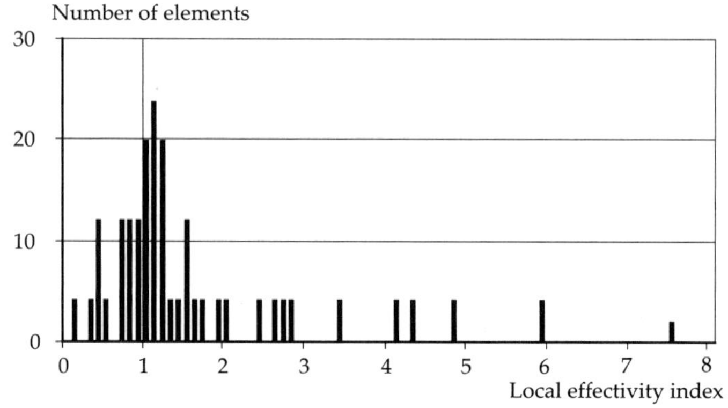

Figure 9.13. *Histogram of the ZZ2 indicator.*

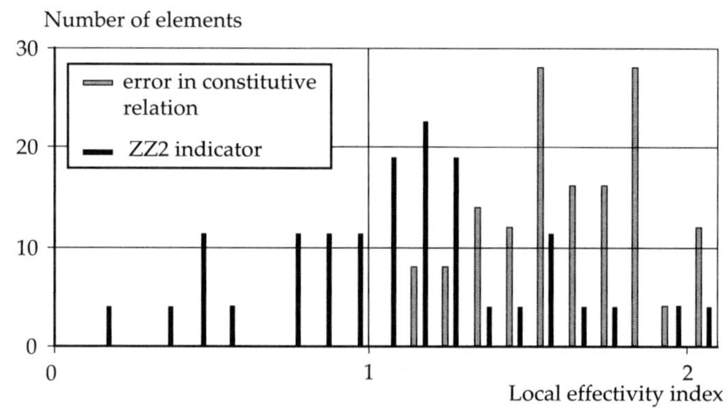

Figure 9.14. *Comparison for values close to 1.*

In this second example, let us examine the quality of the local estimates as a function of the stress intensities.

Let us introduce the norm

$$\| \sigma \|_{loc,E} = \left[\frac{1}{\mathrm{mes}(E)} \int_E \mathrm{Tr}[\sigma\sigma] \, dE \right]^{1/2} \tag{9.9}$$

The stress intensity in an element E can be characterized by

$$i_{\sigma,E} = \frac{\| \sigma \|_{loc,E}}{\underset{E \subset \Omega}{\mathrm{Max}} \| \sigma \|_{loc,E}} \tag{9.10}$$

One obviously has $0 \le i_{\sigma,E} \le 1$.

The quality of the local estimate on E can be characterized by

$$q_E = \text{Max}\left(\gamma_E, \frac{1}{\gamma_E}\right)$$

For any $x \in [0,1]$, let

$$\omega(x) = \left\{E \backslash i_{\sigma,E} \ge x\right\}$$

The quality of an error estimator as a function of the stress intensities can be evidenced by plotting the curve

$$x \mapsto \underset{E \,\in\, \omega(x)}{\text{Max}} \; q_E$$

Figure 9.15 shows the curve drawn for the constitutive relation error estimator and for the ZZ2 indicator.

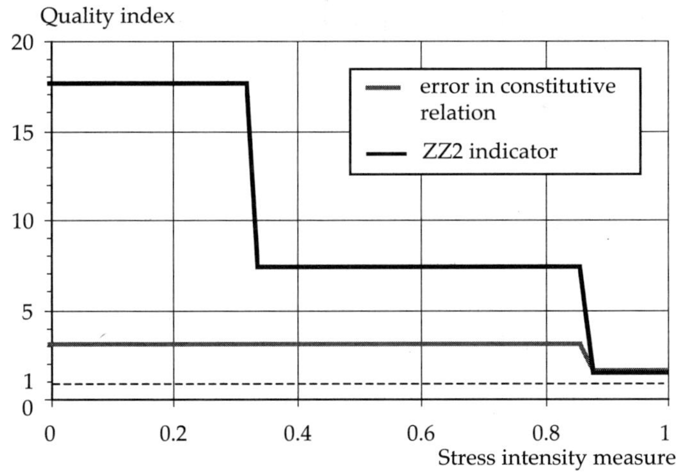

Figure 9.15. *Evolution of the quality as a function of the stress intensity.*

For any stress level in an element, the error in constitutive relation leads to local estimates which are reasonable upper bounds of the actual errors.

Conversely, the ZZ2 indicators give poor local estimates, even in highly loaded elements.

Other tests and practical use

The heuristic property has been studied for 3- and 6-node triangles. More studies on commonly used 3D elements are in progress.

Exceptions to the heuristic upper bound property were found only in a few cases where the local error was extremely small and the stresses were relatively low.

In practice, if one disregards zones where the local error is relatively small (e.g. less than 2% of the mean error) and those where the stresses are low, the heuristic property (9.8) is systematically verified with a constant $\bar{C}$ close to 1.

9.3.3 An estimation of the error in local stresses

Using the heuristic property (9.8), it is easy to construct an estimator of the error in the local stresses whose implementation does not require any additional calculation; in particular, the calculation of a GREEN operator is not necessary.

Property

If the HOOKE's operator $\mathbf{K}$ *is constant on element E, one has*

$$\left[\mathrm{Tr}\left[\left(\langle \sigma_{ex}\rangle - \langle \sigma_h\rangle\right)\mathbf{K}^{-1}\left(\langle \sigma_{ex}\rangle - \langle \sigma_h\rangle\right)\right]\right]^{1/2} \le \frac{\bar{C}\, e_E}{\mathrm{mes}(E)^{1/2}} \tag{9.11}$$

with

$$\langle \mathbb{A}\rangle = \frac{1}{\mathrm{mes}(E)}\int_E \mathbb{A}\, dE$$

Proof

Let $\mathbb{A}$ be an operator. One has

$$\mathrm{Tr}\left[\langle \mathbb{A}\rangle \mathbf{K}^{-1}\langle \mathbb{A}\rangle\right] \le \left\langle \mathrm{Tr}\left[\mathbb{A}\mathbf{K}^{-1}\mathbb{A}\right]\right\rangle$$

Indeed, one has

$$\int_E \mathrm{Tr}\left[\left(\mathbb{A} - \frac{1}{\mathrm{mes}(E)}\int_E \mathbb{A}\, dE\right)\mathbf{K}^{-1}\left(\mathbb{A} - \frac{1}{\mathrm{mes}(E)}\int_E \mathbb{A}\, dE\right)\right] dE$$

$$= \int_E \mathrm{Tr}\left[\mathbb{A}\mathbf{K}^{-1}\mathbb{A}\right] dE + \frac{1}{\mathrm{mes}(E)}\mathrm{Tr}\left[\left(\int_E \mathbb{A}\, dE\right)\mathbf{K}^{-1}\left(\int_E \mathbb{A}\, dE\right)\right]$$

$$- 2\int_E \mathrm{Tr}\left[\mathbb{A}\mathbf{K}^{-1}\left(\frac{1}{\mathrm{mes}(E)}\int_E \mathbb{A}\, dE\right)\right] dE$$

$$= \mathrm{mes}(E)\left[\left\langle \mathrm{Tr}[\mathbb{A}\mathbf{K}^{-1}\mathbb{A}]\right\rangle - \mathrm{Tr}\left[\langle\mathbb{A}\rangle\mathbf{K}^{-1}\langle\mathbb{A}\rangle\right]\right]$$

thus

$$\left\langle \mathrm{Tr}\left[\mathbb{A}\mathbf{K}^{-1}\mathbb{A}\right]\right\rangle - \mathrm{Tr}\left[\langle\mathbb{A}\rangle\mathbf{K}^{-1}\langle\mathbb{A}\rangle\right] \geq 0$$

Then, observing that

$$\left\langle \sigma_{ex} - \sigma_h\right\rangle = \left\langle \sigma_{ex}\right\rangle - \left\langle \sigma_h\right\rangle$$

and using the heuristic upper estimate (9.8), one obtains the desired result.

9.4 Types of local quantities to be estimated

9.4.1 Average variables associated with the stress

As a matter of fact, in practice it is sufficient to seek estimates of the averages of the exact stress: the mean value, the affine part.... The estimation of point values is possible, however it is more delicate and more expensive to obtain.

The mean value is defined by

$$\left\langle \sigma_{ex}\right\rangle = \frac{1}{\mathrm{mes}(E)}\int_E \sigma_{ex}\, dE \tag{9.12}$$

As for the affine part this is, on an element E, the affine stress field $[\sigma_{ex}]_L$ which minimizes

$$\| \sigma_{ex} - \mathbb{A} \|_{\sigma, E} \tag{9.13}$$

on the set of stress fields $\mathbb{A}$ which are affine on the element E.

REMARKS

1. If the operator **K** is constant elementwise, the minimization of (9.13) is equivalent to minimizing $\sigma_{ex} - [\sigma_{ex}]_L$ in the sense of the $L^2(E)$ norm. For simplicity, in the following we will make this assumption.

2. On a sufficiently small element, we can approximate with reasonable accuracy the exact stress by its mean value on element E or even better by its affine part on E.

Determination of the mean value

Let σ be a stress operator.

Property

The determination of the operator $\langle\sigma\rangle$ on an element E is equivalent to the determination of the scalar quantities

$$\int_\Omega \mathrm{Tr}(\sigma\Sigma)\, d\Omega \tag{9.14}$$

for particular operators Σ which are zero outside E and constitute a basis of the space of symmetric constant operators on E.

For example, in two dimensions there are three such operators and are defined on E by

$$\Sigma^1 = \frac{1}{\mathrm{mes}(E)}\begin{bmatrix} 1 & 0 \\ 0 & 0 \end{bmatrix};\ \Sigma^2 = \frac{1}{\mathrm{mes}(E)}\begin{bmatrix} 0 & 0 \\ 0 & 1 \end{bmatrix};\ \Sigma^3 = \frac{1}{2\,\mathrm{mes}(E)}\begin{bmatrix} 0 & 1 \\ 1 & 0 \end{bmatrix}$$

Proof

For the two dimensions, it suffices to observe that

$$\langle\sigma\rangle_{11} = \int_\Omega \mathrm{Tr}[\sigma\,\Sigma^1]\, d\Omega$$

The proof is analogous in three dimensions where we have six operators Σ^i.

Determination of the affine part

Let us put ourselves in the 2D setting in the case where E is a triangle and let σ be a stress operator. We denote by λ_1, λ_2 and λ_3 the barycentrical coordinates.

Property

The determination of the operator $[\sigma]_L$ *in an element E is equivalent to the determination of the three operators*

$$\int_E \sigma \lambda_1 \, dE , \quad \int_E \sigma \lambda_2 \, dE , \quad \int_E \sigma \lambda_3 \, dE$$

and thus of the scalar quantities

$$\int_\Omega \mathrm{Tr}[\sigma \Sigma] \, d\Omega \tag{9.15}$$

for the nine particular Σ *operators, which are zero outside E, and are defined on E by*

$$\Sigma^i \lambda_j , \quad \text{for } i, j = 1, 2, 3$$

Proof

Let

$$[\sigma]_L = \mathbb{A}^1 \lambda_1 + \mathbb{A}^2 \lambda_2 + \mathbb{A}^3 \lambda_3$$

From the definition (9.13), by writing the extremality conditions we obtain

$$\sum_{j=1}^{3} \mathbb{A}^j \int_E \lambda_i \lambda_j \, dE = \int_E \sigma \lambda_i \, dE \tag{9.16}$$

which constitute a regular linear system for the unknowns $\mathbb{A}^i$. Indeed the matrix of the system (9.16) is

$$\frac{\mathrm{mes}(E)}{12} \begin{bmatrix} 2 & 1 & 1 \\ 1 & 2 & 1 \\ 1 & 1 & 2 \end{bmatrix}$$

This proves the first part of the property.

To prove the second part, it suffices to note that the right-hand sides of (9.16) satisfy

$$\left[\int_E \sigma \lambda_i \, dE \right]_{11} = \int_\Omega \mathrm{Tr}[\sigma \Sigma^1 \lambda_i] \, d\Omega$$

$$\left[\int_E \sigma\, \lambda_i \, dE\right]_{22} = \int_\Omega \mathrm{Tr}[\sigma\, \Sigma^2\, \lambda_i]\, d\Omega$$

$$\left[\int_E \sigma\, \lambda_i \, dE\right]_{12} = \int_\Omega \mathrm{Tr}[\sigma\, \Sigma^3\, \lambda_i]\, d\Omega$$

The above property is extended without difficulty to the 3D case and tetrahedral elements. In this case there are simply 24 operators $\Sigma^i \lambda_\varphi$. The above results can also be extended to quadrilateral elements in 2D and to bricks in 3D. Nevertheless, from a practical point of view, for these elements it is preferable not to seek an affine approximation but instead an approximation in terms of the shape functions classically utilized in finite elements for the quadrilateral with four nodes and the brick with eight nodes.

To fix the ideas, let us consider the case of a quadrilateral element E in 2D. Let $\mathbb{A}$ be a stress field zero outside E and defined on E by

$$\mathbb{A} = \mathbb{A}^1 \varphi_1 + \mathbb{A}^2 \varphi_2 + \mathbb{A}^3 \varphi_3 + \mathbb{A}^4 \varphi_4 \tag{9.17}$$

where the $\mathbb{A}^i$ are symmetric constant operators and the φ_i the finite element shape functions of the quadrilateral element with four nodes.

We can approximate σ on E by seeking the field $[\sigma]_Q$ which minimizes

$$\| \sigma - \mathbb{A} \|_{\sigma, E} \tag{9.18}$$

on the set of fields $\mathbb{A}$ of the form (9.17).

Property

The determination of the operator $[\sigma]_Q$ on an element E is equivalent to the determination of the four operators

$$\int_E \sigma\varphi_1 \, dE\,, \quad \int_E \sigma\varphi_2 \, dE\,, \quad \int_E \sigma\varphi_3 \, dE\,, \quad \int_E \sigma\varphi_4 \, dE$$

and hence of the scalar quantities

$$\int_\Omega \mathrm{Tr}[\sigma\, \Sigma]\, d\Omega \tag{9.19}$$

for the 12 particular $\mathbb{\Sigma}$ *operators, which are zero outside* E *and defined on* E *by* $\mathbb{\Sigma}^i \varphi_j$, *for* $i = 1, 2, 3$ *and* $j = 1, 2, 3, 4$.

Proof

It suffices to note that the extremality conditions (9.18) are written

$$\sum_{j=1}^{4} \mathbb{A}^j \int_E \varphi_i \, \varphi_j \, dE = \int_E \mathbb{\sigma} \, \varphi_i \, dE \tag{9.20}$$

and that the matrix of this linear system is not singular.

We can extend the above property to the case of brick elements by an analogous approach. Simply, in this case, there are 48 particular operators $\mathbb{\Sigma}$.

9.4.2 Average variable associated with displacement

For the displacement field, we seek to estimate the mean value on an element E,

$$\langle \underline{U}_{ex} \rangle = \frac{1}{\text{mes}(E)} \int_E \underline{U}_{ex} \, dE \tag{9.21}$$

which for sufficiently small elements is very reasonable.

Let $\underline{U}$ be a displacement field.

Property

To determine $\langle \underline{U} \rangle$, *it suffices to determine*

$$\int_\Omega \underline{U} \cdot \underline{W} \, d\Omega \tag{9.22}$$

where $\underline{W}$ *is constant on* E *and zero outside* E.

Proof

This is immediate by observing that

$$\int_\Omega \underline{U} \cdot \underline{W} \, d\Omega = \langle \underline{U} \rangle \cdot \underline{W} \, \text{mes}(E)$$

REMARK. In practice, it is sufficient to determine (9.22) for the constant and independent vectors (two in 2D and three in 3D).

The following property allows us to reduce the determination of the average of the displacement to the determination of the same type of quantities as the average or the affine part of the stress:

Property

For very field $\underline{U} \in \boldsymbol{U}_0$ *the determination of* $\langle \underline{U} \rangle$ *is equivalent to the determination of the scalar quantities*

$$\int_\Omega \mathrm{Tr}(\sigma\, \Sigma_{\underline{W}})\, d\Omega \tag{9.23}$$

where $\sigma = \mathbf{K}\, \varepsilon(\underline{U})$ *and* $\Sigma_{\underline{W}}$ *is a symmetric operator associated with* $\underline{W}$.

More precisely, this operator can be defined in the following way :

$$\Sigma_{\underline{W}} = \varepsilon(\underline{U}_{\underline{W}})$$

where $\underline{U}_{\underline{W}}$ is the solution of the elasticity problem

Find $\underline{U}_{\underline{W}} \in \boldsymbol{U}_0$ *such that*

$\forall \underline{U}^* \in \boldsymbol{U}_0,$

$$\int_\Omega \mathrm{Tr}[\varepsilon(\underline{U}_{\underline{W}})\mathbf{K}\varepsilon(\underline{U}^*)]\, d\Omega = \int_\Omega \underline{W} \cdot \underline{U}^*\, d\Omega \tag{9.24}$$

Proof

It is sufficient to observe that for $\underline{U}^* = \underline{U}$, we have

$$\int_\Omega \mathrm{Tr}[\varepsilon(\underline{U}_{\underline{W}})\mathbf{K}\varepsilon(\underline{U})]\, d\Omega = \int_\Omega \underline{W} \cdot \underline{U}\, d\Omega \tag{9.25}$$

and hence

$$\int_\Omega \mathrm{Tr}[\sigma\varepsilon(\underline{U})]\, d\Omega = \int_\Omega \underline{W} \cdot \underline{U}\, d\Omega \tag{9.26}$$

In particular, the above property will be used for $\underline{U} = \underline{U}_{ex} - \underline{U}_h$. Then, we have

$$\begin{aligned}\int_\Omega \mathrm{Tr}[\Sigma_{\underline{W}}\mathbf{K}\varepsilon(\underline{U}_{ex} - \underline{U}_h)]\, d\Omega &= \int_\Omega \underline{W} \cdot (\underline{U}_{ex} - \underline{U}_h)\, d\Omega \\ &= \left(\langle \underline{U}_{ex} \rangle - \langle \underline{U}_h \rangle\right) \cdot \underline{W}\, \mathrm{mes}(E)\end{aligned} \tag{9.27}$$

9.4.3 Extraction operators

In summary, in all the considered situations, we are led to estimate the error between and exact quantity and a quantity evaluated by finite elements [(9.15), (9.19) or (9.27)] which can always be put in the form

$$I_{ex} - I_h = \int_\Omega \mathrm{Tr}[(\sigma_{ex} - \sigma_h)\Sigma]\, d\Omega$$

This operator Σ, which depends on the type of quantity that we want to estimate, is called an extraction operator. In the case of mean variables associated with stresses, this operator is explicit. In the case of the average of the displacement, it is implicit. Let us also introduce the elasticity problem associated with the predeformation Σ.

Find a displacment field $\underline{U}_\Sigma$ *and a stress field* σ_Σ *defined on* Ω *such that:*

- *Constraint equations*

$$\underline{U}_\Sigma \in \boldsymbol{\mathcal{U}}_0 \tag{9.28}$$

- *Equations of equilibrium*

$$\sigma_\Sigma \in \boldsymbol{\mathcal{S}} \quad \text{and} \quad \forall \underline{U}^* \in \boldsymbol{\mathcal{U}}_0,$$

$$\int_\Omega \mathrm{Tr}\left[\sigma_\Sigma \varepsilon(\underline{U}^*)\right] d\Omega = \int_\Omega \mathrm{Tr}[\Sigma\, \mathbf{K} \varepsilon(\underline{U}^*)]\, d\Omega \tag{9.29}$$

- *Constitutive relation*

$$\sigma_\Sigma = \mathbf{K}\varepsilon(\underline{U}_\Sigma) \tag{9.30}$$

The exact solution of this problem is denoted by $(\underline{U}_{\Sigma,ex}, \sigma_{\Sigma,ex})$.

REMARKS

1. This problem is central to all the methods of estimation of local errors. It is called the "dual problem" by RANNACHER who uses a terminology stemming from optimization.

The operator $\Sigma \mapsto \underline{U}_\Sigma$ is in fact a generalized GREEN operator well known to mechanicians.

The various approaches proposed up to now differ on two technical aspects:

- the method of approximation of $\underline{U}_\Sigma$;
- the method used for obtaining an upper estimate.

For example, if we employ the initial finite element mesh to approximate $\underline{U}_\Sigma$, the approximation can be mediocre if the studied quantities are very local. Hence, such an approach is limited to the estimation of averages in zones which are relatively large with respect to the dimensions of the elements.

2. In the case where we seek the average magnitudes associated with the stress, the problem [(9.28) - (9.30)] is well defined since Σ is explicit.

3. In the case where we are interested in the average of the displacement, this problem is also completely defined if we note that

$$\forall \underline{U}^* \in \mathcal{U}_0,$$

$$\int_\Omega \mathrm{Tr}[\Sigma_{\underline{W}} \mathbf{K} \varepsilon(\underline{U}^*)]\, d\Omega = \int_\Omega \underline{W} \cdot \underline{U}^*\, d\Omega \qquad (9.31)$$

Equation (9.29) is thus equivalent to the equation

$$\sigma_\Sigma \in \mathcal{S} \quad \textit{and} \quad \forall \underline{U}^* \in \mathcal{U}_0,$$

$$\int_\Omega \mathrm{Tr}\left[\sigma_\Sigma \varepsilon(\underline{U}^*)\right] d\Omega = \int_\Omega \underline{W} \cdot \underline{U}^*\, d\Omega \qquad (9.32)$$

In general, the problem [(9.28) - (9.30)] cannot be solved exactly. On the other hand, it is possible to obtain approximations of its solution by the finite element method. For example, we can employ the finite element discretization associated with the mesh $\mathcal{T}_h$ and the space $\mathcal{U}_h$. The approximate solution is denoted by $(\underline{U}_{\Sigma,h}, \sigma_{\Sigma,h})$.

Similarly, by the techniques of construction of statically admissible fields, we know how to construct a field $\hat{\sigma}_{\Sigma,h}$ such that

$$\forall \underline{U}^* \in \mathcal{U}_0, \quad \int_\Omega \mathrm{Tr}[\hat{\sigma}_{\Sigma,h} \varepsilon(\underline{U}^*)]\, d\Omega = \int_\Omega \mathrm{Tr}[\Sigma \mathbf{K} \varepsilon(\underline{U}^*)]\, d\Omega$$

4. The extractor to be used for estimating the intensity factors is given in [BABUSKA - MILLER, 1984], [LEGUILLON - SANCHEZ-PALENCIA, 1987], [BUI, 1973].

9.4.4 Estimation of the local errors

The estimation of the local errors comes down to the estimation of

$$I_{ex} - I_h = \int_\Omega \mathrm{Tr}(\sigma_{ex}\,\Sigma)\,d\Omega - \int_\Omega \mathrm{Tr}(\sigma_h\,\Sigma)\,d\Omega$$

Estimates employing only the mesh $\mathcal{T}_h$

The estimates of $I_{ex} - I_h$ are built on the following result :

Property

$$I_{ex} - I_h = \int_\Omega \mathrm{Tr}\big[(\varepsilon(\underline{U}_{\Sigma,ex}) - \varepsilon(\underline{U}_{\Sigma,h}))\mathbf{K}(\varepsilon(\underline{U}_{ex}) - \varepsilon(\underline{U}_h))\big]\,d\Omega \tag{9.33}$$

Proof

We have

$$\begin{aligned}\int_\Omega \mathrm{Tr}\big[(\varepsilon(\underline{U}_{\Sigma,ex}) - \varepsilon(\underline{U}_{\Sigma,h}))\mathbf{K}(\varepsilon(\underline{U}_{ex}) - \varepsilon(\underline{U}_h))\big]\,d\Omega \\ = \int_\Omega \mathrm{Tr}\big[(\varepsilon(\underline{U}_{\Sigma,ex}) - \Sigma)\mathbf{K}(\varepsilon(\underline{U}_{ex}) - \varepsilon(\underline{U}_h))\big]\,d\Omega \\ + \int_\Omega \mathrm{Tr}\big[(\Sigma - \varepsilon(\underline{U}_{\Sigma,h})\mathbf{K}(\varepsilon(\underline{U}_{ex}) - \varepsilon(\underline{U}_h))\big]\,d\Omega\end{aligned} \tag{9.34}$$

Let us evaluate the first term of the right-hand side of (9.34) for $\underline{U}^* = \underline{U}_{ex} - \underline{U}_h$,

$$\begin{aligned}\int_\Omega \mathrm{Tr}\big[(\varepsilon(\underline{U}_{\Sigma,ex}) - \Sigma\ )\mathbf{K}(\varepsilon(\underline{U}_{ex}) - \varepsilon(\underline{U}_h))\big]\,d\Omega \\ = \int_\Omega \mathrm{Tr}\big[(\varepsilon(\underline{U}_{\Sigma,ex}) - \Sigma)\mathbf{K}\varepsilon(\underline{U}^*)\big]\,d\Omega\end{aligned}$$

From [(9.28) - (9.30)], we then obtain

$$\int_\Omega \mathrm{Tr}\big[(\varepsilon(\underline{U}_{\Sigma,ex}) - \Sigma)\mathbf{K}(\varepsilon(\underline{U}_{ex}) - \varepsilon(\underline{U}_h))\big]\,d\Omega = 0$$

On the other hand, we have

$$\begin{aligned}\int_\Omega \mathrm{Tr}\big[(\ \Sigma - \varepsilon(\underline{U}_{\Sigma,h})\mathbf{K}(\varepsilon(\underline{U}_{ex}) - \varepsilon(\underline{U}_h))\big]\,d\Omega \\ = I_{ex} - I_h - \int_\Omega \mathrm{Tr}\big[\varepsilon(\underline{U}_{\Sigma,h})(\sigma_{ex} - \sigma_h)\big]\,d\Omega\end{aligned} \tag{9.35}$$

and since $\underline{U}_{\Sigma,h}$ belongs to $\boldsymbol{\mathcal{U}}_{h0}$, the last integral is zero (9.35). Combining all the above results, we obtain (9.33).

A first estimate

Property

The error $|I_{ex} - I_h|$ *is bounded from above by*

$$|I_{ex} - I_h| \le \mathbf{e}_{RdC}(\underline{U}_h, \hat{\sigma}_h) \cdot \mathbf{e}_{RdC}(\underline{U}_{\Sigma,h}, \hat{\sigma}_{\Sigma,h})$$

This upper bound which follows immediately from the result (9.33), is not, in general, very sharp.

A second estimation

A sharper upper estimate of $|I_{ex} - I_h|$ can be obtained by employing the heuristic property (9.8). Let us denote

$$\mathbf{e}_E = \left\| \hat{\sigma}_h - \varepsilon(\underline{U}_h) \right\|_{\sigma,E} \quad \text{and} \quad \mathbf{e}_{\Sigma,E} = \left\| \hat{\sigma}_{\Sigma,h} - \varepsilon(\underline{U}_{\Sigma,h}) \right\|_{\sigma,E}$$

Property

Assuming that the heuristic property (9.8) *holds, we have the upper bound*

$$|I_{ex} - I_h| \le \bar{C} \sum_{E \in \mathcal{T}_h} \mathbf{e}_E \cdot \mathbf{e}_{\Sigma,E} \tag{9.36}$$

Proof

Noting that

$$\int_\Omega \mathrm{Tr}\left[\varepsilon(\underline{U}_{\Sigma,ex})\mathbf{K}(\varepsilon(\underline{U}_{ex}) - \varepsilon(\underline{U}_h))\right] d\Omega = \int_\Omega \mathrm{Tr}\left[\hat{\sigma}_{\Sigma,h}(\varepsilon(\underline{U}_{ex}) - \varepsilon(\underline{U}_h))\right] d\Omega$$

we can replace $\varepsilon(\underline{U}_{\Sigma,ex})$ by $\mathbf{K}^{-1}\hat{\sigma}_{\Sigma,h}$ in (9.33).

Employing the CAUCHY - SCHWARTZ inequality

$$|I_{ex} - I_h| \le \sum_{E \in \mathcal{T}_h} \left\| \sigma_{ex} - \varepsilon(\underline{U}_h) \right\|_{\sigma,E} \cdot \mathbf{e}_{\Sigma,E}$$

Equation (9.36) now follows immediately for the heuristic upper estimate (9.8),

$$\left\| \sigma_{ex} - \varepsilon(\underline{U}_h) \right\|_{\sigma,E} \le \bar{C}\, \mathbf{e}_E$$

Estimates which employ a local reanalysis

In order to improve the estimates given above, we can seek an approximation of $\underline{U}_\Sigma$ on a mesh more refined than $\mathcal{T}_h$. We can also use a local reanalysis on a refined submesh defined in the neighborhood of the element E where we are seeking the estimates. These techniques of local reanalysis are developed in the framework of multiscale approaches and especially for the highly heterogeneous media [FISH - MARKOFELAR, 1994], [FISH - SHEK - PANDHEERAI - SHEPARD, 1997], [ODEN - VEMAGANTE - MOES, 1999], [LADEVEZE - LOISEAU - DUREISSEIX, 2001].

The situation is described in Figure (9.16). A zone ω_E surrounding the element E, where we desire to estimate certain average quantities, is meshed more finely.

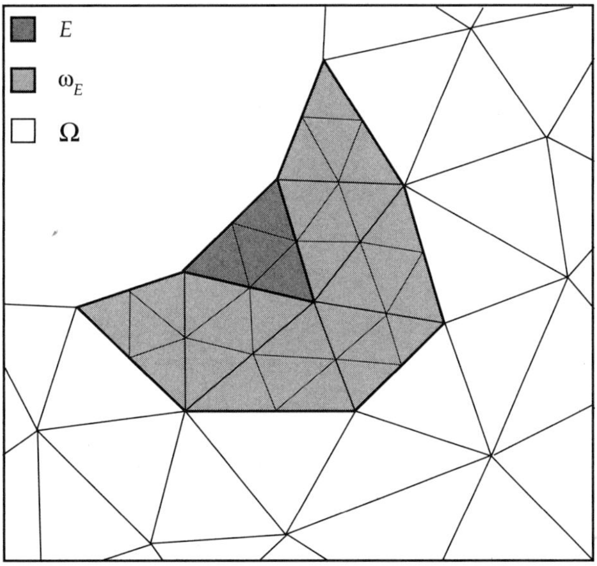

Figure 9.16. *A local submesh.*

We will denote by $\mathbf{t}_{hh}$ the mesh of ω_E. The idea is to improve the estimate

$\underline{U}_{\Sigma,h}$ of $\underline{U}_{\Sigma,ex}$ by performing a new finite element analysis only in ω_E.

More precisely, let $\underline{u}_{\Sigma,hh}$ be the solution field of the following problem posed on ω_E:

Find $\underline{u}_{\Sigma,hh}$ *zero on* $\partial_1\omega_E$ *and such that:*

For every $\underline{u}^*_{hh}$ *of finite element type on* ω_E *which is zero on* $\partial_1\omega_E$,

$$\int_{\omega_E} \mathrm{Tr}\left[\varepsilon(\underline{u}_{\Sigma,hh})\mathbf{K}\varepsilon(\underline{u}^*_{hh})\right] d\omega = \int_{\omega_E} \left[-\underline{\mathrm{div}}\,(\mathbf{K}\,\mathbb{\Sigma})\right]\cdot\underline{u}^*_{hh}\; d\omega$$

with

$$\partial_1\omega_E = \partial\omega_E - (\partial\omega_E \cap \partial_2\Omega)$$

In this way we obtain a new approximation of $\underline{U}_\Sigma$,

$$\underline{V}_{\Sigma,hh} = \underline{U}_{\Sigma,h} + \underline{u}_{\Sigma,hh} \tag{9.37}$$

Property

$$I_{ex} - I_h - \hat{I}_{hh} = \int_\Omega \mathrm{Tr}\left[(\varepsilon(\underline{U}_{\Sigma,ex}) - \varepsilon(\underline{V}_{\Sigma,hh}))\mathbf{K}(\varepsilon(\underline{U}_{ex}) - \varepsilon(\underline{U}_h))\right] d\Omega \tag{9.38}$$

with

$$\hat{I}_{hh} = \int_\Omega \mathrm{Tr}\left[(\hat{\sigma}_h - \sigma_h)\varepsilon(\underline{u}_{\Sigma,hh})\right] d\Omega \tag{9.39}$$

Proof

We have

$$\begin{aligned}\int_\Omega \mathrm{Tr}\left[(\varepsilon(\underline{U}_{\Sigma,ex}) - \varepsilon(\underline{V}_{\Sigma,hh}))\mathbf{K}(\varepsilon(\underline{U}_{ex}) - \varepsilon(\underline{U}_h))\right] d\Omega \\ = \int_\Omega \mathrm{Tr}\left[(\varepsilon(\underline{U}_{\Sigma,ex}) - \mathbb{\Sigma})\mathbf{K}(\varepsilon(\underline{U}_{ex}) - \varepsilon(\underline{U}_h))\right] d\Omega \\ + \int_\Omega \mathrm{Tr}\left[(\mathbb{\Sigma} - \varepsilon(\underline{V}_{\Sigma,hh})\mathbf{K}(\varepsilon(\underline{U}_{ex}) - \varepsilon(\underline{U}_h))\right] d\Omega\end{aligned} \tag{9.40}$$

As in (9.34), the first term of the right-hand side of (9.40) is zero.

On the other hand, we have

$$\int_\Omega \mathrm{Tr}\left[(\Sigma - \varepsilon(\underline{V}_{\Sigma,hh}))\mathbf{K}(\varepsilon(\underline{U}_{ex}) - \varepsilon(\underline{U}_h))\right] d\Omega$$

$$= I_{ex} - I_h - \int_\Omega \mathrm{Tr}\left[\varepsilon(\underline{V}_{\Sigma,hh})\mathbf{K}(\varepsilon(\underline{U}_{ex}) - \varepsilon(\underline{U}_h))\right] d\Omega$$

$$= I_{ex} - I_h - \int_\Omega \mathrm{Tr}\left[\varepsilon(\underline{U}_{\Sigma,h})(\sigma_{ex} - \sigma_h)\right] d\Omega$$

$$- \int_\Omega \mathrm{Tr}\left[\varepsilon(\underline{u}_{\Sigma,hh})(\sigma_{ex} - \sigma_h)\right] d\Omega$$

To obtain the result (9.38), it suffices to note that

$$\int_\Omega \mathrm{Tr}\left[\varepsilon(\underline{U}_{\Sigma,h})(\sigma_{ex} - \sigma_h)\right] d\Omega = 0$$

and that

$$\int_\Omega \mathrm{Tr}\left[\varepsilon(\underline{u}_{\Sigma,hh})\mathbf{K}(\sigma_{ex} - \sigma_h)\right] d\Omega$$

$$= \int_\Omega \mathrm{Tr}\left[\varepsilon(\underline{u}_{\Sigma,hh})\mathbf{K}(\hat{\sigma}_h - \sigma_h)\right] d\Omega = \hat{I}_{hh}$$

This result leads us to retain $I_h + \hat{I}_{hh}$ as an approximation of the exact value I_{ex}. As before, when we estimate I_{ex} by $I_h + \hat{I}_{hh}$, it is possible to obtain two estimates of the exact error.

A first estimate

Property

The error $\left| I_{ex} - I_h - \hat{I}_{hh} \right|$ *is bounded by*

$$\left| I_{ex} - I_h - \hat{I}_{hh} \right| \leq \mathbf{e}_{RdC}(\underline{U}_h, \hat{\sigma}_h) \cdot \mathbf{e}_{RdC}(\underline{V}_{\Sigma,hh}, \hat{\sigma}_{\Sigma,h})$$

This upper estimate which follows immediately from the result (9.38), is not, in general, very sharp numerically.

A second estimate

A sharper upper estimate of $\left| I_{ex} - I_h - \hat{I}_{hh} \right|$ can be obtained by employing the heuristic property (9.8).

Let

$$\mathbf{e}_E = \left\| \hat{\sigma}_h - \varepsilon(\underline{U}_h) \right\|_{\sigma,E} \quad \text{and} \quad \tilde{\mathbf{e}}_{\Sigma,E} = \left\| \hat{\sigma}_{\Sigma,h} - \varepsilon(\underline{V}_{\Sigma,hh}) \right\|_{\sigma,E}$$

Property

Assuming that $\bar{C}\,\mathbf{e}_E$ *is an upper estimate of the exact local error, we have*

$$\left| I_{ex} - I_h - \hat{I}_{hh} \right| \le \bar{C} \sum_{E \in \mathcal{T}_h} \mathbf{e}_E \cdot \tilde{\mathbf{e}}_{\Sigma,E} \tag{9.41}$$

The proof is analogous to that of the property (9.36).

REMARK. The local refinement (of the mesh or of the interpolation) can be obtained at a good cost by employing the method based on the partition of unity associated with finite element basis functions, and the method introduced by [BABUSKA - MELENK, 1999] and developed in [STROUBOULIS - BABUSKA - COPPS, 2000].

Iterative strategy for constructing bounds without heuristics

This strategy follows the approach proposed in [BABUSKA - STROUBOULIS - GANGARAJ, 1999] which has been described in Chapter 3 (Section 3.4).

Let us introduce the solution errors

$$\underline{e}_h = \underline{U}_{ex} - \underline{U}_h \quad \text{and} \quad \underline{e}_{\Sigma,h} = \underline{U}_{\Sigma,ex} - \underline{U}_{\Sigma,h}$$

where $\underline{e}_h$ and $\underline{e}_{\Sigma,h}$ are the respective solutions of the problems

$$\underline{e}_h \in \boldsymbol{\mathcal{U}}_0 \quad \text{and} \quad \forall \underline{U}^* \in \boldsymbol{\mathcal{U}}_0,$$

$$\int_\Omega \mathrm{Tr}\left[\varepsilon(\underline{e}_h)\mathbf{K}\varepsilon(\underline{U}^*)\right] d\Omega = \sum_E \int_E \underline{r}_E \cdot \underline{U}^* dE - \sum_\Gamma \int_\Gamma \underline{t}_\Gamma \cdot \underline{U}^* d\Gamma$$

and

$$\underline{e}_{\Sigma,h} \in \boldsymbol{\mathcal{U}}_0 \quad \text{and} \quad \forall \underline{U}^* \in \boldsymbol{\mathcal{U}}_0,$$

$$\int_\Omega \mathrm{Tr}\left[\varepsilon(\underline{e}_{\Sigma,h})\mathbf{K}\varepsilon(\underline{U}^*)\right] d\Omega = \sum_E \int_E \underline{r}_{\Sigma,E} \cdot \underline{U}^* dE - \sum_\Gamma \int_\Gamma \underline{t}_{\Sigma,\Gamma} \cdot \underline{U}^* d\Gamma$$

As in (3.38) and (3.39) let us then introduce the local problems $\boldsymbol{\mathcal{P}}_E$ and $\boldsymbol{\mathcal{P}}_E^0$ defined on each element E and the analogous problems $\boldsymbol{\mathcal{P}}_{\Sigma,E}$ and $\boldsymbol{\mathcal{P}}_{\Sigma,E}^0$. We will denote the solutions of these problems by

$$\hat{\underline{e}}_{Eh}, \quad \tilde{\underline{e}}_{Eh} \quad \text{and} \quad \hat{\underline{e}}_{\Sigma,Eh}, \quad \tilde{\underline{e}}_{\Sigma,Eh}$$

Here, we will assume that these local problems are solved exactly.

Property

$$\begin{aligned} I_{ex} - I_h = &\sum_E \int_E \mathrm{Tr}\left[\varepsilon(\hat{\underline{e}}_{Eh})\mathbf{K}\varepsilon(\hat{\underline{e}}_{\Sigma,Eh})\right] dE \\ &- \sum_E \int_E \mathrm{Tr}\left[\varepsilon(\hat{\underline{e}}_{Eh} - \tilde{\underline{e}}_{Eh})\mathbf{K}\varepsilon(\hat{\underline{e}}_{\Sigma,Eh} - \tilde{\underline{e}}_{\Sigma,Eh})\right] dE \\ &+ \int_\Omega \mathrm{Tr}\left[\varepsilon(\underline{\delta}_h)\mathbf{K}\varepsilon(\underline{\delta}_{\Sigma,h})\right] d\Omega \end{aligned} \tag{9.42}$$

where $\underline{\delta}_h$ *is always defined by* (3.42) *and* $\underline{\delta}_{\Sigma,h}$ *by the analogous problem :*

Find $\underline{\delta}_{\Sigma,h} \in \boldsymbol{\mathcal{U}}_0$ *such that*

$$\begin{aligned} \forall \underline{U}^* \in \boldsymbol{\mathcal{U}}_0, \quad &\int_\Omega \mathrm{Tr}\left[\varepsilon(\underline{\delta}_{\Sigma,h})\mathbf{K}\varepsilon(\underline{U}^*)\right] d\Omega \\ &= \sum_E \int_E \mathrm{Tr}\left[\varepsilon(\hat{\underline{e}}_{\Sigma,Eh} - \tilde{\underline{e}}_{\Sigma,Eh})\mathbf{K}\varepsilon(\underline{U}^*)\right] dE \end{aligned} \tag{9.43}$$

Proof

This is very similar to that of the property (3.40). First, we show that

$$\begin{aligned} &\int_\Omega \mathrm{Tr}\left[\varepsilon(\underline{\delta}_h)\mathbf{K}\varepsilon(\underline{\delta}_{\Sigma,h})\right] d\Omega \\ &\quad = \int_\Omega \mathrm{Tr}\left[\varepsilon(\underline{e}_h)\mathbf{K}\varepsilon(\underline{e}_{\Sigma,h})\right] d\Omega \\ &\quad - \sum_E \left[\int_E \mathrm{Tr}\left[\varepsilon(\hat{\underline{e}}_{Eh})\mathbf{K}\varepsilon(\hat{\underline{e}}_{\Sigma,Eh})\right] dE\right] \\ &\quad + \sum_E \left[\int_E \mathrm{Tr}\left[\varepsilon(\hat{\underline{e}}_{Eh} - \tilde{\underline{e}}_{Eh})\mathbf{K}\varepsilon(\hat{\underline{e}}_{\Sigma,Eh} - \tilde{\underline{e}}_{\Sigma,Eh})\right] dE\right] \end{aligned} \tag{9.44}$$

The property (9.42) then follows by employing (9.33).

From this we deduce the property :

Property

$$\left| I_{ex} - I_h - \sum_E \alpha_{Eh} \right| \leq \left[\sum_E \left\| \hat{\underline{e}}_{Eh} - \tilde{\underline{e}}_{Eh} \right\|_{u,E}^2 \right]^{1/2} \left[\sum_E \left\| \hat{\underline{e}}_{\Sigma,Eh} - \tilde{\underline{e}}_{\Sigma,Eh} \right\|_{u,E}^2 \right]^{1/2} \qquad (9.45)$$

where we let

$$\alpha_{Eh} = \int_E \mathrm{Tr}\left[\varepsilon(\hat{\underline{e}}_{Eh}) \mathbf{K} \varepsilon(\hat{\underline{e}}_{\Sigma,Eh}) \right] dE$$
$$- \int_E \mathrm{Tr}\left[\varepsilon(\hat{\underline{e}}_{Eh} - \tilde{\underline{e}}_{Eh}) \mathbf{K} \varepsilon(\hat{\underline{e}}_{\Sigma,Eh} - \tilde{\underline{e}}_{\Sigma,Eh}) \right] dE$$

Proof

The proof is immediate by employing (3.43) and an analogous inequality satisfied by $\| \underline{\delta}_{\Sigma,h} \|_{u,\Omega}$.

An iterative strategy can then be set up. It is based on the identity

$$\int_\Omega \mathrm{Tr}\left[\varepsilon(\underline{e}_h) \mathbf{K} \varepsilon(\underline{e}_{\Sigma,h}) \right] d\Omega$$
$$= \sum_E \alpha_{Eh} + \int_\Omega \mathrm{Tr}\left[\varepsilon(\underline{\delta}_h) \mathbf{K} \varepsilon(\underline{\delta}_{\Sigma,h}) \right] d\Omega \qquad (9.46)$$

which follows directly from (9.44).

After the above initialization step, we solve by a standard finite element method the problems defining $\underline{\delta}_h$ and $\underline{\delta}_{\Sigma,h}$; in this way we obtain the approximations

$\underline{\delta}_{hh}$ and $\underline{\delta}_{\Sigma,hh}$

and the corresponding exact errors

$$\underline{e}_{hh}^{\delta} = \underline{\delta}_h - \underline{\delta}_{hh}$$
$$\underline{e}_{hh}^{\delta,\Sigma} = \underline{\delta}_{\Sigma,h} - \underline{\delta}_{\Sigma,hh}$$

The new expression of $I_{ex} - I_h$ is

$$I_{ex} - I_h = \sum_E \alpha_{Eh} + \int_\Omega \mathrm{Tr}\left[\varepsilon(\underline{\delta}_{hh})\mathbf{K}\varepsilon(\underline{\delta}_{\Sigma,hh})\right] d\Omega + \int_\Omega \mathrm{Tr}\left[\varepsilon(\underline{e}\,_{hh}^{\delta})\mathbf{K}\varepsilon(\underline{e}\,_{hh}^{\delta,\Sigma})\right] d\Omega \tag{9.47}$$

After solving the new local problems associated with these errors, we replace the last term in (9.47) by the identity which is analogous to (9.44) but for the errors $\underline{e}\,_{hh}^{\delta}$ and $\underline{e}\,_{hh}^{\delta,\Sigma}$. The bound is obtained by estimating from above the last term as in (9.45). This process can, of course, be iterated ; it is stopped when the obtained bound is considered to be sufficiently sharp.

REMARKS

1. The quality of the bounds depends on the permitted numerical effort. In practice, we will use approximate solutions with an element of degree $p+k$; $k=3$ is sufficient, in general, except in the presence of a zone of singularity of the extractor where we could take a higher value for k, or even introduce functions of an "ad hoc" form.

2. This strategy of constructing bounds can be interpreted as a strategy for multiscale calculation in which we are only interested in quantities calculated in the largest scale.

9.4.5 Example

We will present an example of the evaluation of the local errors in the stresses extracted from the reference [LADEVEZE - ROUGEOT - BLANCHARD - MOREAU, 1999].

The problem is described in the Figure 9.17 as well as the mesh used to determine the finite element approximations. We are interested in the evaluation of the averages of the stresses in the element $\bar{E}$.

With the triangular elements with six nodes, the global error in constitutive relation is 7%.

The averages of the stresses calculated by finite elements are

$$\left\langle \sigma_{h,11} \right\rangle_{\bar{E}} = 0.73; \quad \left\langle \sigma_{h,22} \right\rangle_{\bar{E}} = 55.2; \quad \left\langle \sigma_{h,12} \right\rangle_{\bar{E}} = 0.00$$

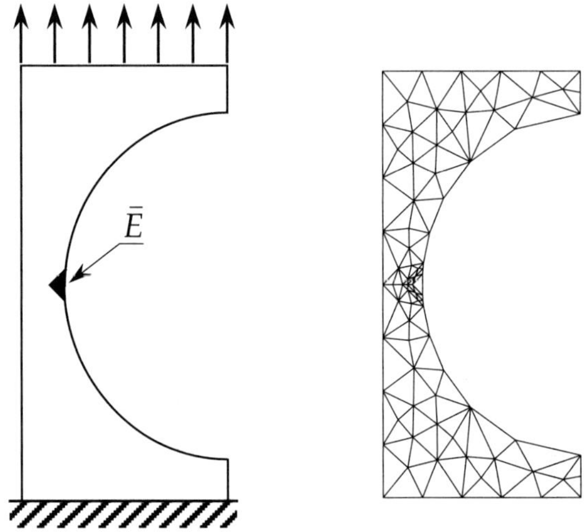

Figure 9.17. *Description of the test and the employed mesh.*

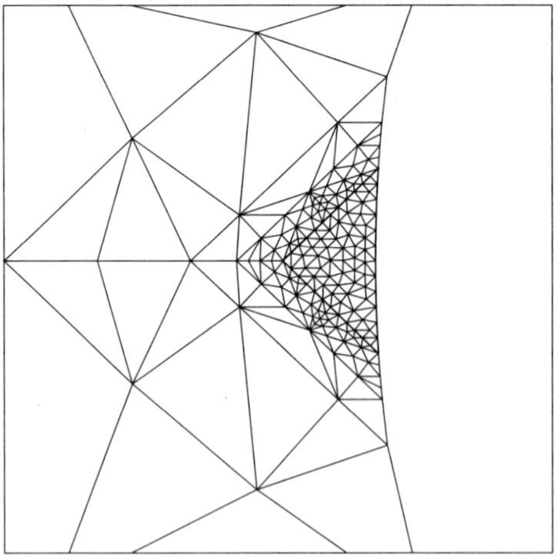

Figure 9.18. *Fine mesh used for the local reanalysis.*

Using (9.36) we obtain the estimates

$$\left|\left\langle \sigma_{h,11}\right\rangle_{\bar{E}} - \left\langle \sigma_{ex,11}\right\rangle_{\bar{E}}\right| \leq 3.61$$
$$\left|\left\langle \sigma_{h,22}\right\rangle_{\bar{E}} - \left\langle \sigma_{ex,22}\right\rangle_{\bar{E}}\right| \leq 6.46$$
$$\left|\left\langle \sigma_{h,12}\right\rangle_{\bar{E}} - \left\langle \sigma_{ex,12}\right\rangle_{\bar{E}}\right| \leq 7.81$$

which gives a relative error of 11.7% for the most significant variable which is here σ_{22}.

Employing a reanalysis on the fine mesh shown in Figure 9.18 and the property (9.41), we obtain the estimates

$$\left|\left\langle \sigma_{h,11}\right\rangle_{\bar{E}} - \left\langle \sigma_{ex,11}\right\rangle_{\bar{E}}\right| \leq 0.45$$
$$\left|\left\langle \sigma_{h,22}\right\rangle_{\bar{E}} - \left\langle \sigma_{ex,22}\right\rangle_{\bar{E}}\right| \leq 0.79$$
$$\left|\left\langle \sigma_{h,12}\right\rangle_{\bar{E}} - \left\langle \sigma_{ex,12}\right\rangle_{\bar{E}}\right| \leq 0.80$$

which now give a relative error of 1.44% for σ_{22}, the most significant magnitude.

9.4.6 Comments

The extension to time-dependent problems calls for the definition of an associated dual problem which is obviously also depending on time. Such a problem is well known to researchers working in mechanics; its treatment for obtaining local estimates is the subject of current work [RANNACHER, 1999].

The extension to nonlinear problems possibly depending on time is more delicate. It follows an idea due to JOHNSON [JOHNSON - HANSBO, 1992]; it consists in using the linearized operator and in considering the associated dual problem [RANNACHER - SUTTMEIER, 1999]. Of course, there is no guaranty for the quality of the estimator obtained in this manner.

Bibliography

ACHCHAB B. - AGOUZAL A. - BARANGER J. - MAITRE J.F. 2000

"A posteriori error estimates in finite element methods for general Friedricks' system." *Comput. Methods Appl. Mech. Eng.* **184**:39-47

ACTIS R. - SZABO B. - SCHWAB C. 1999

"Heriarchic models for laminated plates and shells." *Comput. Methods Appl. Mech. Eng.* **172**:79-107

AINSWORTH M. 1999

"Identification and a posteriori estimation of transported error in finite element analysis." *Comput. Math. Appl.* **176**:3-18

AINSWORTH M. - ODEN J.T. 1993

"A priori error estimators for second-order elliptic systems. Part 2: An optimal order process for calculating self-equilibrating fluxes." *Comput. Math. Appl.* **26**:75-87

AINSWORTH M. - ODEN J.T. 1997

"A posteriori error estimation in finite element analysis." *Comput. Methods Appl. Mech. Eng.* **142** :1-88

ALART P. 1997

"Méthode de NEWTON généralisée en mécanique du contact." *J. Math. Pures Appl.* **76**:83-108

ALLIX O. - LADEVEZE P. 1992

"Interlaminar interface modelling for the prediction of laminates delamination." *Composite Structures* **22**:235-242

ALLIX O. - LADEVEZE P. 1996

"Damage mechanics of interfacial media: Basic aspects identification and application to delamination." in *Damage and Interfacial Debonding in Composites* (VOYIADIS and ALLEN Editors), pp. 167-188

ANTUNEZ H. 1998

"Thermo-mechanical modelling and sensivity analysis for metal-forming operations." *Comput. Methods Appl. Mech. Eng.* **161**:113-125

AUBIN J.P. - BURCHARD H. 1970

"Some aspects of the method of the hypercircle applied to elliptic variational problems." in *Numerical Solution of Partial Differential Equations* (HUBBARD Editor), Academic Press

AUBRY D. - LUCAS D. - TIE B. 1998

"Adaptive strategy for transcient/coupled problems. Applications to thermoelasticity and elastodynamics." in *Advances in Adaptive Computational Methods in Mechanics*, pp. 325-332 (LADEVEZE and ODEN Editors), Elsevier

AUBRY D. - TIE B. 1991

"A posteriori error analysis and *h*-adaptive refinement for the F.E.M. in nonlinear structural computations." *Proc. of the European Conference on New Advances in Computational Structural Mechanics* (LADEVEZE and ZIENKIEWICZ Editors) Giens (France), pp. 1-8

AUBRY D. - TIE B. 1992

"Error estimates *h*-adaptive strategy and hierarchical concept for nonlinear finite element method." *Proc. of the First European Conference on Numerical Methods in Engineering* (HIRSCH, ZIENKIEWICZ and OÑATE Editors) Bruxelles (Belgium), pp. 17-24

AZZIZ K. - BABUSKA I. 1972

The Mathematical Foundations of the Finite Element Method with Applications to Partial Differential Equations. Academic Press, New York

BABUSKA I. 1986

"Feedback adaptivity and a-posteriori estimates in finite elements: Aims, theory and experience." in *Accuracy Estimates and Adaptivity for Finite Element* (BABUSKA, ZIENKIEWICZ, GAGO and OLIVEIRA Editors), Wiley, London, chap. 1, pp. 3-23

BABUSKA I. - MELENK J.M. 1999

"The partition of unity method." *Int. J. Num. Meth. Eng.* **40**:727-758

BABUSKA I. - MILLER A. 1984

"The post-processing approach in the finite element method. Part II: The calculation of stress intensity factors" *Int. J. Num. Meth. Eng.* **20**:1111-1129

BABUSKA I. - RHEINBOLDT W.C. 1978

"Error estimates for adaptive finite element computation." *SIAM J. Num. Anal.* **15**(4):736-754

BABUSKA I. - RHEINBOLDT W.C. 1982

"Computational error estimates and adaptive processes for some nonlinear structural problems." *Comput. Methods Appl. Mech. Eng.* **34**:895-937

BABUSKA I. - STROUBOULIS T. - GANGARAJ S.K. 1997

"A model study of the quality of a posteriori error estimators for finite element solutions of linear elliptic problems with particular reference to the behavior near the boundary." *Int. J. Num. Meth. Eng.* **40**:2521-2577

BABUSKA I. - STROUBOULIS T. - GANGARAJ S.K. 1999

"Guaranteed computable error bounds for the exact error in the finite element solution. Part I: One dimensional model problem." *Comput. Methods Appl. Mech. Eng.* 51-80

BABUSKA I. - STROUBOULIS T. - UPADHYAY C.S. - GANGARAJ S.K. 1994

"A model study of the quality of a posteriori estimators for linear elliptic problems: Error estimation in the interior of patchwise uniform grids of triangles." *Comput. Methods Appl. Mech. Eng.* **114**:307-378

BABUSKA I. - STROUBOULIS T. - UPADHYAY C.S. - GANGARAJ S.K. 1995

"A posterirori estimation and adaptive control of the pollution error in the *h*-version of the finite element method." *Int. J. Num. Meth. Eng.* **38**:4207-4235

BABUSKA I. - STROUBOULIS T. - UPADHYAY C.S. - GANGARAJ S.K. - COPPS K. 1994

"Validation of a posteriori error estimators by numerical approach." *Int. J. Num. Meth. Eng.* **37**:1073-1123

BABUSKA I. - SURI M. 1990

"The *p* and *h* version of the finite element method: an overwiew." *Comput. Methods Appl. Mech. Eng.* **82**:5-26

BABUSKA I. - SZABO B.A. 1982

"On the rate of convergence of the finite element method." *Int. J. for Num. Meth. Eng.* **18**:323-341

BABUSKA I. - SZABO B.A. - KATZ I.N. 1981

"The *p*-version of the finite element method." *SIAM J. Num. Anal.* **18**(3):515-545

BANK R.E. - WEISER A. 1985

"Some a posteriori error estimators for elliptic partial differential equations." *Math. of Comp.* **44**:283-301

BARTHOLD F.J. - SCHMIDT M. - STEIN E. 1997

"Error estimation and mesh adaptivity for elsato-plastic deformations." *Proc. of COMPLAS V Computational Plasticity: Fundamentals and Applications,* Barcelon (Spain), pp. 597-602

BASS J.M. - ODEN J.T. 1987

"Adaptive finite element methods for a class of evolution problems in viscoplasticity." *Int. J. Engrg. Sci.* **25**(6):623-653

BATHE K.J. 1982

Finite Element Procedure in Engineering Analysis. Prentice-Hall, Englewood Cliffs

BATHE K.J. - WILSON E.L. 1976

Numericals Methods in Finite Element Analysis. Prentice-Hall, Englewood Cliffs

BATOZ J.L. 1982

"An explicit formulation for an efficient triangular plate bending element." *Int. J. for Num. Meth. in Eng.* **18**:1077-1089

BATOZ J.L. - DHATT G. 1990

Modélisation des structures par éléments finis, Vols. I, II and III. Hermès, Paris

BAUDRON A.M. - TROMPETTE P. 1986

"Un critère de jugement simple pour un maillage éléments finis." *J. de Mécanique Théor. Appl.* **5**(5):731-746

BAZANT Z.P. - CEDOLIN L. 1991

Stability of Structures. Oxford University Press, p. 168

BAZANT Z.P. - PIJAUDIER-CABOT G. 1988

"Non-local damage: Continuum model and localization instability." *Report 87-2,* Northwestern University

BECKERS P. - DUFEU E. 1996

"Estimation d'erreur basée sur l'équilibre et le lissage des tensions." *Rapport SF - 215 LTAS,* Université de Liège (Belgium)

BECKERS P. - ZHONG H.G. 1991

"Influence of element distorsions on the reliability of some a posteriori error estimators." *Proc. of the European Conference on New Advances in Computational Structural Mechanics* (LADEVEZE and ZIENKIEWICZ Editors) Giens (France), pp. 9-16

BELYTSCHKO T. - HUGHES T.J.R. 1986

Computational Methods for Transcient Analysis. North-Holland

BELYTSCHKO T. - LASRY D. 1989

"Localisation limiters and numerical strategies for strain softening materials." in *Craking and Damage* (MAZARS and BAZANT Editors), pp. 349-362, Elsevier

BELYTSCHKO T. - TABARRA M. 1993

"*h*-Adaptive finite element methods for dynamic problems with emphasis on localization." *Int. J. Num. Meth. Eng.* **36**:4245-4265

BENALAL A. - BILLARDON R. - GEYMONAT G. 1989

"Conditions de bifurcation à l'intérieur et aux frontières pour une classe de matériaux non standards." *C. R. Acad. Sci. Paris* Sér. II, **308**:893-898

BENOIT C. - COOREVITS P. - PELLE J.P. 1999

"Error estimation for plate structures: Application using DKT elements" *Engineering Computations* **16**:584-600

BERGA A. - DE SAXCE G. 1994

"Elastoplastic finite element analysis of soil - Problems with implicit standard material constitutive laws." *Revue Européenne des Eléments Finis* **3**:411-456

BERNADOU M. - HASSAN K. 1981

"Basis functions for general hct triangules complete or reduce." *Int. J. Num. Meth. Eng.* **17**:784-789

BLACKER T. - BELYTSCHKO T. 1994

"Superconvergent patch recovery with equilibrium and conjoint interpolant enhancements." *Int. J. Num. Meth. Eng.* **37**:517-536

BOISSE P. - COFFIGNAL G. 1997

"Indicateur d'erreur en relation de comportement pour les structures composées de poutres et de plaques." *Actes du Troisième Colloque National en Calcul des Structures*, pp. 355-360, Giens (France), Presses Académiques de l'Ouest

BOISSE P. - COFFIGNAL G. 1998

"Error through the constitutive relation for beams or C° plate finite element: Static and vibration analyses." *in Advances in Adaptive Computational Methods in Mechanics*, pp. 493-512 (LADEVEZE and ODEN Editors), Elsevier

BOISSE P. - PERRIN S. 1995

"Calcul de l'erreur en relation de comportement pour les modélisations par éléments finis de plaques C°." *C. R. Acad. Sci. Paris* Sér. IIb, **320**:289-294

BOISSE P. - PERRIN S. - COFFIGNAL G. - HADJEB K. 1999

"Error estimation through the constitutive relation for RESSNER-MINDLIN plate bending element." *Computers and Structures* **73**:615-627

BONNET M. 1999

"Stability of crack front under GRIFFITH criterium: A computational approach using integral equations and domain derivatives of potential energy." *Comput. Methods Appl. Mech. Eng.* **173**:337-364

BOROOMAND B. - ZIENKIEWICZ O.C. 1997

"Recovery by equilibrium in patches (REP)." *Int. J. Num. Meth. Eng.* **40**:137-164

BOROOMAND B. - ZIENKIEWICZ O.C. 1998

"Recovery procedures in error estimation and adaptivity: Adaptivity in nonlinear problems of elasto-plasticity behaviour." in *Advances in Adaptive Computational Methods in Mechanics*, pp. 383-410 (LADEVEZE and ODEN Editors), Elsevier

BOROUCHAKI H. - GEORGE P.L. 1997

Triangulation et maillage de Delaunay. Hermès, Paris

BOUBARCHENE M. - HOCHARD C. - POITOU A. 1997
"Domain optimization using TREFFTZ functions. Application to free boundaries." *Comp. Assisted Mech. Eng. Sci.* **4**:317-326

BOUCARD P.A. - LADEVEZE P. - POSS M. - ROUGEE P. 1997
"A non-incremental approach for large displacement problems." *Computers and Structures* **64**:499-508

BOUDI E.M. - BECKERS P. - ZHONG H.G. 1992
"Estimateurs d'erreurs pour des éléments de plaque basés sur la théorie de KIRCHHOFF." *Proc. of Strucome 92*, Paris (France), pp. 599-611

BOUILLARD P. - IHLENBURG F. 1999
"Error estimation and adaptivity for the finite element method in acoustics: 2D and 3D applications." *Comput. Methods Appl. Mech. Eng.* **176**:147-164

BREBBIA C.A. - TELLES J.C.F.. - WROBEL L.C. 1984
Boundary Integral Techniques. Springer

BREZIS H. 1973
Opérateurs maximaux monotones et semi-groupes de contraction. Mathematics Studies, North-Holland

BRINK U. - STEIN E. 1998
"A posteriori error estimation in large strain elasticity using equilibrated local NEUMANN problems." *Comput. Methods Appl. Mech. Eng.* **161**(1-2) :77-101

BRUNEAU M. 1999
Manuel d'acoustique fondamentale. Hermès, Paris

BUI H.D. 1973
Mécanique de la rupture fragile. Masson, Paris

BUI H.D. 1992
Introduction aux problèmes inverses en mécanique des matériaux. Eyrolles, Paris

BUSSY P. - COFFIGNAL G. - LADEVEZE P. 1985
"Reliability and optimization of nonlinear finite element analyses." *Proc. of NUMETA'85*, Swansea (U.K.), pp. 559-667

CARNET J. 1978
Une méthode heuristique de maillage dans le plan pour la mise en œuvre des Éléments. Thèse de 3ème cycles Université P et M. Curie, Paris

CARNET J. - LADEVEZE P. - LEGUILLON D. 1981
"An optimal mesh procedure in the finite element method." *Proc. of the Third Conference on Mathematics of Finite Element and Application* (MAFELAP), Uxbridge (U.K.)

CARROL W.E. - BARKER R.M. 1973
"A theorem for optimum finite element idealizations." *Int. J. of Solids Struct.* **9**:883-895

CHOUAKI A. - LADEVEZE P. - PROSLIER L. 1998

"Updating structural dynamic models with emphasis on the damping properties." *AIAA J.* **36**:1094-1099

CIARLET P.G. 1978

The Finite Element Method for Elliptic Problems. North-Holland

CIRAK F. - RAMM E. 1998

"A posteriori error estimation and adaptivity for linear elasticity using the reciprocal theorem." *Comput. Methods Appl. Mech. Eng.* **156**:351-362

CLEMENT P. 1975

"Approximation by finite element functions using local regularization." *RAIRO Anal. Num.* 77-84

CLOUGH R.W. - TOCHER J.L. 1965

"Finite element stiffness matrices for analysis of plates in bending." *Proc. Conf. on Matrix Methods in Structural Mechanics*, pp. 515-545, Wright Patterson, Ohio, OH

COCHELIN B. - DAMIL N. - POTIER-FERRY M. 1994

"Asymptotic-numerical method and PADE approximants for nonlinear elastic structures." *Int. J. Num. Meth. Eng.* **37**:1181-1192

COCU M. - PRATT E. - RAOUS M. 1995

"Analysis of an incremental formulation for frictional contact problem:" in *Contact Mechanics*, Plenum Press, New York

COFFIGNAL G. 1987

Optimisation et fiabilité des calculs par éléments finis en élastoplasticité. Thèse d'Etat Université P et M. Curie, Paris

COFFIGNAL G. 1993

"Indicateur d'erreur d'erreur pour les modes propres des modèles éléments finis de poutres." *Actes du Premier Colloque National en Calcul des Structures*, pp. 355-360, Giens (France), Hermès

COFFIGNAL G. - LADEVEZE P. 1983

"Error computation and optimal mesh in elasticity and elastoplasticity." *Proc. SMIRT 7*, North-Holland, pp. 177-182

COMBE J.P. - LADEVEZE P. - PELLE J.P. 1998

"Constitutive relation error estimator for transient finite element analysis." *CDRom of Proc. of the Fourth World Congress on Computational Mechanics*, Buenos Aires (Argentina)

COMBE J.P. - LADEVEZE P. - PELLE J.P. 1999

"Constitutive relation error estimator for transient finite element analysis." *Comput. Methods Appl. Mech. Eng.* **176**(1-4):165-185

COMBE J.P. - PELLE J.P. 1998
"Sur le contrôle des erreurs d'intégration en temps en dynamique." *Rapport Interne n° 207.* LMT-Cachan (France)

COOREVITS P. - DUMEAU J.P. - NOËL F. - PELLE J.P. 1997
"Methods and softwares for the automation of finite element analyses in 3D." in *Integrated Design and Manufacturing in Mechanical Engineering* (CHEDMAIL, BOCQUET and DORNFELD Editors), pp. 193-202, Kluwer Academic

COOREVITS P. - DUMEAU J.P. - PELLE J.P. 1995
"Maillages 3D adaptatifs: Quelques procédures pour respecter une carte de tailles." *Proc. of StruCoMe 95*, Paris (France), pp. 117-127

COOREVITS P. - DUMEAU J.P. - PELLE J.P. 1996
"Analyses éléments finis adaptatives pour les structures tridimensionnelles en élasticité." *Revue Européenne des Eléments Finis* **5**(3):341-374

COOREVITS P. - DUMEAU J.P. - PELLE J.P. 1997
"Error estimator and adaptivity for three-dimensional finite element analyses." in *Advances in Adaptive Computational Methods in Mechanics*, pp. 443-458 (LADEVEZE and ODEN Editors), Elsevier

COOREVITS P. - DUMEAU J.P. - PELLE J.P. 1999
"Control of analyses with isoparametric elements in 2D and in 3D." *Int. J. Num. Meth. Eng.* **46**:157-176

COOREVITS P. - HILD P. - PELLE J.P. 1999a
"Contrôle et adaptation des calculs éléments finis pour les problèmes de contact unilatéral." *Revue Européenne des Eléments Finis* **8**:7-29

COOREVITS P. - HILD P. - PELLE J.P. 1999b
"A posteriori error estimation for unilateral contact with matching and nonmatching meshes." *Comput. Methods Appl. Mech. Eng.* **46**:157-176

COOREVITS P. - LADEVEZE P. - PELLE J.P. 1992
"Anisotropic adaptive mesh: Application to transient problems." *Proc. of First European Conference on Numerical Methods in Engineering* , Bruxelles (Belgium)

COOREVITS P. - LADEVEZE P. - PELLE J.P. 1994
"Mesh optimization for problems with steep gradients." *Engineering Computations* **11**:129-144

COOREVITS P. - LADEVEZE P. - PELLE J.P. - ROUGEOT P. 1992
"Some new applications of a method for the control and optimization of finite element computations." in *New Advances in Computational Structural Mechanics* (LADEVEZE and ZIENKIEWICZ Editors), Elsevier , pp. 205-217

COOREVITS P. - PELLE J.P. - ROUGEOT P. 1994
"Sur l'adaptativité des maillages à base de quadrilatères. Application à l'automatisation de calculs par éléments finis en élasticté 2D." *Revue Européenne des Eléments Finis*. **3**(3):379-409

CORIGLIANO A - PEREGO U. 1990
"Unconditionally stable mid-point time integration in elastic-plastic dynamics." Rend. Acc. Lincei. s. **9**(1) :367-376

COUPEZ T. 1991
Grandes transformations et remaillage automatique. Thèse de l'Ecole Nationale Supérieure des Mines de Paris, CEMEF, Sophia-Antipolis (France)

COURANT R. - HILBERT D. 1953
Method of Mathematical Physics. Interscience, New York

CRISFIELD M.A. 1991
Non-linear Finite Element Analysis of Solids and Structures, Wiley, London, London

DAFALIAS Y. 1983
"Corotational rates for kinematic hardening at large plastic deformation." *J. Appl. Mech.* **50**:561-565

DAUMAS M. - MULLER J.M. 1997
Qualité des calculs sur ordinateurs. Masson , Paris

DAUTRAY R. - LIONS J.L. 1984
Analyse mathématique et calculs numériques pour les sciences et les techniques, Vols. 1 to 9. Masson, Paris

DEBONGNIE J. - BECKERS P. 1999
"On a general decomposition of the error of an approximate stress field in elasticity." *Int. workshop on TREFFTZ methods*, Sintra (Portugal)

DEBONGNIE J. - ZHONG H. - BECKERS P. 1995
"Dual analysis with general boundary conditions." *Comput. Methods Appl. Mech. Eng.* **122**:183-192

DE BORST R. 1988
"Bifurcation in finite element models with non-associated flow law." *Int. J. Num. Analyt. Meth. Geomech.* **12**:99-116

DEMKOWICZ L. - KIM C. 2000
"An implementation of the implicit element residual method for one irregular *h-p* meshes in two dimensions." (to appear)

DENNIS J. - MORE J. 1977
"Quasi-Newton methods: Motivation and theory" *SIAM Rev.* **19**:46-84

DE SAXCE G. 1992
"Une généralisation de l'inégalité de Fenchel et ses applications aux lois constitutives." *C. R. Acad. Sci. Paris* Sér. II, **314**:125-129

DE SAXCE G. - FENG Z.Q. 1998
"The bipotential method: A constructive approach to design the complete contact law with friction and improved numerical algorithms." *Int. J. Math. Comput. Modelling* **28**(4-8) :225-245

DESTUYNDER P. 1991
Modélisation mécanique des milieux continus. Ellipse, Paris

DESTUYNDER P. - COLLOT M. - SALAÛN M. 1997
"A mathematical framework for the Ladeveze a posteriori error bounds in finite element methods." in *Advances in Adaptive Computational Methods in Mechanics*, pp. 65-78 (LADEVEZE and ODEN Editors), Elsevier

DESTUYNDER P. - METIVET B. 1996
"Estimation d'erreur explicite dans une méthode d'éléments finis conformes." *C. R. Acad. Sci. Paris* Sér. I **323**:679-684

DIAZ A.R. - KIKUCHI N. - TAYLOR J.R. 1983
"A method of grid optimization for finite element methods." *Comput. Methods Appl. Mech. Eng.* 29-45

DIEZ P. - EGOZENZ J. - HUERTA A. 1998
"A posteriori error estimation for standard finite element analysis." *Comput. Methods Appl. Mech. Eng.* **163**:141-157

DIEZ P. - HUERTA A. 1999
"A unified approach to remeshing strategies for finite element *h*-adaptivity." *Comput. Methods Appl. Mech. Eng.* **176**:215-229

DRUCKER D.C. 1964
"On the postulate of stability of material in the mechanics of continua." *J. Mécanique* **3**(2):235-249

DUFEU E. 1997
Calcul d'erreur et adaptation de maillages en 3 dimensions. Thèse de l'Université de Liège (Belgium)

DUTTA A. - RAMAKRISHNAN C. 1997
"Error estimation in finite element transient dynamic analysis using modal superposition." *Engineering Computations* **14**(1):135-159

DUVAUT G. 1976
"Analyse fonctionnelle et mécanique des milieux continus." in *Theoretical and Applied Mechanics* (KOITER Editor), North-Holland , pp. 119-132

DUVAUT G. - LIONS J.L. 1972
Les inéquations en mécanique et en physique. Dunod, Paris

EKELAND I. - TEMAM R. 1974
Analyse convexe et problèmes variationnels. Dunod, Paris

EWING R. - LAZAROV R. - VASSILEV P. 1992
"Adaptive technique for time dependent problems." *Comput. Methods Appl. Mech. Eng.* **101**:113-126

FICHERA G. 1965
Numerical and Quantitative Analysis. Surveys and Reference Works in Mathematics, Vol. 3. Pitman, London

FIESTA 1986
User Manual. ISMES

FISH J. - MARKOFELAR S. 1994
"Adaptive global-local refinement strategy based on the interior estimates of the *h*-method" *Int. J. Num. Meth. Eng.* **37**:827-838

FISH J. - SHEK K. - PANDHEERAI M. - SHEPARD M. 1997
"Computational plasticity for composite structures based on mathematical homogenization: Theory and practice." *Comput. Methods Appl. Mech. Eng.* **148**:53-73

FOURMENT L. - CHENOT J.L. 1994
"Adaptive remeshing and error control for forming processes." *Revue Européenne des Eléments Finis* **2**(2):247-280

FOURMENT L. - CHENOT J.L. 1995
"Error estimators for viscoplastic materials: Application to forming processes." *Engineering Computations* **12**(5):469-490

FRAEIJS DE VEUBEKE B.M. 1965
"Displacement and equilibrium models in the finite element method." in *Stress Analysis* (ZIENKIEWICZ and HOLISTER Editors), Wiley, London London, Chap. 9.

FRAEIJS DE VEUBEKE B.M. - HOGGE M.A. 1970
"Dual analysis fot heat conduction problems by finite elements." *Int. J. Num. Meth. Eng.* **5**:65-82

FRAEIJS DE VEUBEKE B.M. - SANDER G. 1968
"An equilibrium model for plate bending." *Int. J. Solids Struct.* **4**:447-468

FRANÇOIS D. - PINEAU A. - ZAOUI A. 1991
Comportement mécanique des matériaux. Hermès, Paris

FREMOND M. 1988
"Contact with adhesion." in *Topics in Nonsmooth Mechanics*, Birkhauser , Berlin, pp. 157-185

GAGO J. - KELLY D.W. - ZIENKIEWICZ O.C. - BABUSKA I. 1983
"A posteriori error analysis and adaptive processes in the finite element method; Part II: Adaptive mesh refinement." *Int. J. Num. Meth. Eng.* **19**:1621-1656

GALLIMARD L. - LADEVEZE P. - PELLE J.P. 1995
"A posteriori error estimator for non linear finite element computations: Application to elastoplasticity." *Proc. of Complas IV 4th International Conference on Computational Plasticity: Fundamentals and Applications*, Barcelona (Spain)

GALLIMARD L. - LADEVEZE P. - PELLE J.P. 1996
"Error estimation and adaptivity in elastoplasticity." *Int. J. Num. Meth. Eng.* **39**:189-217

GALLIMARD L. - LADEVEZE P. - PELLE J.P. 1997a
"Error estimation and time-space parameters optimization for FEM nonlinear computations." *Computers and Structures* **64**(1-4):145-156

GALLIMARD L. - LADEVEZE P. - PELLE J.P. 1997b
"Drucker-type a-posteriori error estimator for elastoplasticity and viscoplasticity finite element computation and an associated adaptivity procedure." *Proc. of COMPLAS V Computational Plasticity: Fundamentals and Applications*, Barcelona (Spain)

GALLIMARD L. - LADEVEZE P. - PELLE J.P. 2000
"An enhanced error estimator on the constitutive relation for plasticity problems." *Computers and Structures* **78**:801-810

GANGARAJ S. 1999
A posteriori estimation of the error in finite element solution by computation of the guaranteed upper and lower bounds. PhD Thesis, Texas A & M University

GASTINE J.L. - LADEVEZE P. - MARIN P. - PELLE J.P. 1992
"Accuracy and optimal meshes in finite element computation for nearly incompressible materials." *Comput. Methods Appl. Mech. Eng.* **94**(3):303-314

GEAR C.W. 1971
Numerical Initial Value Problems in Ordinary Differential Equations. Prentice-Hall

GEORGE P.L. 1991
Génération automatique de maillages: Applications aux méthodes d'éléments finis. Coll. Recherches en Mathématiques Appliquées, Masson, Paris

GERADIN M. - IDELSOHN S - HOGGE M. 1981
"Computational strategies for the solution of large nonlinear problems via quasi-Newton methods." *Computers and Structures* **13**:73-81

GERADIN M. - RIXEN D. 1991
Théorie des vibrations. Presses de l'Université de Liège (Belgium)

GERMAIN P. 1973
Mécanique des milieux continus. Masson, Paris

GIBERT R.J. 1988
Vibrations des structures. Eyrolles, Paris

GUEZEL G. 1982

Sur l'encadrement du taux de restitution d'énergie. Thèse de l'Ecole Normale Supérieure de Cachan (France)

HACKBRICH W. - WITTRUM G. 1994

Adaptive Methods: Algorithms Theory and Applications. Vieweg, Branunschweig

HALPHEN B. - NGUYEN Q.S. 1975

"Sur les matériaux standards généralisés." *J. Mécanique* **14**:39-62

HERAKOVITCH C. 1996

Mechanics of Fibrous Composites. Wiley, London

HERRMANN L.R. 1965

"Elasticity equations for incompressible and nearly incompressible materials by a variational theorem." *AIAA J.* **3**:1896-1900

HILD P. 2000

"Numerical implementation of two nonconforming finite element methods for unilateral contact." *Comput. Methods Appl. Mech. Eng.* **184**:99-123

HINTON E. - CAMPBELL J.S. 1974

"Local and global smoothing of discontinuous finite element functions using a least squares method." *Int. J. Num. Meth. Eng.* **8**:461-480

HJIAI M. 1999

Sur la classe de matériaux standards implicites: Concept aspect discrétisés et estimation de l'erreur a posteriori. Thèse de la faculté polytechnique, Mons (Belgium)

HJIAI M. - DE SAXCE G. - ABRIAK N. 1996

"The bipotential approach for soil plasticity." in *Numerical Methods* (DESIDERI, LE TALLEC, ONATE, PERIAUX and STEIN Editors), Wiley, London, pp. 918-923

HOCHARD C. - LADEVEZE P. - PROSLIER L. 1993

"A simplified analysis of elastic structures." *European J. Mech.* **12**(4):509-535

HOLZER S. - RANK E. - WERNER H. 1990

"An implementation of the *hp*-version of the finite element method for Ressneir-Mindlin plate problems." *Int. J. Num. Meth. Eng.* **37**:2631-2655

HUCHINSON J. 1974

"Plastic buckling." *Adv. Appl. Mech.* **14**:67-144

HUERTA A. - DIEZ P. 2000

"Error estimation including pollution assessment for nonlinear finite element analysis." *Comput. Methods Appl. Mech. Eng.* **181**:21-41

HUERTA A. - DIEZ P. - RODRIGUEZ A. - PIJAUDIER CABOT G. 1998
"Error estimation and adaptive finite element analysis of softening materials." in *Advances in Adaptive Computational Methods in Mechanics*, pp. 333-346 (LADEVEZE and ODEN Editors), Elsevier

HUGHES T.J.R. 1987
The Finite Element Method: Linear Static and Dynamic Finite Element Analysis. Prentice-Hall, Englewood Cliffs

IBRAHIMBEGOVIC A. - GHARZEDDINE F. 1999
"Finite deformation plasticity in principal axes: From a manifold to Euclidian setting." *Comput. Methods Appl. Mech. Eng.* **171**:341-369

IHLENBURG F. - BABUSKA I. 1995
"Finite element solution of the HELMHOLTZ equation with high wave number. Part I: The *h*-version of FEM." *Comput. Math. Appl.* **38**(9):9-37

IHLENBURG F. - BABUSKA I. 1995
"Finite element solution of the HELMHOLTZ equation with high wave number. Part II: The *h-p*-version of FEM." *Int. J. Num. Meth. Eng.* **38**:3745-3774

IMBERT J.F. 1984
Analyse des structures par Eléments Finis (CEPADUES Editions). Toulouse (France)

JIN H. - BERNSPANG L. - LARSSON R. - WIBERG N.E. 1990
"Mesh generation and mesh refinement procedures for computational plasticity." *Proc. of International Conference on Computational Plasticity: Fundamentals and Applications*, Barcelona (Spain)

JIROUSEK J. 1985
"Implementation of local effects into conventional and non-conventional finite element formulations." in *Local effects in the analysis of structures* (LADEVEZE Editor), Elsevier

JIROUSEK J. - WROBLESKI A. 1996
"*T*-elements: State of the art and future trends." *Arch. Comput. Mech. Eng. State of the Art Rev.* **3**:289-314

JOHNSON C. 1976
"Existency theorems for plasticity problems." *J. Math. Pures Appl.* **55**:431-444

JOHNSON C. 1990
"Adaptive finite element methods for diffusion and convection problems." *Comput. Methods Appl. Mech. Eng.* **82**:301-322

JOHNSON C. - HANSBO P. 1991
"Adaptive finite element methods for small strain elasto-plasticity." *Proc. of IUTAM Conf. on Finite Inelastic Deformations Theory and Applications.* Hanover (Germany)

JOHNSON C. - HANSBO P. 1992
"Adaptive finite element methods in computational mechanics." *Comput. Methods Appl. Mech. Eng.* **101**:143-181

KEIBER M. - HEIN T. 1992
The Stochastic Finite Element Method, Wiley, London

KELLY D.W. 1984
"The self-equilibration of residuals and complementary a posteriori error estimates in the finite element method." *Int. J. Num. Meth. Eng.* **20**:1491-1506

KELLY D.W. - GAGO J. - ZIENKIEWICZ O.C. - BABUSKA I. 1983
"A posteriori error analysis and adaptive processes in the finite element method; Part I: Error analysis." *Int. J. Num. Meth. Eng.* **19**:1593-1619

KIKUCHI N. 1986
"Adaptive grid-design methods for finite element analysis." *Comput. Methods Appl. Mech. Eng.* 129-160

KIKUCHI N. - ODEN J.T. 1988
"Contact problems in elasticity: A study of variational inequalities and finite element methods." SIAM, Philadelphia

KOITER W.T. 1967
On the stability of elastic equilibrium. Thèse de l'Univerité de Delft (Holland), NASA TTF 10

KOMPIS V. - FRASTIA L. 1997
"Polynomial representation of hybrid finte elements." *Comput. Assisted Mech. Eng. Sci.* **4**:521-532

KUHN H.W. - TUCKER A.W. 1950
"Non-linear programming." *Proc. of 2nd Berkeley Symp. on Mathematics Statistics and Probability*, Univ. of California Press

LABORDE P. - NGUYEN Q.S. 1990
"Etude de l'équation d'évolution des systèmes dissipatifs standards." *M2AN* **24**(1):67-84

LADEVEZE P. 1975
Comparaison de modèles de milieux continus. Thèse d'Etat. Université P et M. Curie Paris

LADEVEZE P. 1977
"Nouvelles procedures d'estimation d'erreur relative à la méthode des éléments finis et applications." *Proc. Journées Eléments Finis*, Rennes (France)

LADEVEZE P. 1983
"R.D.M tridimensionnelle." *Rapport interrne n° 12*, LMT-Cachan (France)

LADEVEZE P. 1985
"Sur une famille d'algorithmes en mécanique des structures." *C. R. Acad. Sci. Paris*, Sér. II **300**:41-44

LADEVEZE P. 1989
"La méthode à grand incrément de temps pour l'analyse des structures à comportement non linéaire décrit par variables internes." *C. R. Acad. Sci. Paris*, Sér. II **309**:1095-1099

LADEVEZE P. 1990
"Modélisation et calculs des structures: de nouveaux défis." *Recherche Aérospatiale* **5**:39-56

LADEVEZE P. 1992
"La maîtrise des modèles en mécanique des structures." *Revue Européenne des Eléments Finis* **1**(1):9-30

LADEVEZE P. 1993
"Erreur en relation de comportement en dynamique: Théorie et application au recalage de modèles." *Rapport interne n° 150*, LMT-Cachan (France)

LADEVEZE P. 1994
"La méthode d'évaluation d'erreur en relation de comportement appliquée à la MEF: Qualité et amélioration." *Rapport interne n° 153*, LMT-Cachan (France)

LADEVEZE P. 1995a
"A damage computational approach for composites: Basic aspects and micromechanics relations." *Comput. Mech.* **17**:142-150

LADEVEZE P. 1995b
"Les bases de la méthode des erreurs en realtion de comportement pour le contrôle adaptatif des calculs éléments finis: Les travaux des années 1975." *Rapport interne n° 163*, LMT-Cachan (France)

LADEVEZE P. 1996
Mécanique non linéaire des structures. Hermès, Paris (English version, Springer 1998)

LADEVEZE P. 1998
"A modelling error estimator for dynamic structural model updating." in *Advances in Adaptive Computational Methods in Mechanics*, pp. 131-151 (LADEVEZE and ODEN Editors), Elsevier

LADEVEZE P. 2001
"Constitutive relation error estimations for finite element analyses considering (visco) plasticity and damage." *Int. J. Num. Meth. Eng.* **52**(5/6):527-542

LADEVEZE P. - ALLIX O. - DEU J.F. - LEVEQUE D. 2000
"A mesomodel for localisation and damage computation in laminates." *Comput. Methods Appl. Mech. Eng.* **183**:105-122

LADEVEZE P. - ARNAUD L. - ROUCH P. - BLANZE C. 2000
"La théorie variationnelle des rayons complexes pour le calcul des vibrations moyennes fréquences." *Revue Européenne des Eléments Finis* **9**:67-88

LADEVEZE P. - CHOUAKI A. 1999
"Application of a posteriori error estimation for structural model updating." *Inverse Problems* **15**:49-58

LADEVEZE P. - COFFIGNAL G. - PELLE J.P. 1986
"Accuracy of elastoplastic and dynamic analysis." in *Accuracy Estimates and Adaptative Refinements in Finite Element Computations*, Chap. 11, pp. 181-203 (BABUSKA, GAGO, OLIVEIRA and ZIENKIEWICZ Editors), Wiley, London

LADEVEZE J. - LADEVEZE P. 1978
"Majoration de la constante de Poincaré relative au problème de la membrane - Domaines étoilés." *J. Math. Phys. Appl. (ZAMP)* **29**:670-683

LADEVEZE J. - LADEVEZE P. 1979
"Lower bounds for the principal free membrane eigenvalue." *SIAM J. Math. Anal.* **10**(6):1280-1292

LADEVEZE P. - LEGUILLON D. 1981
"Error computation in finite element method and application." *Proc. GAMNI*, Dunod, Paris

LADEVEZE P. - LEGUILLON D. 1983
"Error estimate procedure in finite element method and application" *SIAM J. Num. Anal.* **20**(3):485-509

LADEVEZE P. - LOISEAU O. - DUREISSEIX D. 2001
"A micro-macro and parallel computational strategy for highly heterogeneous structures." *Int. J. Num. Meth. Eng.* **52**(1/2):121-138

LADEVEZE P. - MAUNDER E.A.W. 1996
"A general method for recovering equilibrating element tractions." *Comput. Methods Appl. Mech. Eng.* **137**:111-151

LADEVEZE P. - MAUNDER E.A.W. 1997
"A general methodology for recovering equilibrating finite element tractions and stress fields for plates and solid elements." *Comput. Assisted Mech. Eng. Sci.* **4**:533-548

LADEVEZE P. - MOES N. 1998a
"A new a posteriori error estimation for nonlinear time-dependent finite element analysis." *Comput. Methods Appl. Mech. Eng.* **157**:45-68

LADEVEZE P. - MOES N. 1998b
"A posteriori constitutive relation error estimators for nonlinear finite element analysis and adaptive control." in *Advances in Adaptive Computational Methods in Mechanics*, pp. 231-256 (LADEVEZE and ODEN Editors), Elsevier

LADEVEZE P. - MOES N. 1999
"Adaptive control for time-dependent nonlinear finite element analysis." *Computers and Structures* **73**:45-60

LADEVEZE P. - MOES N. - DOUCHIN B. 1999
"Constitutive relation error estimations for (visco) plasticity finite element analysis with softening." *Comput. Methods Appl. Mech. Eng.* **176**:247-264

LADEVEZE P. - ODEN J.T. 1998
Advances in Adaptive Computational Methods in Mechanics. Elsevier

LADEVEZE P. - PELLE J.P. 1983
"Méthodes de calcul par encadrement des fréquences propres de structures élastiques." *C. R. Acad. Sci. Paris*, Sér. II **296**:1757-1760

LADEVEZE P. - PELLE J.P. 1984
"Sur les méthodes de calcul par ecadrement des fréquences propres de structures élastiques." *J. Méc. Théor. Appl.* **3**(5) :689-715

LADEVEZE P. - PELLE J.P. 1988
"Optimisation des maillages pour les calculs de fréquences propres." *J. Méc. Théor. Appl.* **7**(3) :245-268

LADEVEZE P. - PELLE J.P. 1989a
"Accuracy in finite element computation for eigenfrequencies." *Int. J. Num. Meth. Eng.* **29**:1929-1949

LADEVEZE P. - PELLE J.P. 1989b
"Sur le contrôle adaptatif des calculs par éléments finis." *Bulletin de Liaison de la Recherche en Informatique et Automatique*, INRIA 129, pp. 17-23

LADEVEZE P. - PELLE J.P. - ROUGEOT P. 1991
"Error estimation and mesh optimization for classical finite elements." *Engineering Computations* **8**:69-80

LADEVEZE P. - REYNIER M. - MAÏA N.M.M. 1994
"Error on the constitutive relation in dynamics." in *Inverse Problems in Engineering*, pp. 251-256 (BUI and TANAKA Editors), Balkema

LADEVEZE P. - ROUGEOT P. 1997
"New advances on a posteriori error on constitutive relation in finite element analysis." *Comput. Methods Appl. Mech. Eng.* **150**:239-249

LADEVEZE P. - ROUGEOT P. - BLANCHARD P. - MOREAU J.P. 1999
"Local error estimators for finite element linear analysis." *Comput. Methods Appl. Mech. Eng.* **176**:231-246

LA PORTE M. - VIGNE J. 1974
"Error analysis in computing" in *Information Processing* , North-Holland

LEGUILLON D. - SANCHEZ-PALENCIA E. 1987
Computation of Singular Solutions in Elliptic Problems and Elasticity. RMA 5, Masson, Paris

LEMAITRE J. - CHABOCHE J.L. 1985
Mécanique des Matériaux Solides. Dunod, Paris

LE TALLEC P. 1990
Numerical Analysis of Viscoelastic Problems. RMA 15, Masson-Springer

LEVI - NOUR OMID B. - STANLEY G. - SWENSON L. 1989
"An adaptive mess refinement strategy for analysis of shell structures." in *Computational Structural Mechanics and Multidisciplinary Optimization*, pp. 19-29 (ASME) San Fransisco

LEWIS R. - SCHREFLER B. 1998
The Finite Element Method in the Static and Dynamic Deformation and Consolidation of Porous Media. Wiley, London

LIBAI A. - SIMMONDS J. 1998
The Nonlinear Theory of Elastic Shells. Cambridge University Press

LICHT C. - PRATT E. - RAOUS M. 1991
"Remarks on a numerical method for unilateral contact including friction." *Int. Series of Numerical Mathematics*, Vol. 101, Birkhaüser, pp. 129-144

LIEUSAINT P. - ZLAMAL M. 1979
"Superconvergence of the gradient of finite element solutions." *RAIRO Anal. Numer.* **13**:139-166

MADAY Y. - PATERA A.T. - PERAIRE J. 1999
"A general formulation for a posteriori bounds for output functionals of partial differential equations - Applications to eigenvalue problems." *C. R. Acad. Sci. Paris*, Sér. I **328**(9):823-828

MAIER G. - NAPPI A. 1983
"On the unified framework provided by mathematical programming to plasticity." in *Mechanics of Materials Behavior.* (DVORAK and SHIELD Editors), Elsevier, pp. 253-276

MAIER G. - COMI C. - CORRIGLIANO A. - PEREGO U. 1993
"Internal time general internal variable and multi-yield-surface theories of plasticity and creep: A unification of concept." *Int. J. Plasticity* **2**:37-57

MANDEL J. 1966

Cours de mécanique des milieux continus. Gautiers-Villars, Paris

MANDEL J. 1972

"Director vectors and constitutive equations for plastic and viscoplastic media." *Proc. Sym. on Problems of Plasticity* (SAWCZUK Editor), Nordhoff, pp. 125-141

MELENK J. - BABUSKA I. 1997

"Approximation with harmonic and generalized harmonic polynomials in the partition of unity method." *Comput. Assisted Mech. Eng. Sci.* **4**:521-532

MOES N. 1996

Une méthode de mesure d'erreur a posteriori pour les modèles de matériaux décrits par variables internes. Thèse de l'Ecole Normale Supérieure de Cachan (France)

MORAND H. - OHAYON R. 1992

Interaction fluide-structure. RMA 23, Masson, Paris

MOREAU J.J. 1966

Fonctionnelles convexes. Cours du Collège de France, Paris

MOREAU J.J. 1974

"On unilateral constraints friction and plasticity." in *New Variational Techniques in Mathematical Physics* (CAPRIZ and STAMPACCHIA Editors), edizioni Cremonese, Rome (Italy), pp. 175-322

NAYROLES B. 1973

"Point de vue algébrique - Convexité et intégrales convexes en mécanique des solides. " in *New Variational Techniques in Mathematical Physics*, CISME, pp. 324-404

NECAS J. - HLAVACEK I. 1981

Mathematical Theory of Elastic and Elasto-Plastic Bodies: an Introduction. Elsevier

NGUYEN Q.S. 1984

"Bifurcation et stabilité des systèmes irréversibles obeissant au principe de dissipation maximale." *J. Mécanique* **3**:41-61

NGUYEN Q.S. 2000

Stabilité et mécanique non linéaire. Hermès, Paris

NOËL F. 1994

Mailleur auto-adaptatif pour des surfaces gauches en vue de la conception intégrée. Thèse de l'Institut National Polytechnique de Grenoble (France)

NOOR A. - PETER J. 1991

"Strategies for large scale structural problems on high-performance computers." *Comm. Appl. Num. Meth.* **7**:465-478

ODEN J.T. - DEMKOWICZ L. 1991
"*h-p* adaptive finite element methods in computational fluid dynamics." *Methods Appl. Mech. Eng.* 11-40

ODEN J.T. - DEMKOWICZ L. - RACHOWICZ W. - WESTERMANN T.A. 1989
"Toward a universel *h-p* adaptive finite element strategy - Part 2. A posteriori error estimation." *Comput. Methods Appl. Mech. Eng.* 113-180

ODEN J.T. - MARTINS J. 1985
"Models and computational methods for dynamic friction phenomena." *Comput. Methods Appl. Mech. Eng.* **52**:527-634

ODEN J.T. - PRUDHOMME S. 1998
"A posteriori error estimation of *h-p* approximations of the Stokes equations." in *Advances in Adaptive Computational Methods in Mechanics*, pp. 43-64 (LADEVEZE and ODEN Editors) Elsevier

ODEN J.T. - PRUDHOMME S. 1999
"Goal-oriented error estimation and adaptivity for the finite element method." *TICAM repport* 99-15, Austin, TX

ODEN J.T - VEMAGANTI K. - MOËS N. 1999
"Hierarchical modeling for heterogeneous sollids." *Comput. Methods Appl. Mech. Eng.* **172**:3-25

ODEN J.T. - WU W. - AINSWORTH M. 1994
"A posteriori error estimators for the NAVIER-STOKES problem." *Comput. Methods Appl. Mech. Eng.* **111**:185-202

OKSTAD K.M. - MATHISEN K.M. 1994
"Toward automatic adaptive geometrically nonlinear shell analysis." *Int. J. Num. Meth. Eng.* **37**:2657-2678

ONATE E. - BUGEDA G. 1993
"A study of mesh optimality criteria in adaptive finite element analysis." *Engineering Computations* **10**:307-321

ONATE E. - CASTRO I. - KREINER R. 1991
"Error estimation and mesh adaptivity techniques for plate and shell problems." *Proc. of the 3rd Int. Conf. on Quality Assurance and Standard in F.E. Analysis, p. 281* NAFEMS Stratford (U.S.A.)

ORTIZ M. - QUIGLEY J.J. 1991
"Adaptive mesh refinement in strain localization problems." *Comput. Methods Appl. Mech. Eng.* **90**:781-804

ORTIZ M. - SIMO J.C. 1986
"An analysis of a new class of integration algorithms for elastoplastic constitutive relations." *Int. J. Num. Meth. Eng.* **23**:353-366

OWEN D.R.J. - HINTON E. 1980
Finite Elements in Plasticity. Pineridge Press

PAPADRAKAKIS M. - BABILIS G. - BRAOUZI P. 1997
"Efficiency of refinement procedures for the *p*-version of the adaptive finite element method." *Engineering Computations* **14**(1):98-118

PARASCHIVOIU M. - PERAIRE J. - PATERA A.T. 1997
"A posteriori finite element bounds for linear functional outputs of elliptic partial differential equations." *Comput. Methods Appl. Mech. Eng.* **150**:289-312

PARK K.C. 1975
"An improved stiffly stable method for direct integration of nonlinear structural dynamic equations." *J. Appl. Mech.* (ASME) 464-470

PARK K.C. - FELIPPA C.A. 1983
"Partitioned analysis of coupled systems." in *Computational Methods for Transient Analysis* (BELYSCHKO and HUGUES Editors) North Holland, pp. 157-219

PAYNE L.E. 1967
"Isoparametric inequalities and their applications." *SIAM rev.* **9**:453-488

PELLE J.P. 1985
Fiabilité et optimisation des calculs de fréquences propres par éléments finis. Thèse d'état Université P et M. Curie, Paris

PELLE J.P. 1995
"Sur la maîtrise des calculs Eléments Finis: état actuel et perspectives." *Proc. du 2ème Colloque National en Calcul des Structures,* Giens (France), pp. 43-54

PELLE J.P. - BENOIT C. 1995
"Evaluation des erreurs de discrétisation pour les calculs de plaques." *Rapport interne n° 161*, LMT-Cachan (France)

PELLE J.P. - COOREVITS P. - DUMEAU J.P. 1995
"Contrôle et automatisation de maillages 3D. Application à des structures à symétrie de répétition." *Rapport final MESR-Aerospatiale* (aide n° 92 P 0497)

PELLE J.P. - LADEVEZE P. - ROUGEOT P. 1989
"Post-Processeur de calcul d'erreur." *Rapport interne n° 91*, LMT-Cachan (France)

PELLE J.P. - ROUGEOT P. 1988
"Contrôle de qualité en éléments finis et maillages optimisés." in *Calcul des Stuctures et Intelligence Artificielle* - Tome II (FOUET, LADEVEZE and OHAYON Editors), pp. 209-220 Plurtalis

PELLE J.P. - ROUGEOT P. 1989
"Maillages adaptatifs pour les éléments classiques." in *Calcul des Stuctures et Intelligence Artificielle* - Tome III (FOUET, LADEVEZE and OHAYON Editors), pp. 171-186, Plurtalis

PELLE J.P. - RYCKELYNCK D. 1998a

"About the automatization of nonlinear behavior structure analysis: Piloting and quality control." *Proc. of the Second International Conference on Integrated Design and Manufacturing in Mechanical Engineering,*, pp. 715-724, Compiègne (France)

PELLE J.P. - RYCKELYNCK D. 1998b

"Sur une stratégie adaptative pour les calculs non linéaires. Application aux calculs en viscoplasticité." *Revue Européenne des Eléments Finis* **7**(1-3):89-104

PERAIRE J. - PATERA A.T. 1998

"Bounds for linear-functional outputs of coercive partial differential equations; local indicators and adaptive refinements." in *Advances in Adaptive Computational Methods in Mechanics*, pp. 199-216 (LADEVEZE and ODEN Editors), Elsevier

PEREIRA O.J.B.A. - ALMEIDA J.P.M. - MAUNDER E.W.A. 1999

"Adaptive methods for hybrid equilibrium finite element models." *Comput. Methods Appl. Mech. Eng.* (1-4):19-39

PERIC D. - DUTKO M. - OWEN D.R.J. 1998

"Aspects of adaptive strategies for large scale deformation problems at finite inelastic strains." in *Advances in Adaptive Computational Methods in Mechanics*, pp. 349-363, (LADEVEZE and ODEN Editors), Elsevier

PERIC D. - YU J. - OWEN D.R.J. 1994

"On error estimates and adaptivity in elastoplastic solids: Application to the numerical simulation of strain localization in classical and Cosserat continua." *Int. J. Num. Meth. Eng.* **37**:1351-1379

PERRONNET A. 1993

"Tetrahedrization by the 5-14-OTT tree technique and the DELAUNAY criterion." *8th Conference on Finite Elements in Fluids*, Barcelona (Spain)

PERRONNET A. 1995

"Génération de maillages: Les problèmes, quelques algorithmes, un logiciel." *27ème Congrès National d'Analyse Numérique*, Super-Besse (France)

PIJAUDIER-CABOT G. - BODE L. - HUERTA A. 1995

"Arbitrary Lagrangian Eulerian finite element analysis of strain localization in transcient problems." *Int. J. Num. Meth. Eng* **38**(24):4171-4191

PIJAUDIER-CABOT G. - DUBE J.F. - LABORDERIE C. - BODE L. 1992

"Damage models for concrete in transcient dynamics." in *Fracture and Damage in Quasi-Brittle Structures* (BAZANT Editor), pp. 201-216

POGU M. - TOURNEMINE G. 1992

Modélisation et résolution d'équations de la mécanique des milieux continus. Ellipse, Paris

POTIER-FERRY M. 1987

"Foundation of elastic post-buckling theory." in *Buckling and Post-Buckling.* Lecture Note in Physics, No. 288, pp. 1-82, Springer

PRAGER W. - SYNGE J.L. 1947

"Approximation in elasticity based on the concept of functions spaces." *Quart. Appl. Math.* **5**:261-269

PREDELEANU M. 1987

Computational Methods for Predicting Material Processing Defects. Elsevier

PROBE 1985

Theoretical Manual.

RAMM E. - CIRAK F. 1997

"Adaptivity for nonlinear thin walled structures." *Proc. of International Conference on Computational Plasticity: Fundamentals and Applications*, Barcelona (Spain), pp. 145-163

RANK E. - SCHWEINGRUBER M. - SOMMER M. 1993

"Adaptive mesh generation and transformation of triangular to quadrilateral meshes." *Comm. Num. Meth. Engng.* **9**:121-129

RANNACHER R. 1999

Error control in finite element computations. in *Error Control and Adaptivity in Scientific Computing*, pp. 247-278 (BULGAK and ZENGER Editors), Kluwer

RANNACHER R. - SUTTMEIER F.T. 1998

"A posteriori error control and mesh adaptation for FE models in elasticity and elasto-plasticity." in *Advances in Adaptive Computational Methods in Mechanics*, pp. 275-292 (LADEVEZE and ODEN Editors), Elsevier

RANNACHER R. - SUTTMEIER F.T. 1999

"A posteriori error estimation and mesh adaptation for FE models in elasto-plasticity." *Comput. Methods Appl. Mech. Eng.* **176**:333-361

RAOUS M. - CHABRAND P. - LEBON F. 1988

"Numerical methods for frictional contact problems and applications." *J. Méc. Théor. Appl.* **7**:111-128

RASSINNEUX A. - KROMER V. - GUEURY M. 1997

"Maillage automatique tridimensionnel avec contrôle de taille." *Actes du Troisième Colloque National en Calcul des Structures*, pp. 361-366, Giens (France), Presses Académiques de l'Ouest

RAVIART P.A. - THOMAS J.L. 1975

"A mixed finite element method for second-order elliptic problems." *Proc. of the Symposium on the Mathematical Aspects of the Finite Element Method*, Rome (Italy)

RAVIART P.A. - THOMAS J.L. 1983

Introduction à l'analyse numérique des équations aux dérivées partielles. Masson, Paris

REMACLE J. - DULAR P. - GENON A. - LEGROS W. 1996
"A posterirori error estimator and adaptive meshing using error in constitutive relation." *IEEE Trans. Magnetics* **32**(3):1369-1372

RICE J. 1971
"Inelastic constitutive relations for solids: An internal variable theory and its application to metal plasticity." *J. Mech. Phys. Solids* **9**:433-455

RIKS E. 1991
"On the formulation of path-following techniques for structural stability analysis." in *New Advances in Computational Structural Mechanics* (LADEVEZE and ZIENKIEWICZ Editors), Elsevier, pp. 65-80

RIVARA M.C. 1984
"Algorithms for refining triangular grids suitable for adaptive and multigrid techniques." *Int. J. Num. Meth. Eng.* **20**:745-756

RIVARA M.C. 1989
"Selective refinement/derefinement algorithms for sequences of nested triangulations." *Int. J. Num. Meth. Eng.* **28**:2889-2906

ROCKAFELLAR R.T. 1970
Convex Analysis. Princeton University Press

ROUGEE P. 1991
"A new Lagrangian intrinsic approach of continous media in large deformation." *European J. Mech. A/Solids* **10**(1):15-39

ROUGEE P. 1997
Mécanique des grandes transformations. MA 25, Springer

ROUGEOT P. 1989
Sur le contrôle de la qualité des maillages Eléments Finis. Thèse de l'Ecole Normale Supérieure de Cachan (France)

SALENÇON J. 1988
Mécanique des milieux continus. Ellipse, Paris

SANCHEZ-HUBERT J. - SANCHEZ-PALENCIA E. 1989
Vibration and Coupling of Continuous Systems - Asymptotic Methods. Springer

SANCHEZ-PALENCIA E. 1980
Nonhomogeneous Media and Vibration Theory. Lectures Notes in Physics, No 127, Springer

SELMAN A. - HINTON E. - ATAMAZ S.W. 1990
"Edge effects in Reissner-Mindlin plates using adaptive mesh refinement." *Engineering Computations* **7**(3):217-226

SHEPARD M.S. - GEORGES M.K. 1991

"Automatic three-dimensional mesh generator by the finite octree technique." *Int. J. Num. Meth. Eng.* **32**:709-749

SIMO J. 1988

"A framework for finite strain elastoplasticity based on maximum plastic dissipation and multiplicative decomposition." *Comput. Methods Appl. Mech. Eng.* **66**:199-219

SIMO J. - TAYLOR R. 1985

"Consistent tangent operators for rate independent elastoplasticity." *Comput. Methods Appl. Mech. Eng.* **48**:101-118

STEIN E. - AHMAD R. 1977

"An equilibrium method for stress calculation using finite element displacement models." *Comput. Methods Appl. Mech. Eng.* **10**:175-198

STEIN E. - BARTHOLD F.J. - OHNIMUS S. - SCHMIDT M. 1998

"Adaptive finite element in elastoplasticity with mechanical error indicators and NEUMANN-type indicators." in *Advances in Adaptive Computational Methods in Mechanics*, pp. 81-100 (LADEVEZE and ODEN Editors), Elsevier

STEIN E. - OHNIMUS S. 1996

"Dimensional adaptivity in linear elasticity with herarchical test-spaces for *h*- and *p*-refinement processes." *Engineering with Computers* **12**:107-119

STEIN E. - OHNIMUS S. 1997

"Coupled model and solution adaptivity in the finite element method." *Comput. Methods Appl. Mech. Eng.* **150**:327-350

STEIN E. - OHNIMUS S. 1999

"Anisotropic discretization and model error estimation in solid mechanics by local NEUMANN problems." *Comput. Methods Appl. Mech. Eng.* **176**:363-385

STEPHEN D. - STEVEN G. 1997

"Error estimation for natural frequency analysis using plate elements." *Engineering Computations* **14**(1):119-134

STOLZ C. 1987

Anélasticité et stabilité. Thèse de l'Université P et M. Curie, Paris

STRANG G. - FIX G.J. 1976

An Analysis of the Finite Element Method, Prentice-Hall

STROUBOULIS T. - BABUSKA I. - COPPS K. 2000

"The design and analysis of the generalized finite element method." *Comput. Methods Appl. Mech. Eng.* **181**:43-69

STROUBOULIS T. - BABUSKA I. - DATTA D. - COPPS K. - GANGARAJ S. 2000

"A posteriori estimation and adaptive control of the error in the quantity of interest - Part I: A posteriori estimation of the error in the Von Mises stress and the stress intensity factors." *Comput. Methods Appl. Mech. Eng.* **181**:261-294

STROUBOULIS T. - BABUSKA I. - GANGARAJ S.K. 2000

"Guaranteed computable error bounds for the exact error in the finite element solution - Part II: Bounds for the energy norm of the error in two dmension." *Int. J. Num. Meth. Eng.* **47**:427-475

STROUBOULIS T. - HAQUE K.A. 1992a

"Recent experiences with error estimation and adaptivity. Part I: Review of error estimators for scalar elliptic problems." *Comput. Methods Appl. Mech. Eng.* **97**:339-436

STROUBOULIS T. - HAQUE K.A. 1992b

"Recent experiences with error estimation and adaptivity." Part II: Error estimation for *h*-adaptive approximations on grids of triangles and quadrilaterals." *Comput. Methods Appl. Mech. Eng.* **100**:359-430

SUQUET P. 1981

"Sur les équations de la plasticité: existence et unicité des solutions." *J. Mécanique*, **20**(1):3-39

SZABO B.A. 1986

"Estimation and control of error based on *p* convergence." in *Accuracy Estimates and Adaptive Refinements in Finite Element Computations*, Chap 3, pp. 61-78 (GAGO, OLIVEIRA and ZIENKIEWICZ Editors), Wiley, London

SZABO B.A. 1990

"The *p* and *h-p* versions of the finite element method in solid mechanics." *Comput. Methods Appl. Mech. Eng.* **82**:185-195

TEIXEIRAS DE FREITAS J. 1997

"Hybrid-trefftz displacement and stress elements for elastodynamic analysis in the frequency domain." *Comput. Assisted Mech. Eng. Sci.* **4**:345-368

TEMAM R. 1977

Navier Stokes Equations, Studies in Mathematics and its Applications, North-Holland

TOTTENHAM H. 1970

"Basic principles." in *Finite Element Techniques in Structural Mechanics* (TOTTENHAM and BREBIA Editors), Southampton University Press

TRUSDELL C. 1971

Rational Thermodynamics. Wiley, London

TURCKE D.J. - NEICE G.M. 1974
"Guidelines for selecting finite element grids based on an optimization study." *Computers and Structures* **4**:499-519

VALID R. 1995
The Nonlinear Theory of Shells Through Variational Principle, Wiley, London

VERFÜRTH R. 1996
A Review of a posteriori Error Estimation and Adaptive Mesh-Refinement Techniques. Wiley, London-Teubner Verlag

WASHIZU K. 1975
Variational Method in Elasticity and Plasticity, Oxford Pergamon

WATANABE O. - ALTURI S. 1986
"Internal time, general internal variables and multi-yield-surface theories of plasticity and creep: A unification of concepts." *Int. J. Plasticity*. **2**:37-57

WEINSTEIN A. 1963
"Intermediate problems and the maximun-minimum theory." *J. Math. Mech.* **12**:235-246

WEINSTEIN A. - STENGER W. 1972
Methods of Intermediate Problems for Eigenvalues. Academic Press, New York

WHALBIN L.B. 1995
Superconvergence ingalerkin finite element methods. Lectures Notes in Mathematics, No 1605, Springer

WIBERG N.E. - ABDULWAHAB F. 1992
"An efficient post-processing technique for stress problems based on superconvergent derivatives and equilibrium." in *Numerical Methods in Engineering*, (HIRSCH, PERIAUX and ONATE Editors), Elsevier, pp. 25-32

WIBERG N.E. - ABDULWAHAB F. - ZIUKAS S. 1994
"Enhanced superconvergent patch recovery incorporating equilibrium and boundary conditions." *Int. J. Num. Meth. Eng.* **37**:3417-3440

WIBERG N.E. - BANSYS R. - HAGER P. 1999
"Improved eigenfrequencies and eigenmodes in free vibration analysis." *Computer and Structures* **73**:79-86

WIBERG N.E. - LI X.D. 1993
"A post-processing technique and a posteriori error estimate for the NEWMARK method in dynamic analysis." *Earthquake Eng. Struct. Dyn.* **22**:465-489

WIBERG N.E. - LI X.D. 1994
"A post-processed error estimate and adaptive procedure for the semidiscrete finite elment method in dynamic analysis." *Int. J. Num. Meth. Eng.* **37**:3585-3603

WIBERG N.E. - LI X.D. 1998

"Implementation and adaptivity of a space-time finite element method for structural dynamics." *Comput. Methods Appl. Mech. Eng.* **156**(1/4):211-229

WRIGGERS P. - SCHERF O. 1995

"An adaptive finite element algorithm for contact problems in plasticity." *Comput. Mech.* **17**:88-97

WUNDERLICH W. - CRAMER H. - STEINL G. 1998

"An adaptive finite element approach in associated and non-associated plasticity considering localization phenomena." in *Advances in Adaptive Computational Methods in Mechanics*, pp. 293-308 (LADEVEZE and ODEN Editors), Elsevier

ZENG Q. - COMBESCURE A. 1998

"A new one-point quadrature, general non-linear quadrilateral shell element with physical stabilization." *Int. J. Num. Meth. Eng.* **42**:1307-1338

ZHONG H.G. 1991

Estimateurs d'erreur a posteriori et adaptation de maillages dans la méthode des éléments finis. Thèse de l'Université de Liège (Belgium)

ZIENKIEWICZ O.C. 1997

"Trefftz approximation and the generalized finite element method - History and development." *Comput. Assisted Mech. Eng. Sci.* **4**:305-316

ZIENKIEWICZ O.C. - CRAIG A.W. 1986

"Adaptive refinement, error estimates, multigrid solution and hierarchic finite element method concept." in *Accuracy Estimates and Adaptivity for Finite Element* (BABUSKA, ZIENKIEWICZ, GAGO and OLIVEIRA Editors), Wiley, London, Ch. 2, pp. 25-59

ZIENKIEWICZ O.C. - LIU Y. - HUANG G. 1988

"Error estimation and adaptivity in flow formulation for forming problems." *Int. J. Num. Meth. Eng.* **25**:23-42

ZIENKIEWICZ O.C. - TAYLOR R.L. 1988

The Finite Element Method: Basic Formulation and Linear Problems. McGraw-Hill, London (Traduction française, 1991, AFNOR Techniques)

ZIENKIEWICZ O.C. - WOOD W. - HINE N. - TAYLOR R. 1984

"A unified set of single step algorithms. Part I: General formulation and applications." *Int. J. Num. Meth. Eng.* **20**:1529-1552

ZIENKIEWICZ O.C. - XIE Y. 1991

"A simple error estimator and adaptive time stepping procedure for dynamic analysis." *Earthquake Eng. Struct. Dyn.* **20**:871-887

ZIENKIEWICZ O.C. - ZHU J.Z. 1987
"A simple error estimator and adaptative procedure for practical engineering analysis." *Int. J. Num. Meth. Eng.* **24**:337-357

ZIENKIEWICZ O.C. - ZHU J.Z., 1992
"The superconvergent patch recovery and a posteriori error estimates. Part I: The recovery technique - Part II: Error estimates and adaptivity." *Int. J. Num. Meth. Eng.* **33**:1331-1382

ZIENKIEWICZ O.C. - ZHU J.Z. - GONG N.G., 1989
"Effective and practical *h-p* version adaptive analysis procedures for the finite element method." *Int. J. Num. Meth. Eng.* **28**:879-891

Index

Mechanical Engineering Series *(continued from page ii)*

E.N. Kuznetsov, **Underconstrained Structural Systems**

P. Ladevèze, **Nonlinear Computational Structural Mechanics:** New Approaches and Non-Incremental Methods of Calculation

P. Ladevèze and J.-P. Pelle, **Mastering Calculations in Linear and Nonlinear Mechanics**

A. Lawrence, **Modern Inertial Technology: Navigation, Guidance, and Control, 2nd ed.**

R.A. Layton, **Principles of Analytical System Dynamics**

F.F. Ling, W.M. Lai, D.A. Lucca, **Fundamentals of Surface Mechanics: With Applications, 2nd ed.**

C.V. Madhusudana, **Thermal Contact Conductance**

D.P. Miannay, **Fracture Mechanics**

D.P. Miannay, Time-Dependent Fracture Mechanics

D.K. Miu, **Mechatronics: Electromechanics and Contromechanics**

D. Post, B. Han, and P. Ifju, **High Sensitivity Moiré:** Experimental Analysis for Mechanics and Materials

F.P. Rimrott, **Introductory Attitude Dynamics**

S.S. Sadhal, P.S. Ayyaswamy, and J.N. Chung, **Transport Phenomena with Drops and Bubbles**

A.A. Shabana, **Theory of Vibration: An Introduction, 2nd ed.**

A.A. Shabana, **Theory of Vibration: Discrete and Continuous Systems, 2nd ed.**